ISNM 79:
International Series of Numerical Mathematics
Internationale Schriftenreihe zur Numerischen Mathematik
Série internationale d'Analyse numérique
Vol. 79

Birkhäuser Verlag
Basel · Boston · Stuttgart

Bifurcation: Analysis, Algorithms, Applications

Proceedings of the Conference at the University of Dortmund, August 18–22, 1986

Edited by

T. Küpper
R. Seydel
H. Troger

1987

Birkhäuser Verlag
Basel · Boston · Stuttgart

Editors:

T. Küpper
Universität Hannover
Institut für Angewandte Mathematik
Welfengarten 1
D–3000 Hannover 1

R. Seydel
Universität Würzburg
Institut für Angewandte
Mathematik und Statistik
Am Hubland
D–8700 Würzburg

H. Troger
Technische Universität Wien
Institut für Mechanik
Karlsplatz 13
A–1040 Wien

Library of Congress Cataloging in Publication Data

Bifurcation : analysis, algorithms, applications.
(ISNM ; 79 = International series of numerical mathematics ; vol. 79)
Includes bibliographies.
1. Bifurcation theory -- Congresses. I. Küpper, T. (Tassilo), 1947– . II. Seydel, R. (Rüdiger), 1947– . III. Troger, H. (Hans), 1943– . IV. Series: International series of numerical mathematics ; v. 79.
QA372.B52 1987 515.3'52 87–5111
ISBN 0-8176-1798-1 (U.S.)

CIP-Kurztitelaufnahme der Deutschen Bibliothek

Bifurcation : analysis, algorithms, applications : proceedings of the conference at the Univ. of Dortmund, August 18–22, 1986 / ed. by T. Küpper . . . – Basel ; Boston ; Stuttgart : Birkhäuser, 1987.
(International series of numerical mathematics ; 79)
ISBN 3-7643-1798-1 (Basel . . .)
ISBN 0-8176-1798-1 (Boston)
NE: Küpper, Tassilo [Hrsg.]; GT

Printed in Germany
ISBN 3-7643-1798-1
ISBN 0-8176-1798-1

PREFACE

The conference on BIFURCATIONS: ANALYSIS, ALGORITHMS, APPLICATIONS took place in Dortmund in August 18 - 22, 1986. More then 150 Scientists from 16 countries participated in the meeting, among them mathematicians, engineers, and physicists. A broad spectrum of new results on bifurcation was covered by 49 talks. The diversity of the range of treated topics and of involved fields inspired fruitful discussions.

36 refereed papers are contained in these proceedings. The subjects covered treat bifurcation problems, ranging from theoretical investigations to numerical results, with emphasis placed upon applications. The more theoretical papers include the topics symmetry breaking, delay differential equations, Cornu spirals, homoclinic orbits, and selfsimilarity. Different kinds of bifurcations are treated: Hopf bifurcation, bifurcation from continuous spectrum, complex bifurcation, and bifurcation near tori. Several numerical aspects are discussed, among them continuation, block elimination, and spectral methods. Algorithms are proposed for approximating manifolds, calculating periodic solutions and handling multi-parameter problems. Ample space is devoted to applications. Classical phenomena from fluid mechanics (such as convection rolls and the Taylor vortex problem), buckling, and reaction-diffusion problems are considered. Other applications of bifurcations include railway vehicle dynamics, computer graphics, semiconductors, drilling processes, simulation of oil reservoirs, and rotor dynamics.

The proceedings reflect current research in bifurcation. They are an attempt to bring together researchers from different disciplines to stimulate common effort towards a better understanding and handling of bifurcation problems.

We gratefully acknowledge the support received from Deutsche Forschungsgemeinschaft, Deutscher Akademischer Austauschdienst, Freundesgesellschaft der Universität Dortmund, Ministerium für Wissenschaft und Forschung des Landes NW, the Universität Dortmund, and the Fonds zur Förderung der Wissenschaftlichen Forschung in Österreich.

December 1986

Tassilo Küpper, Rüdiger Seydel, Hans Troger

Contents

International Series of
Numerical Mathematics, Vol. 79

THE EFFECT OF DISCRETIZATION ON HOMOCLINIC ORBITS

Wolf-Jürgen Beyn
Fakultät für Mathematik, Universität Konstanz, FRG

1. Introduction

The numerical analysis of the longtime behavior of a dynamical system

(1) $\dot{z} = f(z,\lambda)$, $z \in \mathbb{R}^m$, $\lambda \in \mathbb{R}$ (a parameter)

usually follows two complementary approaches.

I. Direct methods: Set up and solve defining equations for limit sets of (1), such as steady states and periodic orbits, determine their stability characteristics and find singular points with respect to the parameter.

II. Indirect methods: Integrate (1) numerically for various values of $z(o)$ and λ and analyze the longtime behavior of the numerical trajectories.

Although there has been considerable success with direct methods the integration seems to be indispensable in complicated situations where e.g. global stability properties are of interest or where direct methods are not available. For the integration approach it is important to know in which sense the longtime behavior of continuous trajectories is reflected by the numerical trajectories. For simplicity, we will assume the latter to be sequences z^n generated by some one-step method with constant step size h

(2) $z^{n+1} = \varphi(z^n,\lambda,h) = z^n + hg(z^n,\lambda,h)$

It is the purpose of this paper to comment on both approaches in the case where the system (1) has a homoclinic orbit.

2. A direct method for homoclinic orbits

A homoclinic orbit $\bar{z}(t)$, $t \in \mathbb{R}$ is a solution of (1) at some $\lambda = \bar{\lambda}$ which satisfies

(3) $\bar{z}(t) \to z_\infty$ as $t \to \pm\infty$ and $f(z_\infty,\bar{\lambda}) = 0$

Homoclinic orbits are typical one parameter phenomena and - as with Hopf points - they mark the begin and end of branches of periodic orbits (see [5,10] for a theoretical and a numerical illustration of this fact). We consider the following two-dimensional model example (see [8],Ch.6.1)

(4) $\dot{x} = y, \ \dot{y} = x - x^2 + \lambda y + axy$

Fig. 1 gives the numerical bifurcation picture for $a = 0.5$ as abtained with the Code AUTO by Doedel [5].

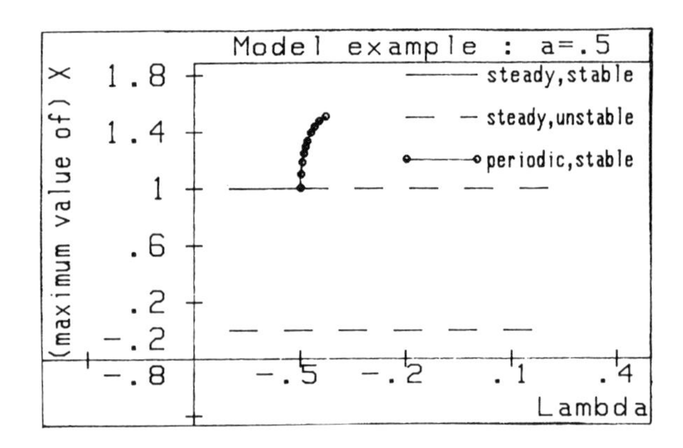

Figure 1

The periodic orbits created at the Hopf point $\lambda = -a$ vanish through a homoclinic orbit at $\bar{\lambda} = -0.429505$ (see [8] for the typical changes in the phase plane near this point). In [5] homoclinic orbits are simply computed as periodic orbits with large period, nevertheless it seems attractive (and probably more efficient) to have a direct method for homoclinic orbits which make use of the base point z_∞ and of the linearization at z_∞ (see below).

We return to the general system (1) and assume that it has some smooth branch $(z_\infty(\lambda),\lambda)$ of steady states such that $A(\lambda) = \frac{\partial f}{\partial z}(z_\infty(\lambda),\lambda)$ has no eigenvalues on the imaginary axis. The computation of a homoclinic orbit with a base point on this branch would then require to find λ and $z(t)\,(t \in \mathbb{R})$ satisfying (1) and

(5a) $\lim_{t\to\infty} z(t) = z_\infty(\lambda),\ \lim_{t\to-\infty} z(t) = z_\infty(\lambda)$

As for periodic orbits we have to add a phase fixing condition which we take in the simple form

(5b) $\Psi^T z(o) - \alpha = 0$ where $\Psi \in \mathbb{R}^m,\ \alpha \in \mathbb{R}$.

For numerical purposes we have to replace the infinite interval $(-\infty,\infty)$ by a finite one, say $[-T,T]$, and we have to replace (5a) by appropriate boundary conditions at $-T,T$. For boundary value problems on semi-infinite intervals there is a well developed theory of how to do this [9,11,12] and how to estimate the resulting error. This theory easily carries over to the present case. Let $P(\lambda)$ and $Q(\lambda)$ be the projectors onto the invariant subspaces of $A(\lambda)$ associated with the unstable and stable eigenvalues. Then we replace (1),(5a,b) by (5b) and

(6a) $\dot{z} = f(z,\lambda),\ -T \le t \le T$

(6b) $P(\lambda)(z(T) - z_\infty(\lambda)) = 0,\ Q(\lambda)(z(-T) - z_\infty(\lambda)) = 0.$

The boundary condition (6b) forces the orbit to leave $z_\infty(\lambda)$ close to the unstable manifold and to approach it again close to the stable manifold. There are many more possible choices for (6b), however (cf.[9]).

Finally, by setting $y(t) = z(-t)$ we obtain a two-point boundary value problem of dimension 2m+1 on [0,T] to which we can apply a standard code

$$(7a)\quad \dot{z} - f(z,\lambda) = 0,\ \dot{y} + f(y,\lambda) = 0,\ \dot{\lambda} = 0\ (0 \le t \le T)$$

$$(7b)\quad \begin{array}{l} P(\lambda(T))(z(T) - z_\infty(\lambda(T))) = 0,\ Q(\lambda(T))(y(T) - z_\infty(\lambda(T))) = 0 \\ \psi^T z(o) - \alpha = 0,\ z(o) = y(o) \end{array}$$

The choice of ψ and α is relatively easy if we are following a branch of periodic solutions with increasing period. If $\hat{z}(t)$ is such a large period solution we could use $\psi = \hat{z}'(o)$, $\alpha = \psi^T\hat{z}(o)$. Using the theory from [9] we can show the following result. Suppose that $(\bar{z}(t)(t\in\mathbb{R}), \bar{\lambda})$ is a nondegenerate homoclinic pair, i.e. it solves (1), (5a,b) and the linearized problem

$$\dot{y} - \frac{\partial f}{\partial z}(\bar{z},\bar{\lambda})y - \frac{\partial f}{\partial \lambda}(\bar{z},\bar{\lambda})\mu = 0,\ \lim_{t\to\pm\infty} y(t) = 0,\ \psi^T y(o) = 0$$

has only the trivial solution (y,μ). Then (5b), (6a,b) has a unique solution (z_T,λ_T) for T sufficiently large and

$$\max_{|t|\le T} \| \bar{z}(t) - z_T(t)\| + |\bar{\lambda} - \lambda_T| \le C\{\| Q(\bar{\lambda})(\bar{z}(-T) - z_\infty(\bar{\lambda}))\| + \|P(\bar{\lambda})(\bar{z}(T) - z_\infty(\bar{\lambda}))\|\}$$

In fact, due to the hyperbolicity of $z_\infty(\bar{\lambda})$ the right hand side decays like $\exp(-\beta T)$ for some $\beta > 0$.
The defining system (7a,b) can be modified in an obvious way for heteroclinic orbits connecting two hyperbolic points at which the Jacobian of f has stable subspaces of equal dimension.

Finally, it is interesting to note that in the two-dimensional case the nondegeneracy of the homoclinic pair $(\bar{z},\bar{\lambda})$ is equivalent to one of the assumptions in the homoclinic bifurcation theorem ([8],Th.6.1.1.(2)). Together with trace $A(\bar{\lambda}) \neq 0$ this theorem ensures that in a one sided neigborhood of $\bar{\lambda}$ there exists a branch of periodic orbits which turn into the homoclinic orbit at $\bar{\lambda}$.

2. Numerical integration of systems with homoclinic orbits

We now consider the asymptotic behavior $n\to\infty$ of a one-step method

(2) if h is small and if λ is close to the value $\bar{\lambda}$, for which the system (1) has a homoclinic orbit.
First of all, it is instructive to review the results on (2) if the system (1), for a fixed λ, has a hyperbolic, periodic orbit $z(t)$ with period T. Apart from some smoothness we assume p-th order accuracy of the method (2) in the form

$$(8)\quad \frac{\partial^i \varphi}{\partial h^i}(z,\lambda,o) = \frac{\partial^i \phi}{\partial h^i}(z,\lambda,o),\ i = 0,\ldots p$$

where $\phi(z,\lambda,h)$ denotes the flow of the system (1) with time step h. It was shown in [1, 2, 4, 6, 7] that, for h sufficiently small, the one step method (2) has a closed invariant curve $\{z_h(t) : t \in \mathbb{R}\}$ where z_h is T-periodic and satisfies

$$(9)\quad \mathrm{Max}\{\| z(t) - z_h(t) \| : 0 \le t \le T\} = O(h^p)$$

$$(10)\quad \varphi(z_h(t),\lambda,h)=z_h(\sigma_h(t))\ (t\in\mathbb{R}) \text{ where } \sigma_h(t)=t+h+O(h^{p+1})$$

These results may be generalized to compact branches of hyperbolic periodic orbits, but it is clear that the critical value of h, below which the invariant curve exists, tends to zero if we approach a Hopf point or a homoclinic point. The situation near a Hopf bifurcation was successfully analyzed in [3]. But - as far as we know - there is no precise answer to the question, what happens to the invariant curves of (2) if λ passes the homoclinic point.
Fig. 2 shows some numerical experiments with Euler's method for the model example (4) with $h = 0.4$, $a = 0.5$.

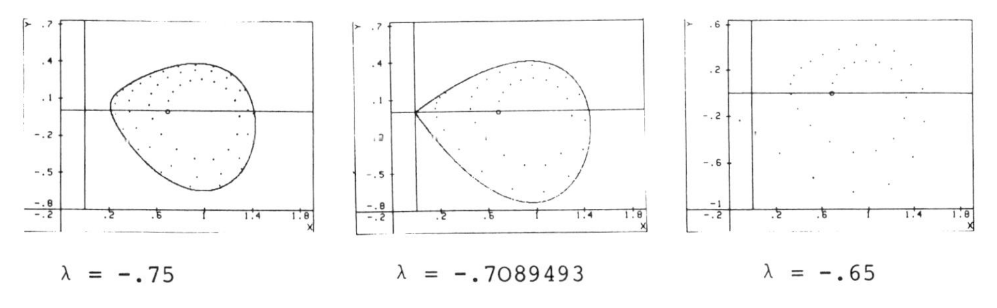

$\lambda = -.75$ $\qquad$ $\lambda = -.7089493$ $\qquad$ $\lambda = -.65$

Figure 2

For $\lambda < \bar{\lambda}_h \cong -0.7089493$ the points of the iteration starting at $(0.7,0)$ filled an invariant curve after some time whereas for $\lambda > \bar{\lambda}_h$ the sequence became unbounded. It seems that, exactly as for the continuous system, there is a critical value $\bar{\lambda}_h$ at which we have an "invariant homoclinic curve". From a generic point of view this is quite surprising (see below) but it has been observed also for smaller step sizes h and different Runge Kutta methods. What we can actually prove is the following. Let the branch $(z_\infty(\lambda),\lambda)$ from section 1 be the trivial one and let $(\bar{z},\bar{\lambda})$ be a nondegenerate homoclinic pair. Moreover, assume (8) and $\varphi(o,\lambda,h) \equiv 0$ for all λ and h as well as

(11) $g(z,\lambda,o) = f(z,\lambda)$ (compare (1) and (2)), $g \in C^1(\mathbb{R}^m \times \mathbb{R} \times [0,1], \mathbb{R}^m)$

Then, for h sufficiently small, the iteration (2) has a discrete homoclinic pair $(z_h^n (n \in \mathbb{Z}),\ \lambda_h)$, i.e.

(12) $z_h^{n+1} = \varphi(z_h^n, \lambda_h, h)\ (n \in \mathbb{Z})$ and $z_h^n \to 0$ as $n \to \pm\infty$, $\psi^T\ z_h^o - \alpha = 0$.

Moreover, we have the estimate

$$\sup\{\| z_h^n - \bar{z}(nh) \| : n \in \mathbb{Z}\} + |\lambda_h - \bar{\lambda}| = O(h^p).$$

The following table shows a few values of λ_h for our model example (Euler's method) which were obtained by truncating the boundary conditions in (12) in a way analogous to (6b) and using $\psi^T = (0,1), \alpha = 0$.

h	0.8	0.4	0.2	0.1	0.05
λ_h	-0.9645	-0.7089	-0.5716	-0.5010	-0.4653

, $a = 0.5$

In all cases we found that λ_h coincided with the value at which the invariant curve vanishes.

In the two dimensional case, under our assumptions above, we have that 0 is a hyperbolic point of the mapping $\varphi(\cdot,\ \lambda_h, h)$. The stable and unstable manifolds M_s^h and M_u^h of this point both contain the discrete homoclinic orbit so that it is crucial to decide whether these manifolds (actually curves) intersect transversely or not.

A transverse intersection would exclude the possibility of a homoclinic curve and would also imply certain chaotic features for the one-step method (2) (e.g. infinitely many discrete periodic orbits and horseshoes [8]).
In fact, following a suggestion by B. Fiedler during this conference, a transversal intersection was observed after adding an artificial perturbation to Euler's method for (4) (actually we added to the first component $10h^3 \phi((x-1.5)/h)\,\phi((y-\frac{h}{4})\frac{5}{h})$ where $\phi(x) = \exp(-1/(1-x)^2)$ for $|x| < 1$ and 0 otherwise) which did not disturb the discrete homoclinic orbit and which still satisfies our smoothness and consistency requirements. Fig. 3 shows the resulting unstable manifold and the typical oscillation effect due to the transversal intersection. The picture was produced by plotting the points of the iteration (2) by starting randomly on a straight line which was very close to the local unstable manifold. Similar observations were made for smaller step-sizes and different one-step methods.

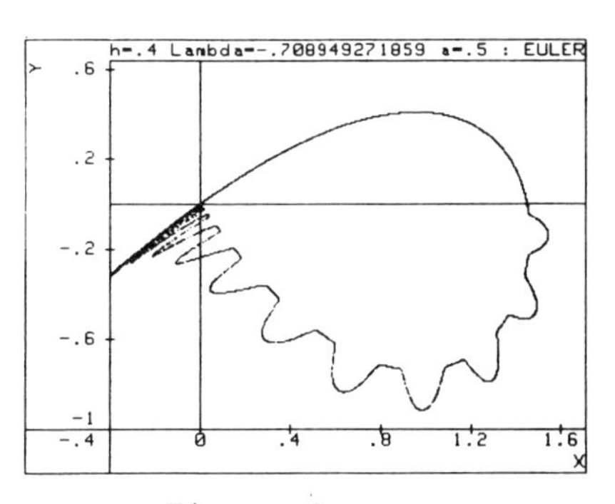

Figure 3

Let us finally relate our observations for one-step methods to the behavior of the dynamical system (1) under perturbations. It is well known that the nondegenerate homoclinic bifurcation in $\mathbb{R}^2$ is stable with respect to autonomous perturbations. However, nonautonomous perturbations introduce a third dimension and hence may produce transverse homoclinic points. For a periodic perturbation, for example, the stable and unstable manifolds of the stationary point now appear as stable and unstable manifolds of a Poincaré map. Hence they may - and in a generic sense also will - intersect transversely after perturbation (see [13]).
In view of our results above we are therefore led to the following question. Do standard one-step methods - without any artificial perturbation - have some intrinsic property which forces them to

act rather as an autonomous than as a nonautonomous perturbation of the dynamical system itself?

References

[1] Beyn, W.-J. (submitted 1985) On invariant closed curves for one-step methods.

[2] Braun, M., Hershenov, J. (1977) Periodic solutions of finite difference equations. Quart. Appl. Math. 35, 139 - 147.

[3] Brezzi, F., Ushiki, S., Fujii, H. (1984) Real and ghost bifurcation dynamics in difference schemes for ODE's. Numerical Methods for Bifurcation Problems (T. Küpper, H.D. Mittelmann, H. Weber Eds.), 79 - 104. Birkhäuser Verlag.

[4] Doan, H.T. (1985) Invariant curves for numerical methods. Quart. Appl. Math. 43, 385 - 393.

[5] Doedel, E.J., Kernevez, J.P. (1986) Software for continuation problems in ordinary differential equations with applications. Preprint

[6] Eirola, T.: Two concepts for numerical periodic solutions of ODE's, to appear in Appl. Math. & Comp.

[7] Eirola, T.: Invariant circles of one-step methods, submitted to BIT

[8] Guckenheimer, J., Holmes, Ph. (1983) Nonlinear oscillations, dynamical systems, and bifurcations of vector fields. Applied Mathematical Sciences Vol. 42, Springer Verlag.

[9] de Hoog, F.R., Weiss, R. (1980) An approximation theory for boundary value problems on infinite intervals. Computing 24, 227 - 239.

[10] Keener, J.P. (1981) Infinite period bifurcation and global bifurcation branches. SIAM J. Appl. Math. 41, 127 - 144.

[11] Lentini, J., Keller, H.B. (1980) Boundary value problems over semi-infinite intervals and their numerical solution. SIAM J. Numer. Anal. 17, 577 - 604.

[12] Markowich, P.A. (1982) A theory for the approximation of solutions of boundary value problems on infinite intervals. SIAM J. Math. Anal. 13, 484 - 513.

[13] Palmer, K.J. (1984) Exponential dichotomies and transversal homoclinic points. J. Diff. Equ. 55, 225 - 256.

Dr. Wolf-Jürgen Beyn, Fakultät für Mathematik der Universität Konstanz, Postfach 55 60, D-7750 Konstanz, FRG.

International Series of
Numerical Mathematics, Vol. 79

BIFURCATION IN ELLIPTIC SYSTEMS

Chris Budd,
Oxford University,
England.

1. Introduction

In this paper we shall examine bifurcations which occur in the semilinear elliptic system

$$\Delta u + \lambda(u + u|u|^{p-1}) = 0 , \quad u|\partial B = 0, \tag{1.1}$$

where B is the unit ball in the space $\mathbb{R}^3$. The system (1.1) has recently been the subject of much research stimulated by the paper by BREZIS & NIRENBERG [2] which studied the special case $p = 5$: the critical Sobolev exponent for $\mathbb{R}^3$. We shall study in this paper the two regimes $p > 5$ (supercritical) and $p \approx 5$ (near critical) and will show that in this parameter region problem (1.1) has a very rich bifurcation structure. This includes an infinity of fold bifurcations, and , for special values of $p > 5$ an infinity of symmetry breaking bifurcations from non positive solution branches. These properties of the solution structure are intimately connected with a reformulation of problem (1.1) as a dynamical system which, for the particular case $p = 5$, is a perturbed Hamiltonian system . We thus will demonstrate that as p passes through the value 5 the uniqueness properties of the solutions of problem (1.1) change significantly.

2. Results on existence and uniqueness.

If $p < 5$ the imbedding of the Sobolev space $H_0^1(B)$ into the space $L^{p+1}(B)$ is compact and the solutions of problem (1.1) are critical points of the functional

$$I(u) = \int_B [\ |\nabla u|^2/2 - \lambda(\ u^2/2 + |u|^{p+1}/(p+1))]\ d\Omega\ . \quad (2.1)$$

An application of standard theory from the calculus of variations given , for example , in AMBROSETTI & RABINOWITZ [1] ensures the existence of a minimising sequence for the functional I(u) and from this we may deduce the existence of non zero solutions for problem (1.1) for each positive value of λ . In Fig. 1 we give a bifurcation diagram for the radially symmetric solution branches.

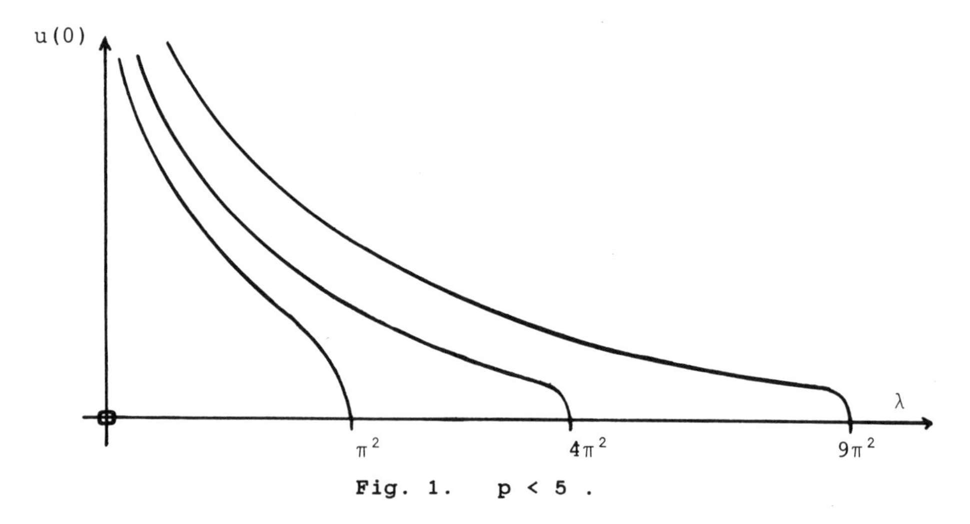

Fig. 1. $p < 5$.

When $p = 5$ however, the above imbedding is not compact and when $p > 5$ it is not even continuous, thus the existence results given above need no longer apply. Information about the solution structure may however be obtained by studying radially symmetric solutions. These satisfy the ordinary differential equation

$$u_{rr} + \frac{2}{r} u_r + \lambda(u + u|u|^{p-1}) = 0, \qquad (2.2)$$
$$u_r(0) = 0 \ , \ u(1) = 0.$$

When p = 5 the following result has recently been proved by BREZIS & NIRENBERG [2] and by BUDD [3].

Theorem 2.1 (i) There exists a positive solution of problem (2.2) iff $\pi^2/4 < \lambda < \pi^2$. This solution satisfies

$$u(0) \to \infty \quad \text{as} \quad \lambda \to \pi^2/4 \ .$$

(ii) There is a value $\delta > 0$ such that any solution of problem (2.2) which is non positive has $\lambda > \pi^2 + \delta$.

Theorem 2.1, together with numerical and asymptotic studies predicts the following bifurcation diagram for u(r) in this case.

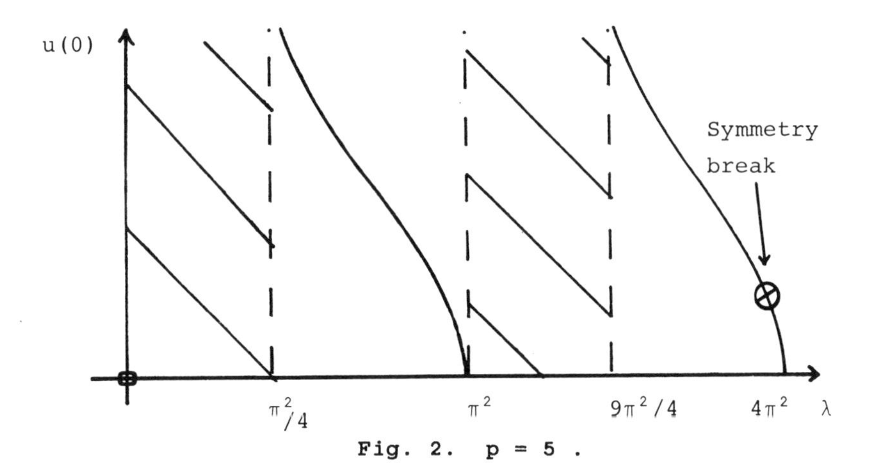

Fig. 2. p = 5 .

Here the shaded areas are regions for the parameter λ in which no solution exists. We indicate in this diagram the location of a symmetry breaking bifurcation point the existence of which is proven in BUDD & NORBURY [8] . We shall discuss these points further in section 4.

When $p > 5$ there exist values $\lambda_{\infty,k}$ of λ at which problem (1.1) has an infinity of solutions and an infinity of fold bifurcations as $\lambda \to \lambda_{\infty,k}$. The corresponding bifurcation diagram for these solutions is given in Fig. 3.

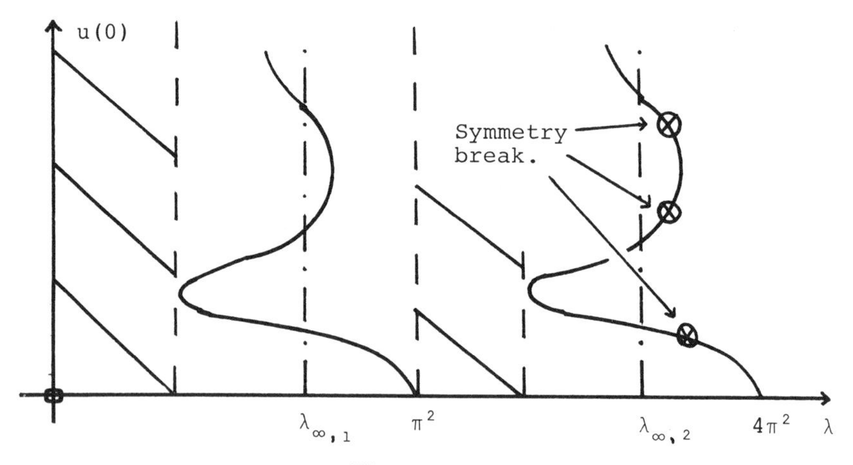

Fig. 3. $p > 5$.

Here we have indicated again the location of the symmetry breaking bifurcation points and observe that only the lowest of these may be obtained from a continuous deformation of the corresponding symmetry break when $p = 5$. The precise form of the bifurcation diagram is discussed in BUDD & NORBURY [7] , where it is further shown that for $\lambda = \lambda_{\infty,k}$, problem (1.1) has a weak singular solution $M(r)$ such that

$$M(r) \sim \kappa / (\lambda_{\infty,k}^{1/2}\, r)^{\alpha} \quad \text{as } r \to 0 , \qquad (2.3)$$
$$\text{where} \quad \alpha = 2/(p-1) \quad \text{and} \quad \kappa^{p-1} = \alpha(1-\alpha).$$

3. Numerical methods, fold bifurcations and dynamical systems.

When numerically solving problem (2.2) it is

convenient to use a variable other than λ as a continuation parameter to avoid problems with computing fold bifurcations. To do this we rescale problem (2.2) in the following manner. Let $s = \lambda^{1/2}r$, $v(s) = u(\lambda^{-1/2}s)$ and for each value of N let v(s) be a solution of the following initial value problem.

$$v_{ss} + \frac{2}{s} v_s + v + v|v|^{p-1} = 0, \qquad (3.1)$$
$$v(0) = N \ , \ v_s(0) = 0 \ .$$

We now use N as a continuation parameter and, if $v(\mu) = 0$ we may set $\lambda = \mu^2$. It is of interest to study the location of the fold bifurcations indicated in Fig. 3 when $p > 5$ as this helps us to understand how Fig. 3 deforms into Fig. 2 as $p \to 5$. The location of these bifurcations may be calculated by solving the following problem.

Find N such that $\varphi(\mu) = 0$ where $\varphi(s)$ is the solution of the differential equation problem

$$\varphi_{ss} + \frac{2}{s} \varphi_s + (1 + p|v|^{p-1})\varphi = 0 \ , \qquad (3.2)$$
$$\varphi(0) = 1 \ , \ \varphi_s(0) = 0 \ .$$

Here v(s) is the solution of problem (3.1) and $v(\mu) = 0$. The problem (3.2) was solved by using a rapidly convergent Newton-Raphson algorithm. It may be shown BUDD [4] that if the mth. fold in problem (2.2) occurs at the value $N = N_m$, then, as $p \to 5$, there exists a value A_m such that

$$N_m^4 \sim A_m \ (p-5)^{-(2m-1)}, \qquad (3.3)$$

where, in particular, $A_1 = 96$. This asymptotic formula is verified by the numerical calculations made using (3.2). To understand how these fold bifurcations arise we make use of the following change of variables described in JONES & KÜPPER [9].

Let $a = s^{\alpha} v$, $b = s^{\alpha+1} v_s$ and $t = \ln s$

where $\alpha = 2/(p-1)$. The problem (3.1) transforms into the following dynamical system.

$$da/dt = \alpha a + b ,$$
$$db/dt = (\alpha-1)b - a|a|^{p-1} - s^2 a , \qquad (3.4)$$
$$ds/dt = s .$$

We observe that this system has an invariant plane $s = 0$ and that when $s << 1$ the restriction of it to this plane is a small perturbation of (3.4). The system (3.4) has a singular point at the origin which is a saddle when $p > 3$ and a further singular point $\underset{\sim}{P}$ at $(a,b,s) = (\kappa,-\alpha\kappa,0)$. This point has an unstable manifold $W_u(t)$ corresponding directly to the singular function $M(s)$ defined in (2.3). This manifold leaves the plane $s = 0$ parallel to the s axis and it intersects the plane $a = 0$ at the points $s = \lambda_{\infty,k}^{1/2}$. Further, if we examine the restriction of the system (3.4) to the plane $s = 0$ we find that the point $\underset{\sim}{P}$ is a spiral repeller when $p < 5$, a centre when $p = 5$ and a spiral attractor when $p > 5$. If $p > 5$ then a solution of problem (3.1) corresponds to a solution trajectory of problem (3.4) which leaves the origin at time $t = -\infty$ and which then spirals around the manifold $W_u(t)$ before intersecting the plane $a = 0$ when $s = \mu$. The spiralling nature of this trajectory causes μ to oscillate about $\lambda_{\infty,k}^{1/2}$ and this behaviour gives rise to the structure observed in Fig. 3. The above argument may be proved rigorously using ideas from Shilnikov theory and details are given in BUDD [6]. If $p = 5$ we use different arguments based upon the Hamiltonian structure of (3.4) in this case and we may obtain the asymptotic relation (3.3).

4. Symmetry breaking bifurcations.

It is of considerable interest to study the location of symmetry breaking bifurcations from radially symmetric solution branches of problem (1.1) as these points are associated with a change in the stability of the solution and may also be used as an initial point in a numerical search for a nonsymmetric solution branch. If we define a symmetry breaking bifurcation point to be a point where a nonsymmetric solution branch intersects a symmetric one, then a necessary condition for the existence of such a point is that the following eigenvalue problem has a nontrivial solution.

$$\Delta\varphi + \lambda(1 + p|u|^{p-1})\varphi = 0 , \qquad (4.1)$$

$$\varphi|\partial B = 0 ,$$

where $u(r)$ is a solution of problem (2.2). To study this problem we set $\varphi(r,\theta,\phi) = R_\ell(r)\, Y_{\ell m}(\theta,\phi)$, where $Y_{\ell m}(\theta,\phi)$ is a Spherical Harmonic of degree ℓ . We introduce the scaling given in (3.1) and set $S_\ell(s) = R_\ell(\lambda^{-1/2}s)$. The following theorem then characterises the location of the symmetry breaking bifurcation points.

Theorem 4.1. Let $S_\ell(s)$ be the solution of the following ordinary differential equation problem

$$(S_\ell)_{ss} + \frac{2}{s}(S_\ell)_s - \ell(\ell+1)(S_\ell)/s^2 + (1+p|v|^{p-1})S_\ell = 0 ,$$

$$S_\ell(s) \to 0 \quad \text{as} \quad s \to 0 . \qquad (4.2)$$

Suppose now that $\mu_k(N)$ is the $k^{th.}$ zero of the function $v(s)$ and that $\alpha_{\ell j}(N)$ is the $j^{th.}$ zero of $S_\ell(s)$. Then

(i) a necessary condition for a symmetry breaking bifurcation is

$$\alpha_{\ell j}(N) = \mu_k(N) \quad , \qquad (4.3)$$

(ii) a sufficient condition for such a bifurcation is

$$d\,\alpha_{\ell j} / dN \neq d\,\mu_k / dN . \qquad (4.4).$$

This result is established in BUDD [5].

To numerically locate the existence of symmetry breaking bifurcation points we solve, for each triple (ℓ,j,k)

the problem (4.3) to find a value of N at which bifurcation may occur. The condition (4.4) then tests whether bifurcation does indeed occur. (We note that the results of SMOLLER & WASSERMAN [10] show that no such bifurcation may occur if $k = 1$ i.e if the function $v(s)$ is positive in the interval of interest.) In Fig. 4. we present a graph of the values of N for which bifurcation occurs for the triple $(\ell, j, k) = (2,1,2)$ and study how N varies as we alter the value of p.

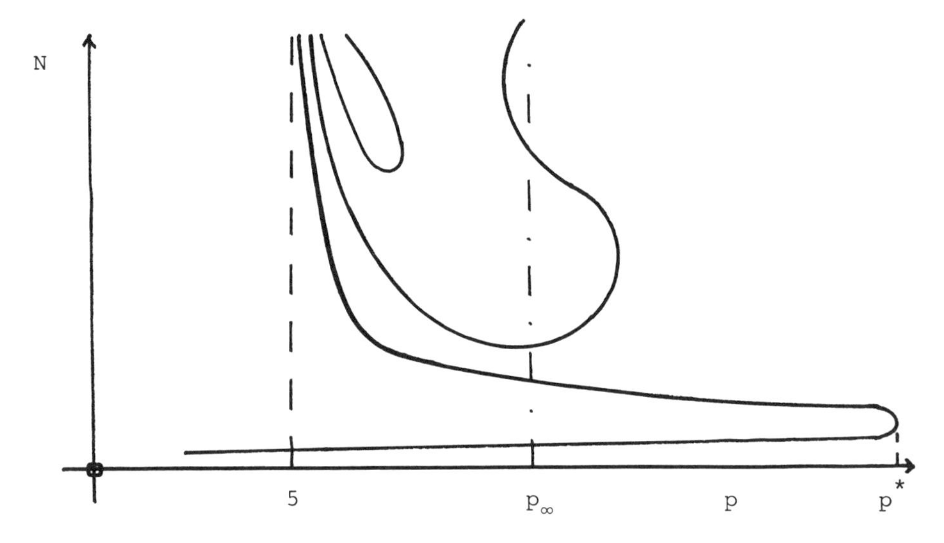

Fig. 4. The locus of symmetry breaking bifurcations.

Three important features are evident from this graph. Firstly, there exist points p^* at which two symmetry breaking bifurcation points coalesce. At these points the condition (4.4) fails and problem (1.1) may not have a branch of nonsymmetric solutions bifurcating from this point. Secondly, when $p = p_\infty$ there are an infinite number of symmetry breaking bifurcations. This point is closely associated with changes in the structural stability of the function $M(r)$ defined by (2.3). Thirdly, there exist a series of branches of symmetry breaking points which tend to infinity as p tends to 5 from above. The numerical evidence indicates that there exist values B_m such that if

(N_m, p) lies on the mth such branch then

$$N_m^4 \sim B_m \, (p-5)^{-(2m-2)} \quad . \tag{4.5}$$

By comparing the formula (4.6) with (3.3) we may see that the location of the symmetry breaking bifurcation points interlaces that of the fold points as $p \to 5$.

We conclude by noting that when $p > 5$ there exist a large number of nonsymmetric solutions which bifurcate from symmetric solution branches of problem (1.1).

5. References

[1] Ambrosetti,A. & Rabinowitz,P., (1973) Dual variational methods in critical point theory.J.Funct.Anal.14,349-381

[2] Brezis,H. & Nirenberg,L., (1983) Positive solutions of nonlinear elliptic equations involving critical Sobolev exponents. Comm. Pure Appl. Math., 36, 436-478.

[3] Budd,C. (1985) Comparison theorems for radially symmetric solutions of semilinear elliptic equations. Submitted to J. Diff. Eqns.

[4] Budd,C. (1986) Semilinear elliptic equations with near critical growth rates.Submitted to Proc.Roy.Soc.Edin.

[5] Budd,C. (1986) Symmetry breaking in elliptic systems. OUCL Nag report 86/7.

[6] Budd,C. (1986) Applications of Shilnikov methods to semilinear elliptic equations. OUCL Nag report 86/13.

[7] Budd,C. & Norbury,J. (1985) Semilinear elliptic equations with supercritical growth rates. Submitted to J.Diff.Eqns

[8] Budd,C. & Norbury,J. Symmetry breaking in semilinear elliptic equations with critical growth rates. In preparation.

[9] Jones,C. & Küpper,T. (1986) On the infinitely many solutions of a semilinear elliptic equation. SIAM Anal.

[10] Smoller,J. & Wasserman,A. (1984) Symmetry breaking from positive solutions of semilinear elliptic equations, Proc. 1984 Dundee conference on differential eqns. Springer.

Dr. Chris Budd, Numerical analysis group, Oxford University computing laboratory, 8-11 Keble Rd., Oxford, OX1 3QD , ENGLAND.

International Series of
Numerical Mathematics, Vol. 79

TRANSITION TO ASYMMETRIC CONVECTION ROLLS

by F. H. Busse
Theoretische Physik IV, Universität Bayreuth, FRG

1. Introduction

Hydrodynamic instabilities characterized by supercritical bifurcations often exhibit the property that the physically realized solution possesses a maximum of symmetries. The axisymmetric Taylor vortex and convection in the form of rolls in Rayleigh-Bénard layer are well known examples. Only the approximate translational symmetry of the axially extended coaxial cylinders or of the extended convection layer is broken by the onset of instability. Bifurcation theory has shown, however, that typically all symmetries of the equations describing the secondary flow can eventually be broken depending on the signs of terms introduced by the nonlinearities of the basic equations. Physical constraints usually prevent the occurrence of all possible signs of those terms even if there exist many possibilities for varying external parameters of the problem. But there remain some interesting cases in which additional symmetries are broken leading to unusual bifurcation phenomena.

The bifurcating solution with the highest degree of symmetry are usually those which can be calculated most easily. It is thus not surprising that solutions of lesser symmetry have often been overlooked, especially when they do not seem to be realized in experiments. In recent years the solutions of lesser symmetry have received increased attention, however, because it has become evident that they might be observed in parameter regimes not yet attained by the experiments. Convection in a layer heated from below offers particularly good opportunities because the Prandtl number and various deviations from the Boussinesq approximation provide a variety of parameters as a function of which terms in the bifurcation equations could change sign. In an early paper SEGEL (1962) investigated the interaction of two parallel convection rolls one of which is the subharmonic of the other and found that a low Prandtl number is benificial for the stability of the composite solution. This result was confirmed by the independent analysis of KNOBLOCH and GUCKENHEIMER (1983) who approached the problem from a different angle. In a recent paper (BUSSE and OR, 1986) another type of composite solution was found, which does not even exhibit a vertical plane of symmetry as Segel's solution does. The present paper presents an extension of the paper by Busse and Or which will be referred to by BO in the following and focusses the attention on the effect of asymmetries of the convection layer.

From the physical point of view solutions of lesser symmetry are of special interest since they may be associated with physical properties which are absent in the case of solutions of higher symmetry. An example is the strong differential rotation that develops when a transition from symmetric convection columns to asymmetric columns occurs in a cylindrical fluid annulus heated from the outside and cooled from within (OR and BUSSE, 1987). The centrifugal force acts as gravity in this case and the Coriolis force enforces an approximate two-dimensionality of the convection flow. Convection driven by centrifugal buoyancy in a rotating annulus is actually an ideal system for the experimental realisation of two-dimensional convection rolls. When the limit of a small annular gap is assumed and when the end plates bounding the annular fluid region in the axial direction are perpendicular to the axis, the equations governing convection rolls parallel to the axis are identical to those for rolls in Rayleigh-Bénard layer (BUSSE, 1970). The thin Ekman layers at the ends of the convection rolls have a lesser effect than the side walls in an ordinary convection layer. The problem studied by OR and BUSSE (1987) differs in that the end boundaries have conical shape and thus are inclined with respect to the equatorial plane of symmetry of the annulus. This configuration is of interest to planetary physics since it exhibits the major features governing convection in rotating spherical shells. According to the theory of Busse (1976) the zonal flows observed in the equatorial regions of the major planets are caused by the same mechanism responsible for the differential rotation in the convecting rotating annulus.

In the present paper the complications introduced by the end boundaries will not be analyzed. Instead we shall focus the attention on the effect of a small quadratic dependence of the density on the temperature.

2. Mathematical Analysis

The conventional treatments of Rayleigh-Bénard convection in a layer heated from below assumes a linear dependence of the density on temperature. In many applications the quadratic term in the relationship

$$\rho = \rho_0\{1-\alpha(T-T_0)-\beta(T-T_0)^2+\ldots\} \qquad (1)$$

must be considered, however. As in an earlier paper (BUSSE, 1967) we use the small parameter β as a representation for all variations of material properties which may cause an asymmetry of the convection layer. Using the same non-dimensional variables as in BO, we arrive at the following equation of motion for the stream function ψ,

$$\nabla^4\psi+R\partial_x\theta-P^{-1}(\frac{\partial}{\partial t}-\partial_z\psi\partial_x+\partial_x\psi\partial_z)\nabla^2\psi = \gamma\partial_x[2z\theta-\theta^2] \tag{2}$$

which differs from (2.1a) of BO only through the addition of the terms on the right hand side. The dimensionless parameter γ is defined by

$$\gamma = \frac{\beta(T_2-T_1)^2gd^3}{\nu\kappa} \tag{3}$$

where T_2-T_1 is the temperature difference accross the layer and the other symbols have their usual meaning. The equation for the deviation θ of the temperature from the static state is the same as in BO,

$$\nabla^2\theta+\partial_x\psi-(\frac{\partial}{\partial t}-\partial_z\psi\partial_x+\partial_x\psi\partial_z)\theta = 0 \tag{4}$$

and the boundary conditions also remain unchanged,

$$\psi = \partial^2_{zz}\psi = \theta = 0 \quad \text{at} \quad z = \pm\frac{1}{2} \tag{5}$$

We look for steady solutions of equations (2), (4) by considering γ as a small parameter. Instead of the single expansion in powers of the amplitude of convection used in BO we introduce a double expansion with γ as the second expansion parameter,

$$\begin{aligned} \psi &= \psi_0+\gamma\psi_{01}+\ldots+\psi_1+\gamma\psi_{11}+\ldots+\psi_2+\ldots \\ \theta &= \theta_0+\gamma\theta_{01}+\ldots+\theta_1+\gamma\theta_{11}+\ldots+\theta_2+\ldots \\ R &= R_0+\gamma R_{01}+\ldots+R_1+\gamma R_{11}+\ldots+R_2+\ldots \end{aligned} \tag{6}$$

Starting with expression (2.4) of BO for ψ_0,

$$\psi_0 = (A\sin\alpha x+B\sin(\beta x+\chi))\cos\pi z \equiv A\psi^{(1)}+B\psi^{(2)} \tag{7}$$

where α,β satisfy approximately the relationships

$$R_A \equiv (\pi^2+\alpha^2)^3\alpha^{-2} \simeq R_B \equiv (\pi^2+\beta^2)^3\beta^{-2} \simeq R_O \tag{8}$$

we find as equation for ψ_{01},

$$[\nabla^6 - R_O \partial^2_{xx}]\psi_{01} = -2z\partial^2_{xx}\psi_O - 4R_O^{-1}\partial_z\nabla^4\psi_O + R_{01}\partial^2_{xx}\psi_O \tag{9}$$

The solvability condition for this equation is satisfied with $R_{01} = 0$ and the solution can be written in the form

$$\psi_{01} = Af_A(z)\sin\alpha x + Bf_B(z)\sin(\beta x+\chi) \tag{10}$$

where f_A and f_B are antisymmetric functions of z. A non-vanishing contribution in the expansion for R is obtained when the solvability condition for the equations for ψ_{11}, θ_{11} is considered and the special case $\beta = 2\alpha$ is assumed,

$$\begin{aligned} R_{11}\langle\psi^{(n)}\partial_x\theta_O\rangle = -R_O\langle\theta^{(n)},(\partial_x\psi_{01}\partial_z-\partial_z\psi_{01}\partial_x)\theta_O+(\partial_x\psi_O\partial_z-\partial_z\psi_O\partial_x)\theta_{01}\rangle \\ +\langle\psi^{(n)},(\partial_x\psi_{01}\partial_z-\partial_z\psi_{01}\partial_x)\nabla^2\psi_O+(\partial_x\psi_O\partial_z-\partial_z\psi_O\partial_x)\nabla^2\psi_{01}\rangle \\ +\langle\psi^{(n)},\partial_x(2z\theta_1-\theta_O\theta_O)\rangle \end{aligned} \tag{11}$$

The angular brackets indicate the average over the fluid layer and the index n assumes the value 1 and 2. An evaluation of the left hand side of (11) indicates that it is proportional to $\cos\chi\cdot\delta(\beta-2\alpha)$ where δ denote the Dirac δ-function. Assuming that γ is of the order of the amplitude of convection, $(A^2+B^2)^{1/2}$, or smaller we combine the contributions R_1, and R_2 to the Rayleigh number and neglect higher order terms. We thus obtain equations (2.11) of BO supplemented by terms proportional to γ,

$$\{R-R_A-f_OA^2-(g_O+f_1(P))B^2+\gamma_1B\cos\chi\cdot\delta(\beta-2\alpha)\}A = 0 \tag{12.a}$$

$$\{R-R_B-g_OB^2-(f_O+g_1(P))A^2\}B = \gamma_2A^2\cos\chi\cdot\delta(\beta-2\alpha) \tag{12.b}$$

The coefficients γ_1 and γ_2 which include γ as factor have not been computed, but rough estimates indicate that they have the same sign as γ.

In contrast to the case of a symmetric layer treated in BO there exists only a single pure mode solution when $\beta = 2\alpha$ and $\chi = 0$,

$$B^2 = (R-R_B)/g_0, \qquad A = 0 \tag{13.a}$$

The other pure mode solution now corresponds to the limit when B is of the order γ which is assumed to be small,

$$A^2 = (R-R_A+\gamma_1 B)/f_0, \qquad B = \gamma_2 A^2 (R-R_B-(f_0+g_1)A^2)^{-1} \tag{13.b}$$

At the point, however, at which the denominator in the expression for B tends to vanish the expression for B loses its validity. Instead of solving equations (12) numerically it is more illuminating to see the changes introduced in the bifurcation diagram by a positive γ as sketched in figure 1. While the pure mode

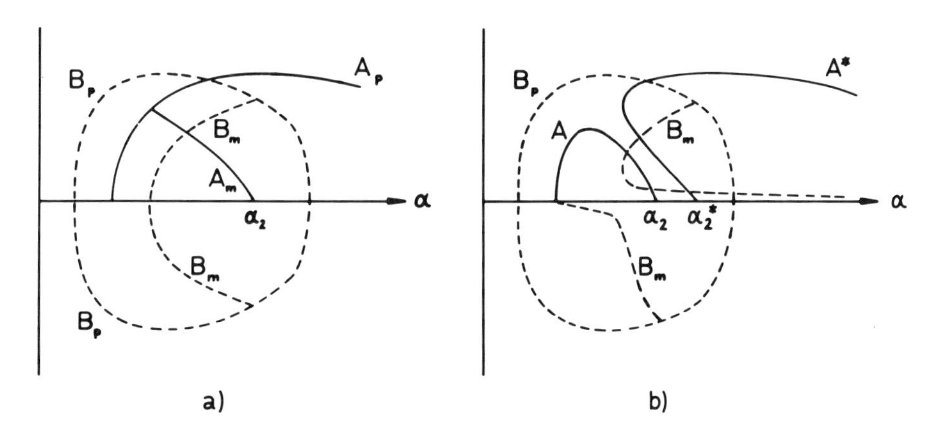

Fig. 1a. Bifurcation diagrams for pure solutions (A_p, B_p) and mixed solutions (A_m, B_m) in the case of a symmetric layer, $\gamma=0$. Figure b indicates the changes in the bifurcation diagram introduced by a finite asymmetry, $\gamma>0$.

(13.a) is changed little from the symmetric case despicted in figure 1a, the B-component of the mixed solution experiences a change analogous to the effect of imperfections on a pitchfork bifurcation. Instead of the pure solution for A and the mixed solution bifurcating from it in the symmetric case we find two branches for the A-component of the solutions. The changes introduced by the asymmetry of the problem have already been suggested by BO. But the variation of the amplitude B has not been indicated correctly in figure 8 of that paper.

3. Relationship to the Taylor Vortex Problem

The close relationship between Taylor vortices occuring between two coaxial differentially rotating cylinders and convection rolls in a Rayleigh-Bénard layer has long been noticed (see, for example, CHANDRASEKHAR, 1961). The Taylor vortex problem becomes symmetric with respect to the median plane of the cylindrical fluid layer in the limit when the gap width is small in comparison to the radius of the cylinders and when the angular velocities of the cylinders are nearly the same. When the difference of the angular velocities becomes of the same order of magnitude as the average angular velocity, asymmetric terms become important in the equations for the axisymmetric Taylor vortex which are analogous to the terms proportional to γ in equation (2) with positive γ. We therefore expect that the bifurcation properties discussed in this paper will also be apparent in the Taylor vortex case. The computations by Meyer-Spasche and Keller (1985) do indeed indicate the features of figure 1b. In their figure 1 the bifurcation points (4) and (5) correspond to the bifurcation points at α_2 and α_2^*, respectively, in figure 1b. The point at which $dA_*/d\alpha$ becomes infinite corresponds to the folds (6), (6') of Meyer-Spasche and Keller. Since they use the wavelength λ instead of α as abscissa and plot the radial velocity, the correspondence of the two bifurcation diagrams is not directly evident. But the comparison of the solution (7) with their numerical results indicates a qualitative correspondence in all respects except for the connection with the solution proportional to $\sin 3\alpha x$ which enters their computations at large λ because of the relatively high value chosen for the Reynolds number.

So far the discussion has been restricted to the case $\chi = 0$. As has been mentioned in BO, steady solutions in the interval $0<\chi<\pi/2$ do not obey the solvability conditions in the higher orders of the problem, as is evident from the numerical computations and the analytical study by Armbruster (private communication, 1986). A solution exists, however, for $\chi = \frac{\pi}{2}$ which does not seem to be affected by asymmetries of the layer to the order to which the problem has been considered here. It will be interesting to analyse the bifurcation properties of the analogous solution in the Taylor vortex case.

References

Busse, F.H. (1967) The stability of finite amplitude cellular convection and its relation to an extremum principle. J. Fluid Mech. 30, 625-649.

Busse, F.H. (1970) Thermal instabilities in rapidly rotating systems. J. Fluid Mech. 44, 441-460.

Busse, F.H. (1976) A simple model of convection in the Jovian atmosphere. Icarus 20, 255-260.

Busse, F.H., and Or, A.C. (1986) Subharmonic and asymmetric convection rolls. J. Appl. Math. Phys. (ZAMP) 37, 608-623.

Chandrasekhar, S. (1961) Hydrodynamic and hydromagnetic stability. Oxford Clarendon Press.

Knobloch, E., and Guckenheimer, J. (1983) Convective Transitions induced by a varying aspect ratio. Phys. Rev. A27, 408-417.

Meyer-Spasche, R., and Keller, H.B. (1985) Some bifurcation diagrams for Taylor vortex flows. Phys. Fluids 28, 1248-1252.

Or, A.C., and Busse, F.H. (1987) Convection in a rotating cylindrical annulus, II. Transitions to asymmetric and vacillating flow. J. Fluid Mech., 174, 313-326.

Segel, L.A. (1962) The nonlinear interaction of two disturbances in the thermal convection problem. J. Fluid Mech. 14, 97-114.

Prof. Friedrich H. Busse, Theoretische Physik IV, Universität Bayreuth, 8580 Bayreuth, FRG

International Series of
Numerical Mathematics, Vol. 79

WAVE PATTERNS IN COUPLED STATIONARY BIFURCATIONS WITH O(2)-SYMMETRY

Gerhard Dangelmayr
Universität Tübingen, FRG

Abstract

In this paper interactions between different steady states are discussed for systems that are equivariant under the symmetry group O(2). In a previous paper it was shown that the normal forms underlying these interactions are capable of producing primary and secondary bifurcations to single mode and mixed mode steady states, respectively, and tertiary bifurcations to travelling waves. By using averaging methods, it is shown here that also tertiary bifurcations to standing waves and quarternary bifurcations to modulated waves can appear.

1. Introduction

The bifurcations leading to complicated behaviour in physical systems are often accessible to a large extent analytically in a neighbourhood of multiple or degenerate bifurcations [6]. This is because near such a bifurcation the system can be described by a low-dimensional set of ordinary differential equations, resulting in a high reduction of the number of degrees of freedom, in particular if the system is governed by partial differential equations. If symmetries are present, still more complicated behaviour is possible [5], however, the presence of the symmetry usually imposes restrictions on the normal form associated with a multiple bifurcation, and therefore makes the study of equivariant systems particularly rewarding.

The fundamental codimension-1 (or primary) bifurcations are steady state and Hopf bifurcations. Codimension-2 degeneracies, then, occur in the case of certain mode interactions. In this paper, interactions between two different steady state modes in the presence of an O(2)-symmetry are considered. This problem has been discussed in [1] from a singularity theory point of view, and in [3] first steps towards an understanding of the dynamics have been made, showing that for certain ranges of the parameters the normal form is capable of producing tertiary bifurcations to travelling waves. The purpose of this paper is to review and to extend the results of [3]; in particular, it will be demonstrated that tertiary bifurcations to standing waves and even quarternary bifurcations to modulated waves can appear in the vicinity of such a degeneracy.

In Section 2 a simple model is introduced which is used as a guide throughout the paper, that is, all the results presented in subsequent sections will find their interpretation in terms of that model. In Section 3 some of the results of [3] concerning weak resonances are reviewed, including a description of the normal form and of the basic solutions -- trivial state, single mode and mixed mode steady states and travelling waves. Section 4 is devoted to a discussion of standing waves and of quarternary bifurcations to modulated waves. In the concluding Section 5 some remarks are made concerning problems that require further investigation.

2. A Model Equation with O(2)-Symmetry

The model I consider consists of a system of reaction diffusion equations for two chemical concentrations $U=(U_1,U_2)$. It is assumed that the spatial domain for the system is a circle so that, in a shorthand notation, the equation can be written in the form

$$\partial_t U = L(\partial_\Phi^2)U + N(U) \ , \tag{1}$$

where Φ is an angular variable, L is a linear operator containing the diffusion part and $N:\mathbb{R}^2\to\mathbb{R}^2$ is a nonlinear smooth mapping. The presence of the angular variable may be either due to a real circular geometry (a ring), or may be caused by imposing periodic boundary conditions on an infinitely extended system. Since L and N do not depend explicitly on Φ, we have equivariance against translations $\Phi\to\Phi+\Psi$, whereas the fact that L depends on ∂_Φ^2 implies equivariance against the reflection $\Phi\to-\Phi$. In other words, the equation (1) is equivariant under an action of the circle group O(2). Assume now that L and N depend also on some externally controllable parameters and that one of these parameters is varied. Typically, then, for a special value of this parameter, L possesses a nontrivial null space ($\det L(-m^2)=0$ for some integer m), giving rise to a steady state bifurcation. In virtue of the O(2)-symmetry, a generic steady state bifurcation leads here to a 2-dimensional null space V_m which is most conveniently parametrized in terms of a complex variable $z = x + iy \in \mathbb{C}$ in the form

$$U = U_m \mathrm{Re}(z e^{im\Phi}) \in V_m \,, \tag{2}$$

where $L(-m^2)U_m = 0$. The standard approach to studying the behaviour of (1) near the degeneracy is to perform a center manifold reduction [7] yielding a simple o.d.e. for z which is equivariant under the action $z \to e^{im\Phi}z$, $z \to \bar{z}$ of O(2), i.e., it has the form $\dot{z} = zf(|z|^2) = z(\alpha - a|z|^2 + \ldots)$. Here, α is an unfolding parameter that describes deviations of the control parameter from criticality and a is the leading coefficient determined by the nonlinearities.

Assume now that also another parameter is varied. Then, for certain critical values of this parameter the singularity described above may become higher degenerate and a codimension-2 bifurcation occurs. One possibility is that the null space is still 2-dimensional, but the coefficient a vanishes. This is a fairly simple situation and requires just the inclusion of a 5th-order term in the normal form. More interesting cases occur if the linear part encounters a higher degeneracy, giving rise to mode interactions. There are three cases to be distinguished: (i) $\det L(-m^2) = \mathrm{Tr}L(-m^2) = 0$, (ii) $\det L(-m^2) = \mathrm{Tr}L(-n^2) = 0$ with $\det L(-n^2) > 0$, $m \neq n$, and (iii) $\det L(-m^2) = \det L(-n^2) = 0$, $m \neq n$. The first and second cases correspond, respectively, to a double zero eigenvalue with a nilpotent linear part and to the coalescence of a Hopf and a steady state bifurcation. These singularities are discussed in [4] and [2]. The third case leads to interactions between two different stationary modes and is the subject of the subsequent sections.

3. Equivariant Vector Fields and Basic Solutions

In the case of steady state-steady state interactions, an arbitrary element of the null space of L can be parametrized by two complex variables $(z_1, z_2) \in \mathbb{C}^2$ via

$$U = \mathrm{Re}\{z_1 e^{im\Phi} U_m + z_2 e^{in\Phi} U_n\} \,, \tag{3}$$

where $L(-m^2)U_m = L(-n^2)U_n = 0$. It is assumed that m and n are relatively prime (otherwise, factor out the g.c.d.) and that $m < n$. The action of O(2) in the function space, where the solutions of (1) are defined, induces an action on

$\mathbb{C}^2$, given by

$$(z_1,z_2) \to (e^{im\Phi}z_1, e^{in\Phi}z_2), \quad (z_1,z_2) \to (\bar{z}_1,\bar{z}_2) . \tag{4}$$

As in the codimension-1 case discussed in Section 2, a center manifold reduction of (1) near the codimension-2 singularity yields a system of o.d.e.'s for (z_1,z_2) that is equivariant under the action (4) and whose linear part vanishes at the degeneracy. It is easy to see that a general smooth equivariant vector field has the form

$$\dot{z}_1 = p_1z_1 + q_1\bar{z}_1^{n-1}z_2^m , \quad \dot{z}_2 = p_2z_2 + q_2z_1^n\bar{z}_2^{m-1} . \tag{5}$$

Here, p_1,q_1,p_2,q_2 are general invariant functions, i.e., real smooth functions of the basic invariants $(|z_1|^2,|z_2|^2,\mathrm{Re}(z_1^n\bar{z}_2^m))$. For the problem of steady state-steady state interactions one performs a Taylor expansion of p_1,p_2,q_1,q_2 about the origin,

$$\begin{aligned} p_1 &= a|z_1|^2 + b|z_2|^2 - \alpha + \ldots , \quad q_1 = \mu_1 + \ldots \\ p_2 &= c|z_1|^2 + d|z_2|^2 - \beta + \ldots , \quad q_2 = \mu_2 + \ldots , \end{aligned} \tag{6}$$

where the coefficients a,b,c,d and μ_1,μ_2 have to satisfy certain non-degeneracy conditions [3], in particular both μ_1 and μ_2 must be nonzero. α and β are the unfolding parameters making the linear part of (5) non-degenerate. There are certain simplifications possible in (6) by means of appropriate rescalings and near-identity transformations, especially for (m,n)=(1,2), but I will not discuss the details here (see [3]).

Before presenting specific results, I first describe the basic solutions in terms of the general form of (5). Clearly, there is always the trivial solution $T: z_1 = z_2 = 0$, which is preserved under all group operations. This solution has two pairs of equal eigenvalues, i.e., there are two signs in the stability assignment. Two other steady states follow also immediately from (5), namely,

$$S_1 : |z_1|^2 \neq 0, \quad z_2 = 0 \quad (\text{if } m>1); \quad S_2 : |z_2|^2 \neq 0, \quad z_1 = 0 ,$$

with equations $p_1(|z_1|^2,0,0) = 0$ and $p_2(0,|z_2|^2,0) = 0$, respectively. These steady states are determined up to an arbitrary phase as a consequence of the group equivariance. They are refered to as single mode solutions because for the model introduced in the preceeding section they correspond to a spatial pattern associated with one of the two critical wave numbers. In order to obtain information about other special solutions of (5) satisfying $z_1 z_2 \neq 0$, it is convenient to introduce polar coordinates,

$$z_1 = xe^{i\phi}, \quad z_2 = ye^{i\psi}; \quad x,y > 0 ,$$

and a "mixed phase"

$$\theta = n\phi - m\psi$$

which is fixed by the rotations of O(2). Noting that $\mathrm{Re}(z_1^n \bar{z}_2^m) = x^n y^m \cos\theta$, (5) reduces to a closed system for (x,y,θ), given by

$$\begin{aligned} \dot{x} &= p_1 x + x^{n-1} y^m q_1 \cos\theta \\ \dot{y} &= p_2 y + x^n y^{m-1} q_2 \cos\theta \\ \dot{\theta} &= -(nq_1 x^2 + mq_2 y^2) x^{n-2} y^{m-2} \sin\theta , \end{aligned} \tag{7}$$

whereas the rates of changes of the individual phases are

$$\dot{\phi} = -q_1 x^{n-2} y^m \sin\theta , \quad \dot{\psi} = q_2 x^n y^{m-2} \sin\theta .$$

Observe that the planes $\theta = 0$ and $\theta = \pi$ are invariant under the flow of (7). In particular, if q_1 and q_2 have equal signs, the flow of the vector field (7) evolves always towards one of these planes. Within these planes there are steady states, denoted by S_o and S_π if $\theta = 0$ and $\theta = \pi$, respectively, which are determined by the equations

$$p_1 \pm q_1 x^{n-2} y^m = 0 , \quad p_2 \pm q_2 x^n y^{m-2} = 0 .$$

Both amplitudes x and y are nonzero and $\dot{\phi} = \dot{\psi} = 0$ along S_o, S_π so that we have true steady states of (5). I refer to S_o and S_π as mixed mode steady states because they correspond to a superposition of the single mode states. The reduced system (7) possesses also another type of steady state, determined by

$$nq_1 y^2 + mq_2 x^2 = 0$$

$$p_1 + q_1 x^{n-2} y^m \cos\theta = 0 , \quad p_2 + q_2 x^n y^{m-2} \cos\theta = 0 ,$$

which is denoted by TW. Actually, TW occurs in pairs $(\theta = \pm\theta_o)$ because (7) possesses a reflection symmetry $\theta \to -\theta$. Along TW, $\dot{\phi}$ and $\dot{\psi}$ are constant but nonzero, so that we have oscillations of z_1 and z_2 around the origin with constant frequencies. The frequency ratio has the simple form $\dot{\phi}/\dot{\psi}=m/n$, i.e., TW corresponds to a periodic solution. For the model of Section 2, TW corresponds to a rotating mixed mode pattern. In other words, TW is a rotating or travelling wave. The two copies of TW correspond to opposite rotation directions.

For the vector field (5) with the expansion (6), a huge number of cases has to be distinguished concerning different choices of (m,n) and of the range of the coefficients in (6). I will not go into details here (see [3]), but mention that the cases m = 1 (strong resonance) and m > 1 (weak resonances) are radically different. For each case one has to construct a stability diagram, that is, the plane of unfolding parameters is divided into a number of regions separated by lines emanating from the origin, where subordinate codimension-1 bifurcations take place. In each region a certain configuration of solutions with specific stability assignments occurs. A complete discussion for the solutions described above is given in [3] for (m,n)=(1,2) and $2 \leq m < n$. To illustrate the main features of the stability diagrams, I focus from now on on the case $4 \leq m < n$ with $a=\mu_1=-\nu$, $b=-\nu\rho$, $c=\kappa$, $d=\mu_2=1$ where ν,ρ,κ are positive and satisfy $\rho\kappa>1$ and $\nu^2>m/n$. The stability diagram for this case is shown in Fig. 1. In this figure, the existence domain of a particular solution is

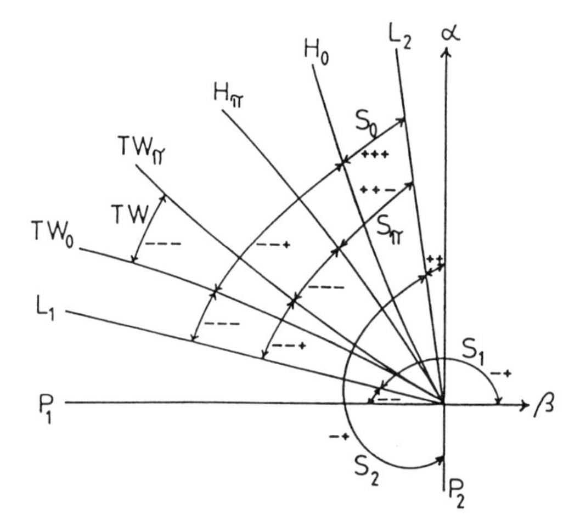

Fig. 1. Stability diagram for a particular subcase of $4 \leq m < n$. The diagram includes existence and stabilities of the solutions S_1, S_2; S_o, S_π and TW.

marked by a curve passing through it and connecting the two bifurcation lines where the solution is created. Each curve is divided into pieces, separated by arrows, on which the branch possesses a particular stability assignment. The lines P_j (j=1,2) are primary bifurcation lines where S_j branches off T. Along the secondary bifurcation lines L_1 and L_2, both mixed mode branches S_o, S_π bifurcate from S_1 and S_2, respectively. The two lines TW_o and TW are tangent at the origin and denote tertiary bifurcation lines where TW is created. Finally, along the lines H_o and H_π, the mixed mode solutions S_o and S_π undergo a Hopf bifurcation leading to limit cycles in the invariant planes $\theta = 0$ and $\theta = \pi$, respectively. The existence domains of the limit cycles -- denoted by SW_o and SW_π -- are not included in Fig. 1, but will be discussed in Section 4. The basic state T has assignments (-,-) in $\{\alpha < 0, \beta < 0\}$, (+,+) in $\{\alpha > 0, \beta > 0\}$ and (+,-) in $\{\alpha > 0, \beta < 0\}$ and $\{\alpha < 0, \beta > 0\}$.

4. Bifurcations to Standing and Modulated Waves

The limit cycles SW_o, SW_π in the invariant planes are pure amplitude oscillations ($\dot\phi, \dot\psi = 0$), i.e., in the model of Section 2 they correspond to pulsating mixed mode patterns which are commonly called standing waves. The behaviour of the SW-solutions is more difficult to analyze than that of the solutions described in the preceeding section. First, one sets $\beta = -\varepsilon^2(\nu + \kappa)$, $\alpha = (1 + \nu\rho)\varepsilon^2 + \gamma\varepsilon^4$, γ being a yet undetermined parameter, and rescales the variables according to $(x,y) \to (\varepsilon x, \varepsilon y)$, $t \to t/\varepsilon^2$, considering ε as a small parameter. The system (7), then, becomes

$$\begin{aligned} \dot x/\nu &= -x^3 - \rho x^2 y + (1+\nu\rho)x + \varepsilon^2 x(\gamma - e x^4) - \varepsilon^{n+m-4} x^{n-1} y^m \cos\theta + \ldots \\ \dot y &= \kappa x^2 y + y^3 - (\nu+\kappa)y + \varepsilon^{n+m-4} x^n y^{m-1} \cos\theta + \ldots \\ \dot\theta &= \varepsilon^{n+m-4}(n\nu y^2 - m x^2) x^{n-2} y^{m-2} \sin\theta + O(\varepsilon^{n+m-3}) \ , \end{aligned} \tag{9}$$

where the dots in the $\dot x$-equation represent terms of the form $xf(x^2,y^2)O(\varepsilon^4) + O(\varepsilon^{n+m-3})$, and analogously for the $\dot y$-equation. The term $e\varepsilon^2 x^5$ results from taking a fourth order term $e|z_1|^4$ in the expansion of p_1, Eq. (6), into account. Some inspection shows that all other quartic terms in p_1 and p_2 can be removed by means of a local coordinate transformation and a proper time rescaling. The important feature of (9) is that for $\varepsilon = 0$ (unperturbed system), the (x,y)-equations are integrable and can be written in the form

$$\dot x = -(\nu/2\sigma)x^{1-r}y^{1-s}H_y, \quad \dot y = (\nu/2\sigma)x^{1-r}y^{1-s}H_x \tag{10}$$

(subscripts denote partial derivatives), where $1/\sigma = \rho\kappa - 1$, $r = (2\sigma/\nu)(\kappa + \nu)$, $s = 2\sigma(1 + \rho\nu)$, and the Hamiltonian H is given by

$$H(x,y) = x^r y^s \{x^2/(1+\rho\nu) + y^2/(\kappa+\nu) - 1\} . \tag{11}$$

The form (10) for the unperturbed (x,y)-system of (9) has been obtained by Guckenheimer [6] in the context of Hopf-Hopf interactions without symmetry. The integrability of the unperturbed system is a consequence of the fact that, if only cubic terms are taken into account, the system (7) encounters a Hopf bifurcation along the straight half-line $(\kappa + \nu)\alpha + (1 + \rho\nu)\beta = 0$, $\beta < 0$. The presence of the $\cos\theta$-terms in (7) splits this single Hopf line into the two lines H_o and H_π of Fig. 1. The phase portrait of the unperturbed system contains a family of closed orbits parametrized by a constant energy, $H = E$, and bounded by a heteroclinic orbit (see [6,7]). Thus, in order to determine closed orbits of the unperturbed system ($\varepsilon \neq 0$), one can apply the averaging method [7], yielding an averaged system of o.d.e.'s for the energy E and θ. This system has the form

$$\begin{aligned} \dot{E} &= \varepsilon^2\{\gamma P(E) - eR(E)\} - \varepsilon^{n+m-4}Q(E)\cos\theta + \ldots \\ \dot{\theta} &= \varepsilon^{n+m-4}\{n\nu A(E) - mB(E)\}\sin\theta + O(\varepsilon^{n+m-3}) , \end{aligned} \tag{12}$$

where the dots represent here terms of the form $g(E)O(\varepsilon^4) + O(\varepsilon^{n+m-3})$. The functions R(E) etc. in (12) are all nonnegative. They are defined in terms of certain integrals over the level curves $H = E$, which will not be written down explicitly. Recall that γ describes deviations from the Hopf line in the (α,β)-plane so that (12) contains one free parameter. The averaged system (12) has not yet been analyzed in detail, however, the structure of (12) allows one to draw some general conclusions. First, one observes that the lines $\theta = 0$ and $\theta = \pi$ are invariant under the flow. Within each of these lines there is a steady state, given to leading order by $\gamma = eR(E)/P(E)$. This steady state takes values of E in the range $E_{min} \leqq E \leqq E_{max}$, $P(E_{min}) = 0$, where E_{min} corresponds to the minimum of the Hamiltonian and E_{max} is the value of H along the heteroclinic orbit of the unperturbed system. For the perturbed (x,y,θ)-system (9), this steady state corresponds to a standing wave solution. Hence, when γ is varied, a SW-branch is created in each invariant plane at a Hopf bifurcation from the mixed mode steady state and disappears in a heteroclinic connection. The phase portraits for this sequence are shown in Fig. 2 (see also [6,7]).

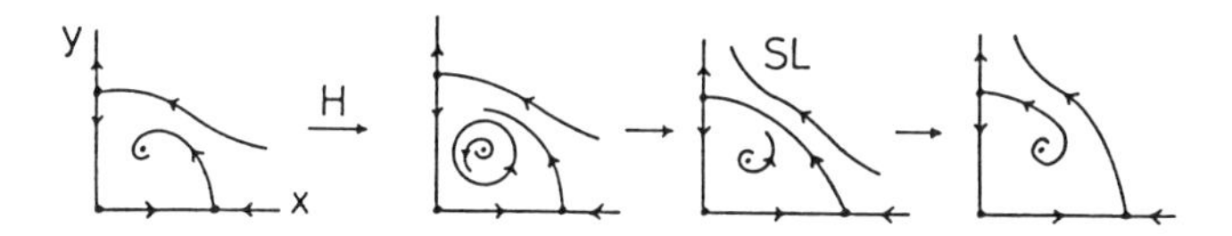

Fig. 2. Creation and disappearance of a limit cycle in an invariant plane.

If there exists E_o such that $E_{min} < E_o < E_{max}$ and $n\nu A(E_o) = mB(E_o)$, then the SW-solution becomes unstable with respect to phase variations and a pitchfork bifurcation for periodic orbits occurs where a pair of periodic orbits with $\cos^2\theta < 1$ is created from a limit cycle in an invariant plane. These new periodic orbits are denoted by MW because for the model of Section 2 they correspond to modulated waves. There are three frequencies associated with MW, namely, $\dot{\phi},\dot{\psi}$ and the frequency of the amplitude oscillations. However, to leading order in ε, one obtains $\dot{\phi}/\dot{\psi} = m/n$ so that MW corresponds to a 3-torus foliated by a family of 2-tori in the original (z_1,z_2)-system.

The discussion above shows that the stability diagram of Fig. 1 has to be completed, for a certain range of the coefficients ν,ρ,κ, by the subdiagram of Fig. 3. In this figure, PP and SL denote the pitchfork bifurcation of periodic orbits and the heteroclinic bifurcation of the standing

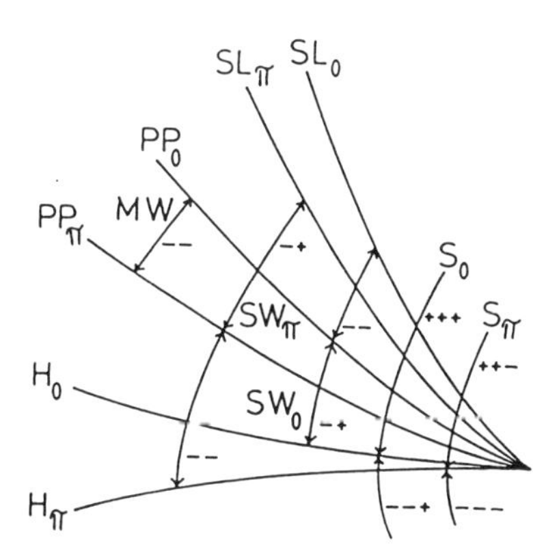

Fig. 3. Completion of the stability diagram of Fig. 1.

waves, respectively. The subscripts refer to the invariant planes in which these bifurcations take place. To leading order, they are given by $\gamma = eR(\bar{E})/P(\bar{E})$ where $\bar{E} = E_o$ for PP and $\bar{E} = E_{max}$ for SL. The presence of the $\cos\theta$-term in (12) splits PP and SL into two lines PP_π, PP_o and SL_π, SL_o, respectively, which are all tangent to H_o and H_π at the origin. The MW bifurcation

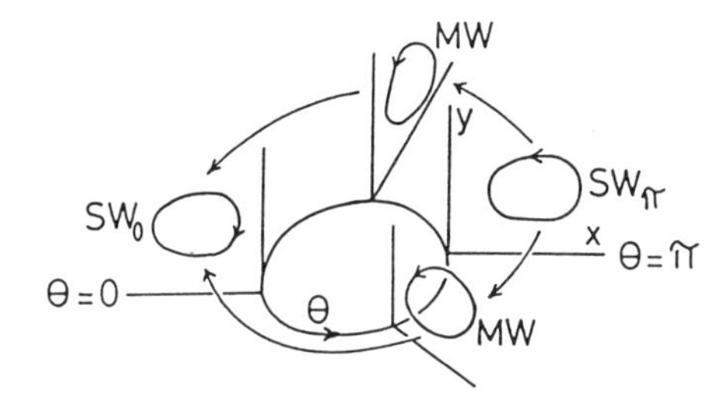

Fig. 4. A pair of MW-solutions branches off SW_π and collapses onto SW_o when the region between PP_π and PP_o in Fig. 3 is traversed from below.

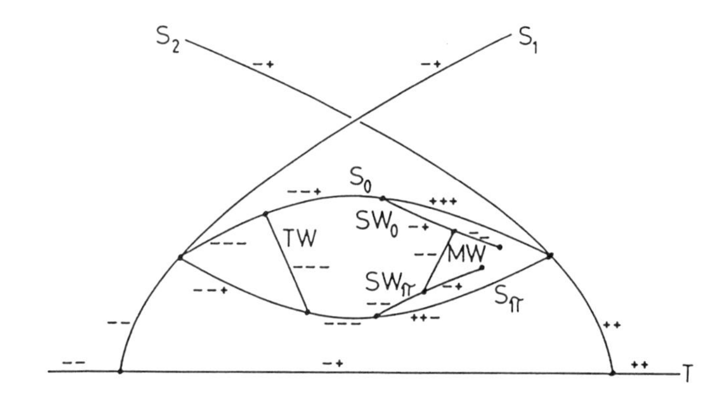

Fig. 5. A bifurcation diagram corresponding to the stability diagram of Fig. 1 completed by Fig. 3.

is illustrated in Fig. 4. In Fig. 5 the bifurcation behaviour corresponding to Figures 1 and 3 is sketched schematically in terms of a bifurcation diagram that is obtained by traversing the (α,β)-plane along a path passing from the region $\{\alpha < 0,\beta < 0\}$ through $\{\alpha > 0,\beta < 0\}$ into the region $\{\alpha > 0,\beta > 0\}$.

5. Conclusions

In this paper a qualitative analysis was presented for steady state-steady state interactions in the presence of O(2)-symmetry which extends the results of [3] to include also bifurcations to standing and modulated waves. The precise regime of the coefficients ν,ρ,κ where the stability diagram of Fig. 3 holds still remains to be determined. Also, the dynamics associated with the modulated waves for the full 4-dimensional phase space have not yet been analyzed. Because MW corresponds to a 3-torus foliated by 2-tori, there is the possibility of lock-in behaviour on a 2-torus. These problems will be discussed at another occasion.

Acknowledgements. I am grateful to R. Cushman, J. Guckenheimer and J. Sanders for helpful discussions. This work was supported in part by the Stiftung Volkswagenwerk, F.R.G.

References

[1] E. Buzano & A. Russo, "Non-axisymmetric post-buckling behaviour of a complete thin cylindrical shell under axial compression", to appear in Dynamics and Stability of Systems (1986).

[2] P. Chossat, M. Golubitsky & B. Keyfitz, "Hopf-Hopf mode interactions with O(2)-symmetry", preprint (1986).

[3] G. Dangelmayr, "Steady-state mode interactions in the presence of O(2)-symmetry", to appear in Dynamics and Stability of Systems (1986).

[4] G. Dangelmayr & E. Knobloch, "The Takens-Bogdanov bifurcation with O(2)-symmetry", to appear in Phil. Trans. Roy. Soc. Lond. A (1986).

[5] M. Golubitsky, I. Stewart & D. Schaeffer, "Singularities and groups in bifurcation theory", Springer-Verlag, to appear 1987.

[6] J. Guckenheimer, "Multiple bifurcation of codimension two", SIAM J. Math. Anal. **15** (1984), 1-49.

[7] J. Guckenheimer & P. Holmes, "Non-linear oscillations, dynamical systems and bifurcations of vector fields", Springer (1983).

Institut für Informationsverarbeitung
Köstlinstr. 6
D-7400 Tübingen
Fed. Rep. of Germany

International Series of
Numerical Mathematics, Vol. 79

DETERMINATION OF BIFURCATION POINTS AND CATASTROPHES FOR THE BRUSSELATOR MODEL WITH TWO PARAMETERS.

Bart De Dier and Dirk Roose

Dept. of Chemical Engineering
Katholieke Universiteit Leuven.
Leuven - Belgium.

1. Introduction.

In this paper we will present an investigation on some of the features of the Brusselator model. The early origin of this scheme returns to the begin of 1950 when Turing was investigating whether the behaviour of biological systems, in casu an embryological system, could be translated into a mathematical model. He detected that chemical reaction in combination with diffusion, ensuring the transport, are principal mechanisms to explain for example self organization, cell formation, cell division and other phenomena in morphogenetics. The proposed mathematical model was quite complex [1]; 10 reactants were involved together in 8 reaction steps; however it possessed some physical inconveniencies as for example negative concentrations in some cases. Later, Prigogine and his coworkers [2,3] were investigating systems based on nonlinear interactions, because it is possible that such systems can produce highly symmetric structures due to their self-organizing capacities. They claimed that a better and functional organization of a reacting system increases the intensity of the related process: in chemical reaction systems for example reaction rates increase. Prigogine and Nicolis compare herefore the non-equilibrium system with a company, where a spacial or temporal reorganization of the work leads to a higher efficiency. Since they knew that Turings original model possessed such features, they corrected it and derived a new reaction-scheme which is known in litterature as the Brusselator [3]. It is based on the following four chemical reaction steps between the initial components A and B, the intermediates X and Y, and the reaction products D and E.

$$
\begin{aligned}
A &\Longleftrightarrow X \\
2X + Y &\Longleftrightarrow 3X \\
X + B &\Longleftrightarrow Y + D \\
X &\Longleftrightarrow E
\end{aligned}
$$

which gives the following overall reaction

$$A + B \Longleftrightarrow C + D$$

X and Y are involved in an auto- and crosscatalytic reaction, which are responsible for the nonlinear interactions in this scheme.

In the second part of this paper we will detail on the numerical aspect of the continuation of turning points for the Brusselator model. Results of these calculations will be presented in a third part.

2. Numerical aspects of the continuation of turning points.

2.1 Computation and continuation of regular solutions.

For the four reaction model of Prigogine, taking into account that only diffusion ensures mass transport and dropping the reverse reactions, the following one-dimensional steady-state reaction-diffusion equations can be derived:

$$\frac{d^2A}{dz^2} = \frac{L^2}{D_A} A \tag{1}$$

$$\frac{d^2X}{dz^2} = -\frac{L^2}{D_X} ((B+1)X + X^2Y - A) \tag{2}$$

$$\frac{d^2Y}{dz^2} = -\frac{L^2}{D_Y} (X^2Y - BX) \tag{3}$$

$$\frac{d^2B}{dz^2} = -\frac{L^2}{D_B} BX \tag{4}$$

Here D_A, D_B, D_X and D_Y represent the respective diffusion-coefficients of the components, while A, B, X and Y stand for their concentrations; L is the parameter of the scheme and indicates the length of the one-dimensional reactor; z is the dimensionless space coordinate and varies between 0 and 1.

Still some boundary conditions have to be defined in order to solve the equations (1) - (4). Since A and B are initial components, their supply to the reacting system has to go on continuously; so Dirichlet boundary conditions can be chosen for A and B, while for X and Y as well Dirichlet as von Neumann boundary conditions can be used. We have chosen von Neumann conditions here, more specifically zero flux boundary conditions.

While the solution of (1) can be obtained analytically, the differential equations for X, Y and B, can be solved by using a discretization of the finite difference type. Both second and fourth order discretizations are available in literature [5,6]; here a fourth-order method has been employed since in this case the number of mesh points can be kept lower which reduces computer-time expenditure; while according to Hildebrandt [5] a fourth order discretization can be given for the zero flux boundary conditions of X and Y.

This set of nonlinear algebraic equations, accompanied with the boundary conditions , which will be denoted further on as F(u,L) = 0, has been solved by an iterative technique, i.e. the Newton-Raphson method, to calculate the regular solutions. Afterwards a continuation was carried out in function of the parameter L. The continuation package used herefore was the Pitcon-code of Rheinboldt and Burkardt [7]. In this way the bifurcation diagrams of figs. 1 and 2 have been computed [4,8]. There X(0), the concentration of X at the left border point, is depicted as a function of the reactor length L. Only these parts of the figs. which are relevant here are depicted.

2.2 Computation and continuation of turnig points.

Starting from the diagram of fig. 1, limit points have been calculated. The determining system for a turning point in a one-parameter problem, according to Seydel [9] and Moore and Spence [10] is given by

$$
\begin{aligned}
F(u,L) &= 0 \\
F_u(u,L)\mu &= 0 \\
\mu_{i,r} &= 1
\end{aligned}
\tag{5}
$$

Moore and Spence proved that a turning point is an isolated solution of (5), and however the dimension of the system equals 2m + 1 (where m is the dimension of F = 0) , they also proved that (5) can be solved by 1 LU decomposition of an m*m matrix and 4 backsubstitutions.

These limit points can be continued in a two parameter space. As second parameter a choice can be made of the diffusioncoefficients, the border concentrations of A and B etc. We have chosen the diffusioncoefficient of A as the second parameter since decreasing its value would lead to a tremendous drop in possible solutions because the minimum in the A-profile quickly becomes smaller for decreasing values of D_A. In this way we expected to detect a lot of high order singularities.

As for the continuation of regular solutions, the continuation of turning points also was percieved using the Pitcon-code. In each continuation step, a set of linear equations of the following type has to be solved to compute the Newton-correction:

$$\begin{bmatrix} A & b \\ c^t & d \end{bmatrix} \begin{bmatrix} f \\ g \end{bmatrix} = -R \qquad (6)$$

where A is a (2m+1)*(2m+1) matrix, c, b and f are arrays of dimension (2m+1) and d and g are scalars. To solve this system in an economical way, a combination of two numerical methods has been used, as has been suggested by Spence & Werner [14]. At first, with the block elimination algorithm of Keller [10], it is possible to eliminate the last row and column of the Jacobian (6) and hence only work with the matrix A. The system then can be solved by 1 LU decomposition of A and by two backsubstitutions. Then, using the algorithm of Moore and Spence we already referred to above, the remaining system can be solved by 1 LU decomposition of an m*m system and 4 backsubstitutions. The overall operation count then is 1 LU decomposition of an m*m system and 8 backsubstitutions. More information on this can be found in [10,11,12,13]. This technique implies that only a small amount of extra work has to be carried out, to continue turning points in comparison with the continuation of regular solutions.

3. Results.

Observing the diagrams, depicted in fig. 1 for D_A = 0.5 and fig. 2 for D_A = 0.02, it is obvious that for the larger D_A the number of branches and limit points is a lot larger as when D_A = 0.02; which means that either a high number of singularities can be detected. Therefore we started first by calculating the turning points shown on fig. 1, using the method present in the Pittsburgh Continuation code; afterwards a continuation of these limit points was percieved employing the techniques explained in par. 2.

3.1 The symmetric branches.

Observing the trivial branch T for D_A = 0.5 and for D_A = 0.02 learns that for a smaller diffusioncoefficient this branch has no turning point at all, while for the larger value of D_A several limit points (12) occur in the area $0 < L < 1.1$. This means that all these turning points should end and disappear in a singularity of higher order, i.e. a bifurcation point, an isola formation point or a cusp.

At first, let us consider the first two turning points T_1 and T_2. From the shape of the branch T at the two different values of D_A it is expectable that they will coincide in the same cusp and in fact form a classic couple of turning points. This, however, did not happen. As can be seen on fig. 3,where the continuation curves for T_1 and T_2 are displayed, the two limit points drift away from one another until T_1 reaches a singularity at $D_A = 0.0316$ after which curve T follows a shape as branch T for $D_A = 0.02$ without any turning points. The shape of the continuation curves starting at other turning points of the trivial branch T is of the same kind. There are however still two things to remark: the continuation branch from T_6 contains two bifurcation points and two neighbour-cusps while the curves starting at T_8 and T_9 end in the same cusp as well for increasing as for decreasing value of D_A.

What concerns the branch M, we could report that the curve starting at M3 also contains a pair of neighbour-cusps for which a continuation in a three-parameter space certainly would be interesting.

The continuation curves for branches T and M can be found on figs. 3 and 4.

3.2 The asymmetric branches.

Let us first note that for this model and the applied boundary conditions, turning points of curves with asymmetric profiles always have to appear in pairs, except in the case of a bifurcation-turning point or in other words a pitchfork bifurcation point. The reason for this is that the equations (1) - (4) and the zero flux boundary conditions are invariant for a substitution of z by (1 - z) and so an asymmetric profile which is rotated around a vertical axis through $z = 0.5$ also will give a valid solution of (1) - (4). This implies that the continuation curves in a two-parameter space of such a pair of turning points will coincide.

The continuation curves of some of the asymmetric branches are depicted in fig. 5; the following remarks can be made here:

- The turning points of branch A for $D_A = 0.5$ and of branch A for $D_A = 0.02$ correspond to each other. As an example the limit point A'_2 corresponds to the limit point A''_2 (see fig. 5a).
- On the other hand no analogue existed for branch C of fig. 1 in case of $D_A = 0.02$. This is illustrated, for example by the two continuation curves through C_2 and C_4 (fig. 5a) which have a high order singularity before

D_A reaches 0.02.

- Branch K gives an example of limit points lying on the same branch which coincide also in the same singularity; nl. the turning points K_2 and K_1 are connected by means of a continuation curve through the part of the (D_A, L) plane where D_A is larger than 0.5 (see fig. 5a).
- Also for asymmetric branches, some interesting continuation curves were calculated. Consider for example the curves through J_4 (fig. 5b), and R_3 (fig. 5c). There we find a few pairs of neighbour-cusps which probably will lead to swallow-tail singularities.
- Finally, a very complex continuation curve was calculated starting from the limit point N_1 on branch N (D_A = 0.5, fig. 5d). First it passes a single cusp singularity, then for decreasing length a pair of neighbour-cusps after which length increases again to pass another pair of neighbour-cusps, while finally another three singularities were detected. This set of cusps certainly has to lead to some higher order singularities; a further continuation in a three- parameter space will be necessary to point this out.

Ir. B. De Dier, Department of Chemical Engineering, Katholieke Universiteit Leuven,De Croylaan 2, 3030 Leuven, Belgium.

Dr. Ir. D. Roose, Department of Computer Science, Katholieke Universiteit Leuven, Celestijnenlaan 200A, 3030 Leuven, Belgium.

References.

[1] Turing, A., Phil. Trans. Roy. Soc. London, B.237, 37, 1952

[2] Erneux, T., Hiernaux, J., Nicolis, G., Bull. of Math. Biology, 40, 771, 1978

[3] Nicolis, G., Prigogine, I.,Selfoganization in non-equilibrium systems, J. Wiley & Sons, New York, 1977

[4] De Dier, B., Engineering Thesis, Dept. of Chem. Eng., Katholieke Universiteit Leuven, Belgium, 1984

[5] Hildebrandt, F.B., Finite difference equations and simulations, Prentice Hall, New Jersey, 1968

[6] Kantorovich, L.V., Krylov, V.I., Approximate methods of higher analysis, P.Noordhoff, Groningen,The Nederlands, 1964

[7] Rheinboldt, W.C., Burkhardt, J.V., TR ICMA 81-30, Inst. for Comp. and Appl. Math. and Applic., University of Pittsburgh, 1981
[8] Walraven, F., Engineering Thesis, Dept. of Chem. Eng.,Katholieke Universiteit Leuven, Belgium, 1984
[9] Seydel, R., Numer. Math., 33, 339, 1979
[10]Moore, G., Spence, A., SIAM J. Numer. Anal., 17, 567, 1980
[11]Roose, D., Piessens, R., Numer. Math., 46, 189, 1985
[12]Jansen, R., Ph. D. dissertation, Dept. Chem. Eng., Katholieke Universiteit Leuven, 1984
[13]Jansen, R., De Dier, B., Z. Fur Naturforsch., in preparation, 1986
[14]Spence, A., Werner, B., IMA J. Numer. Anal., 2, 413, 1982.

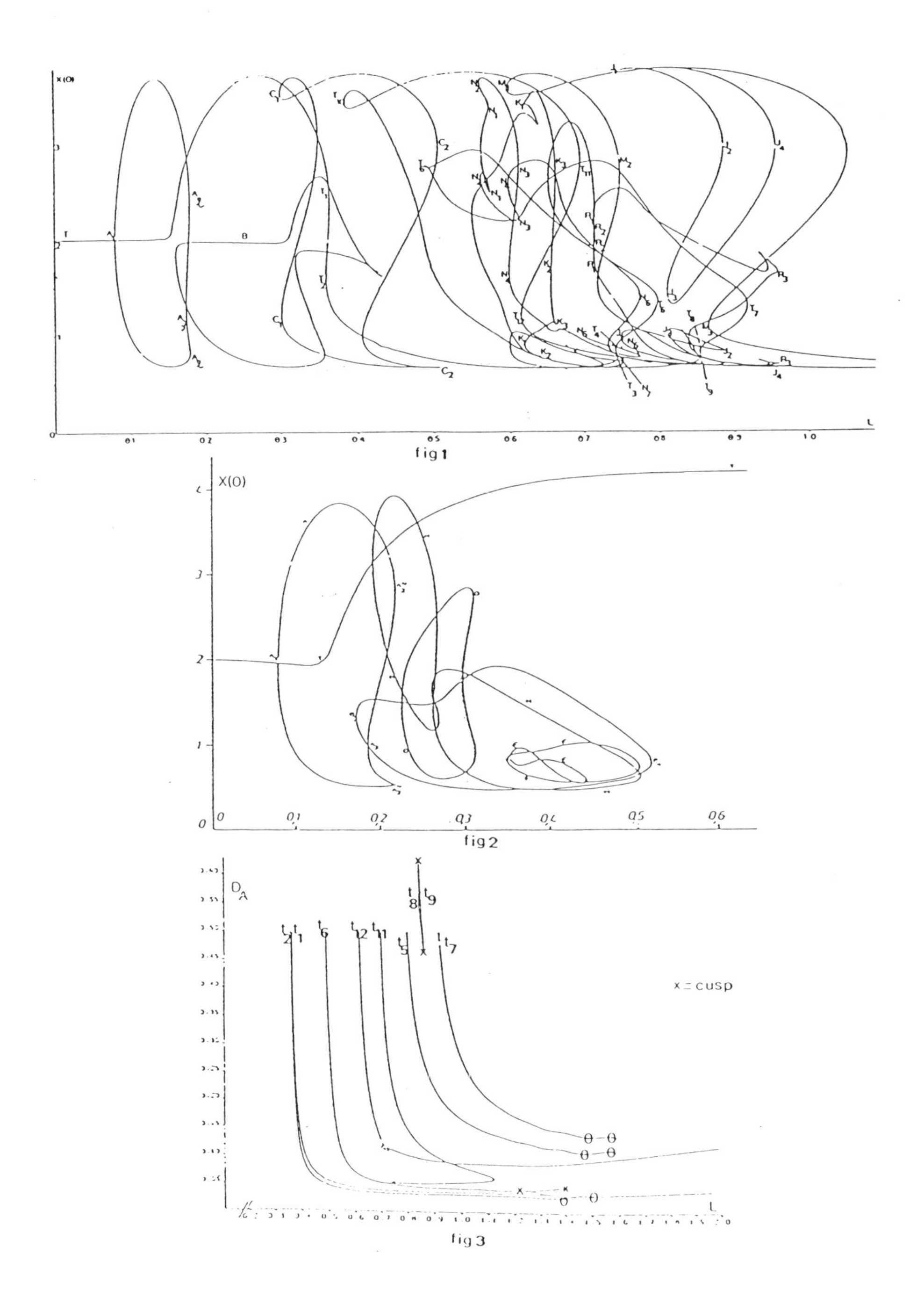
x(O)
L
fig 1
X(O)
fig 2
D_A
x = cusp
L
fig 3

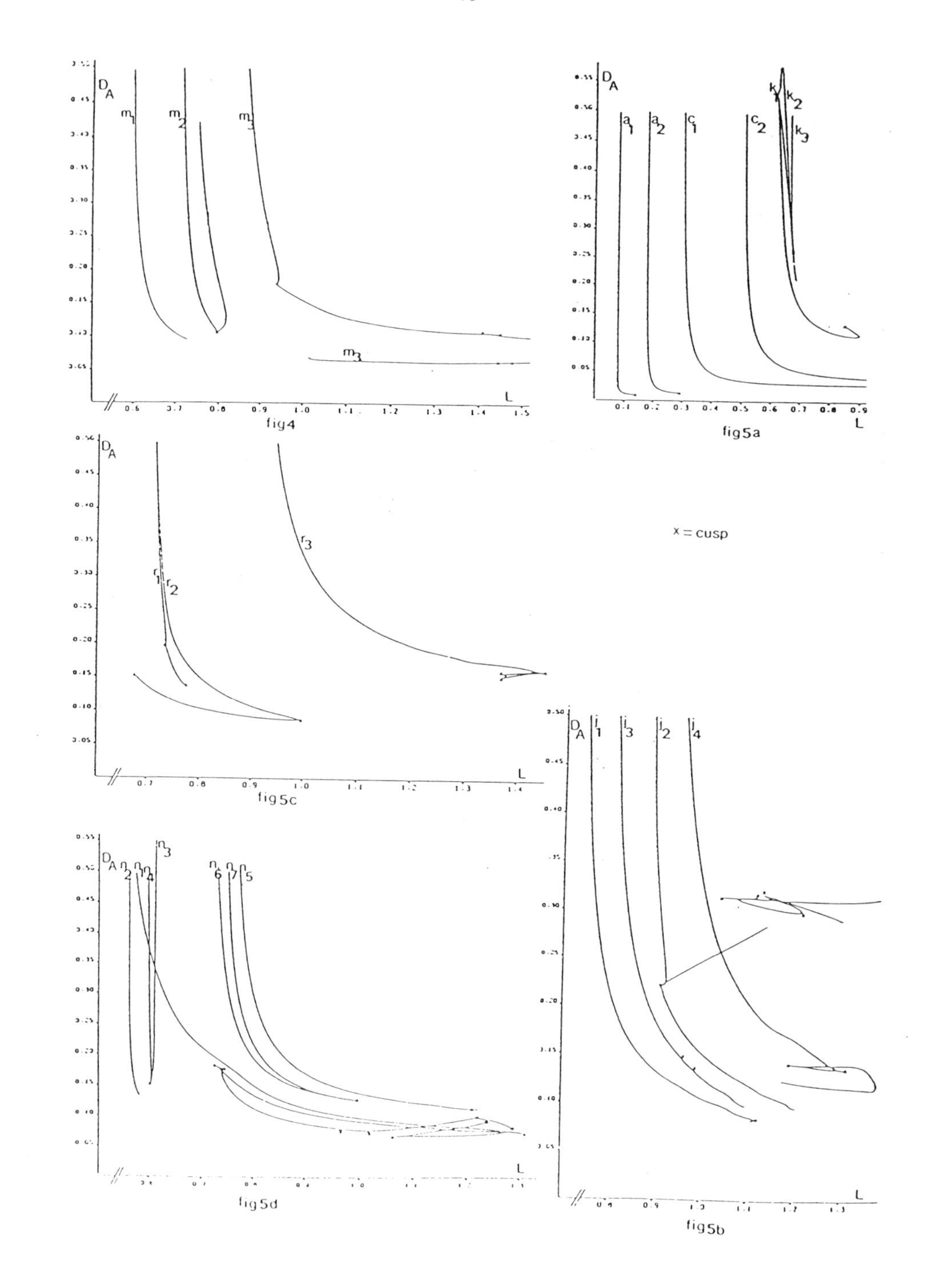
fig4
fig5a
fig5c
fig5d
fig5b
x = cusp

International Series of
Numerical Mathematics, Vol. 79

NUMERICAL STUDIES OF BIFURCATION IN REACTION-DIFFUSION MODELS USING PSEUDO-SPECTRAL AND PATH-FOLLOWING METHODS.

J.C. EILBECK

Heriot-Watt University, Department of Mathematics, Edinburgh, Scotland.

1. Introduction

A number of authors have studied the numerical calculation of time-dependent and steady-state solutions of reaction-diffusion equations, i.e. equations of the form (c.f. De Dier and Roose 1986, Fijii et al. 1982, Manoranjan 1984, Eilbeck 1986),

$$\underline{u}_t = d\Delta\underline{u} + G(\underline{u},\alpha_1,\ldots,\alpha_m) \qquad (1.1)$$

where $\underline{u}(x,t) \in \mathbb{R}^k$; $x \in \Omega \subset \mathbb{R}^n$, $n = 1,2$ or 3; d is a $k \times k$ (usually diagonal) matrix of diffusion coefficients; $\alpha_1,\ldots\alpha_m$ are parameters, and F is a nonlinear function $F: \mathbb{R}^k \times \mathbb{R}^m \to \mathbb{R}^k$. Usually these studies involve discretizing the space dependent part of the equations in some way; in the time-dependent case the resulting differential-difference equations are solved using standard o.d.e. solver packages or more specialized techniques. In the time-independent case $(\underline{u}_t = 0)$ the resulting algebraic equations are often studied using path-following techniques, to investigate the variation and bifurcation of solutions as one or more of the parameters $\alpha_1,\ldots,\alpha_m$ vary. Sometimes these two possibilities are combined to study the bifurcation of time-dependent solutions. In this paper we will consider only the study of steady-state solutions, although the tehniques discussed are applicable to the time-dependent case also. The main purpose of the paper is to point out that the pseudo spectral method is a particularly efficient way of carrying out the spatial discretization process in many cases. Although all the examples discussed here involve only one space dimension $(n = 1)$, pseudo-spectral methods are particularly efficient in two and three space dimensions compared to more standard finite difference and finite element methods (Orszag, 1980; Eilbeck 1983).

The paper is laid out as follows. In Section 2 we review briefly the application of pseudo-spectral methods to reaction-diffusion problems. The accuracy of such methods is discussed in Section 3, and some recent work on boundary condition correction terms for the pseudo-spectral method with trigometric polynomials is discussed. In Section 4 we apply these methods to study fold-following problems in the simple model problem, which gives insight into the unfolding of a pitchfork bifurcation in this system. Finally in Section 5 we summarise some recent work on a coupled reaction-diffusion enzyme model ($k = 2$).

2. The pseudo-spectral method

We follow the same notation and approach as Eilbeck (1986). For simplicity in this section we consider only a scalar reaction-diffusion equation in one space dimension with one parameter

$$v_t(x,t) = v_{xx}(x,t) + g(v,\alpha) \qquad t \geqslant 0, \quad x \in [0,1] \tag{2.1}$$

and with Dirichlet boundary conditions

$$v(0,t) = v(1,t) = 0. \tag{2.2}$$

The vector case is treated in Eilbeck (1983) in one, two and three space dimensions in some spherical geometries. We approximate the exact solution by a finite expansion in terms of eigenfunctions of some Sturm-Louiville problem

$$v(x,t) \cong \tilde{v}(x,t) = \sum_{j=1}^{N} c_j(t)\phi_j(x). \tag{2.3}$$

Two common choices for the $\phi_j(x)$ are trigometric polynomials and Chebyshev polynomials. Each has some advantages and disadvantages, as discussed below. Inserting (2.3) into (2.1) gives

$$\sum_{j=1}^{N} \dot{c}_j(t)\phi_j(x) = - \sum_{j=1}^{N} \pi^2 j^2 c_j(t)\phi_j(x) + g(\tilde{v},\alpha) + r(x,t) \tag{2.4}$$

The residual function $r(x,t)$ is to be made small in some way: in the pure spectral method a Galerkin choice would be adopted, i.e. $r(x,t)$ would be chosen to be orthogonal to the $\phi_j(x)$ for $j = 1,\dots,N$. For nonlinear $g(\tilde{v})$ it is more convenient to adopt a collocation approach, i.e. the pseudo-spectral method: $r(x,t)$ is set zero on a set of discrete points, the collocation points. If the $\phi_j(x)$ satisfy the boundary conditions (2.2) we choose N internal points $\{x_i;\ i = 1,\dots,N\}$ and impose the condition that

$$r(x_i,t) = 0, \qquad i = 1,\dots,N. \tag{2.5a}$$

If the $\phi_j(x,t)$ do not satisfy the boundary conditions, we choose $N-2$ internal points and add the two boundary conditions

$$r(x_i,t) = 0, \quad i = 2,\dots,N-1$$
$$\sum_{j=1}^{N} c_j(t)\phi_j(0) = \sum_{j=1}^{N} c_j(t)\phi_j(1) = 0. \tag{2.5b}$$

Either way leads to a set of N o.d.e.'s or coupled o.d.e.'s and algebraic equations for the coefficients $c_j(t)$. Since we are concerned in this paper only with steady-state solutions, we assume from this point that all time derivatives vanish. For $\phi_j(x)$ satisfying the b.c.'s. (2.2) this gives N algebraic equations for the c_j

$$F(\underline{c},\alpha) \equiv B\underline{c} + \underline{g}(\tilde{v},\alpha) = 0 \tag{2.6}$$

where $B = \{b_{ij}\} = \{-\pi^2 j^2 \phi_j(x_i)\}$ is an $N \times N$ matrix and $\underline{c} = (c_1,c_2,\dots,c_N)^T$. The i th component of $\underline{g}(\tilde{v},\alpha)$ is $g(\tilde{v}(x_i),\alpha)$ where $\tilde{v}(x_i) = \Sigma c_j\phi_j(x_i)$. A similar set of equations for the c_j can be written down in the case where the $\phi_j(x)$ do not satisfy the b.c.'s. Once (2.6) is solved for the c_j, the approximate solution $\tilde{v}(x)$ can be reconstructed from (2.3). Standard path-following techniques can be used to solve (2.6) as α varies, and to calculate turning points, bifurcation points, etc. (see Küpper et al. (1984), Jepson and Spence (1986) and the proceedings of this conference for an up-to-date list of references on path-following methods).

Although stability of solutions is not discussed here, linear stability of the steady-state solutions can easily be calculated (Eilbeck 1986).

3. Accuracy of the pseudo-spectral method

We do not attempt here to derive any rigorous results on error bounds for the nonlinear problems discussed here: our arguments are purely heuristic, based on numerical experience and extrapolation of results from linear error analysis. In particular we consider only the comparison between trigometric and Chebyshev polynomials, applied to a simple test problem, discussed in more detail by Eilbeck and Manoranjam (1986). We consider a simple model problem in one space dimension

$$u_t(x,t) = u_{xx}(x,t) + u(1-u)(u-a) \tag{3.1a}$$

with boundary conditions

$$u(-L,t) = u(L,t) = b. \tag{3.1b}$$

A simple change of variable transforms this to a problem defined on [0,1] with zero boundary conditions

$$v_t = v_{xx} + \alpha f(v) \tag{3.2a}$$

$$v = u - b, \quad v(0,t) = v(1,t) = 0 \tag{3.2b}$$

where

$$f(v) = -v^3 + (1 + a - 3b)v^2 + (2ab - 3b^2 + 2b - a)v + b(1 - b)(b - a)$$

and $\alpha = 4L^2$. As the parameter α varies, the different steady states ($v_t = 0$) of (3.2) trace out paths in solution space. We project these paths down onto the (v^*,α) plane, where v^* is the extremum value of $v(x)$, approximated by $v(x_i)$ over the collocation points i.e. $v^* = \{\tilde{v}_i : |\tilde{v}_i| \geqslant |\tilde{v}_j|, j = 1,\ldots,N\}$. Figure 1 shows schematically some typical results for fixed $a < 0.5$ and various values of b. (See Eilbeck (1986) for some detailed examples). In this and the following section, we consider only bifurcations of the first nonconstant solution, corresponding to $\sin \pi x$ in the $a = b$ linear case. Corresponding pictures for $a > 0.5$ can be formed by reflection in the α axis. Various turning points (fold points, simple limit points) are marked by dots. One of the turning points in Figure 1b is a point of inflexion formed by two nearby turning points (c.f. Fig. 1c) coalescing. In Figure 1d, corresponding to $a = b$, two turning points have joined to form a bifurcation point. Only simple turning points are structurally stable. The positions of these turning points as functions of α and b can be calculated directly by solving the set of equations

$$\begin{aligned} F(\underline{c},\alpha,b) &= 0 \\ F_c^T\psi &= 0 \\ \psi . F_\alpha &= 0. \end{aligned} \tag{3.3}$$

In order to test the accuracy of the pseudo-spectral method for various values of N and different choices of basis functions, Eilbeck and Manoranjan (1986) calculated the position of two different simple turning points corresponding to those illustrated in Figure 1d and 1e respectively. In case 1d we have $a = b$ and hence $f(0) = 0$, whereas in case 1e we have $f(0) \neq 0$. It is not difficult to show that for this Dirichlet boundary value problem with zero

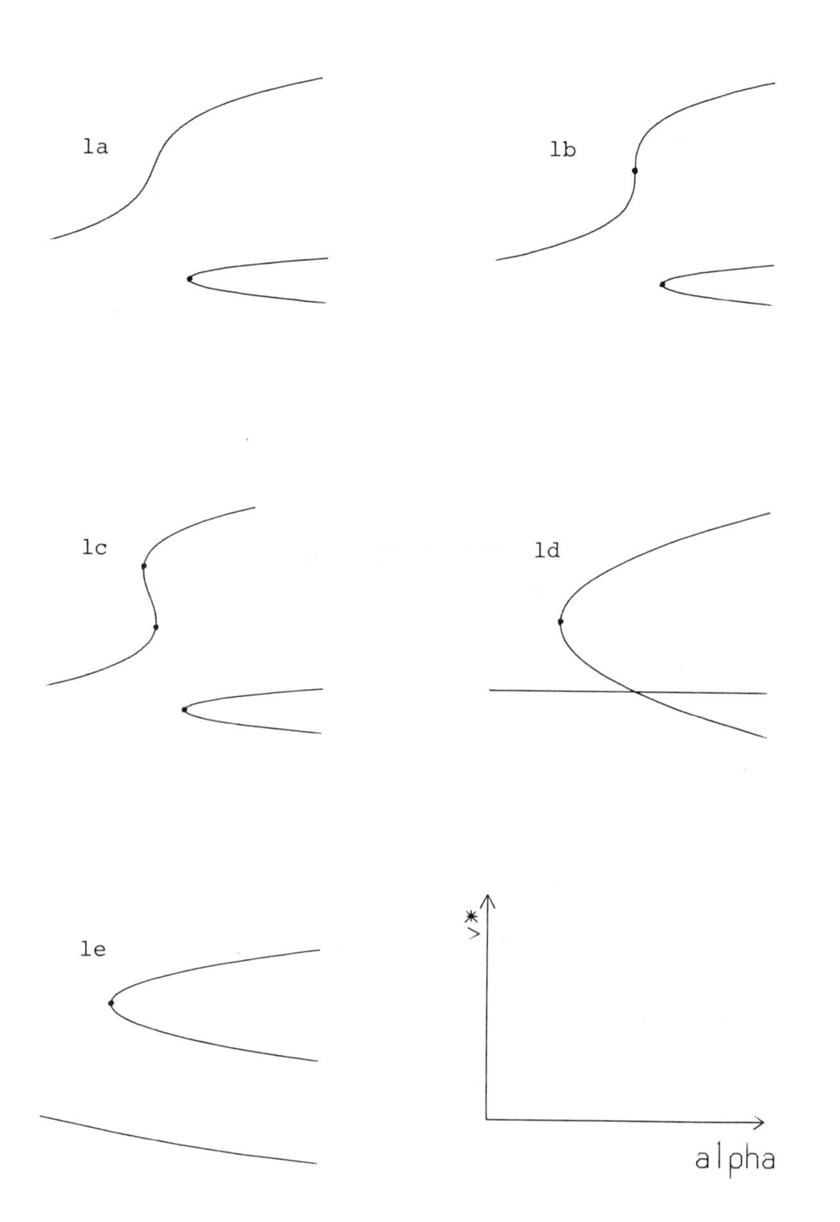

Figure 1. Some bifurcation diagrams for steady-state solutions of (3.2).

boundary values, the odd periodic extension to v(x) will have a discontinuity in its fourth and higher even powers in case 1d, and in its second and higher even powers in case 1e. Arguing from approximation theory, this suggests that the pseudo-spectral method using trigometric polynomials should give fourth order accuracy in case 1d and only second order accuracy in case 1e, whereas the method with Chebyshev polynomials may perhaps still give superalgebraic convergence (Orszag 1980). Figure 2 shows the numerical results for cases 1e. A calculation of the slopes of the asymptotic lines in this case and in case 1d show a good agreement with these heuristic estimates. (Eilbeck and Manoranjan 1986).

Also shown in this figure is a modified calculation for case 1(e) for the choice of trigometric polynomials. Since the discontinuity in the periodic extension of v_{xx} at $x = 0,1$ is simply $g(0,\alpha) = \alpha f(0)$, we can

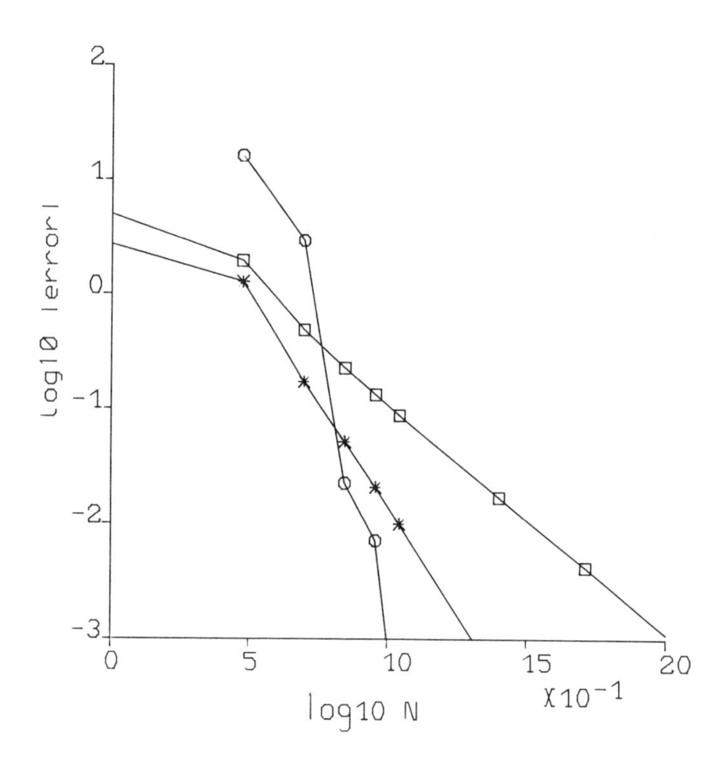

Figure 2: Error curves for the example discussed in the text:
-□-□-□- trigometric polynomials; -0-0-0- Chebyshev polynomials;
-*-*-*- trigometric polynomials with boundary correction term.

add another (known) basis function to the expansion (2.3) to cancel out this discontinuity. The simplest choice, and the one which gave the best results in practice, was to take

$$v(x) = \frac{1}{2}\alpha f(0)x(1-x) + \sum_{j=1}^{N} c_j \sin_j \pi x. \tag{3.4}$$

With this additional term the method is changed in the general case from second order to fourth order at no extra cost! However the Chebyshev expansion is still more accurate than the trigometric expansion for $N \gtrsim 7$. For small values of N, as discussed by Eilbeck and Manoranjan, the trigometric expansion is more accurate. Eilbeck (1986) showed that we can get useful results in the trigometric case even for $N = 1$. Corrections for higher order discontinuities could also be made, but this would be more complicated since these derivatives are not given explicitly as in the second derivative case.

These results apply to the Dirichlet problem with zero b.c.'s. However for the Neumann problem it is straightforward to show that the even periodic extension for $u(x)$ is C^∞ and hence the trigometric polynomials may exhibit superalgebraic convergence in this case.

4. Fold Following

We can use path-following techniques to follow the fold points (v^*,α) given by the solution of (3.3) for fixed a as b varies, and the results are given in Figure 3 for the case $a = 0.25$. The interpretation of this graph is as follows. For $b < 0.25 = a$ we have the case corresponding to Figure 1e, i.e. only one simple turning point on the upper branch of the curve. As we pass through the bifurcation point corresponding to Figure 1d at $a = b = 0.25$ two new fold points appear, as shown schematically in Figure 1(c). At $b \approx 0.265$ the two fold points on the upper branch come together at the point of inflexion shown in Figure 1b, giving a cusp in the (α,b) graph. Finally for $b \gtrsim 0.265$ we have the case shown schematically in Figure 1a, with only one turning point present. The slope of the graph in this region shows that the turning point moves to large values of α as b increases, which explains why previous searches for this branch of solutions for $b \gtrsim 0.32$ in the region $0 < \alpha < 400$ have failed (Eilbeck 1986).

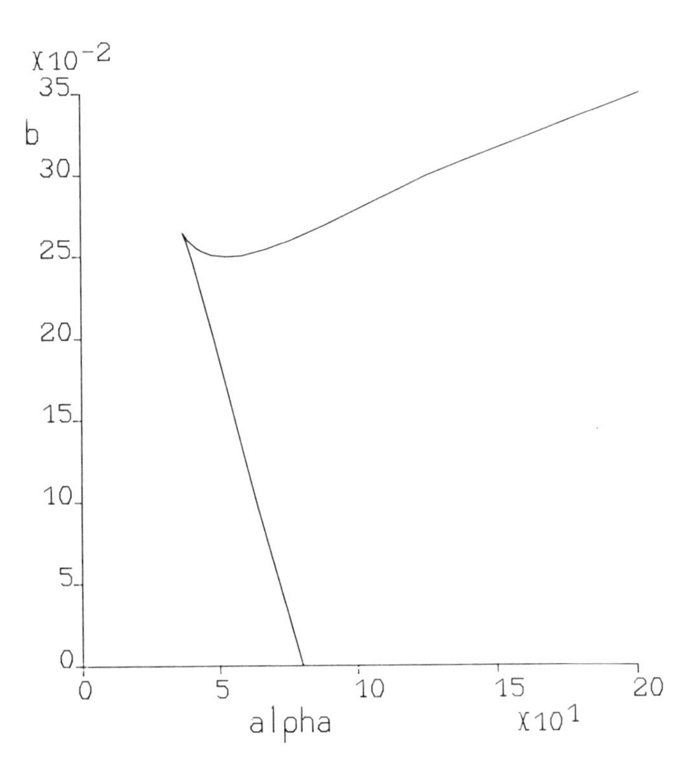

Figure 3: Path of fold points of (3.3)

Fürter (1986) has recently used Liapunov-Schmidt reduction techniques to analyse the unfolding of the pitchfork bifurcation in this system at $a = b = 0.5$. The results of his investigations are summarised schematically in Figure 4. The broken line is the line $a = b$ on which

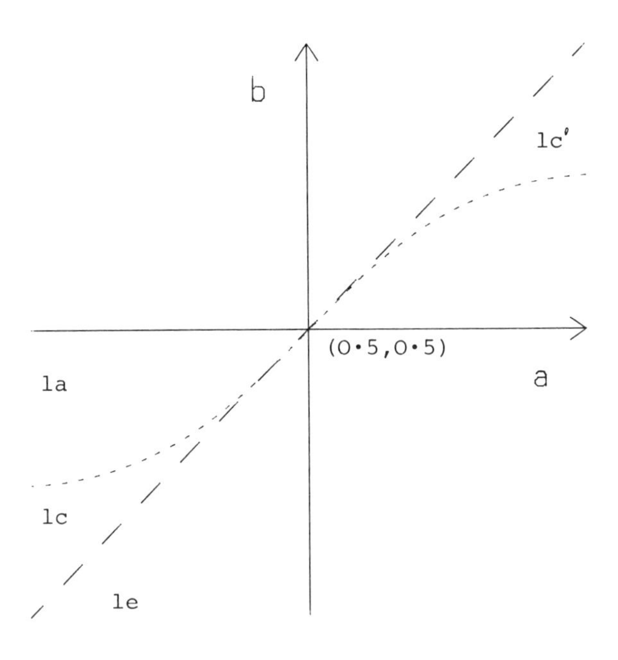

Figure 4

bifurcation points of the type shown in Figure 1d appear, with the limit point lying above the bifurcation point for $a < 0.5$ and below for $a > 0.5$. The dotted line represents the path of the cusp point shown in Figure 3, corresponding to the point of inflexion shown in Figure 1b occurring on the upper (lower) branch for $a < 0.5$ ($a > 0.5$). Investigations are under way to trace numerically the path of this cusp points as a function of a. The regions 1a, 1c, and 1e in Figure 4 represent bifurcation diagrams of the types shown in Figure 1, whereas the region 1c' is the same as 1c except that the curves are reflected through the α axis.

Since the unfolding of a pitchfork bifurcation occurs in many different systems, it is not surprising to find figures similar to Figure 3 elsewhere. A good example, taken from studies of Taylor-vortex flows in a short annulus, is given by Mullin (1982).

5. Path-following for coupled reaction-diffusion systems

Recently Duncan and Eilbeck (1986) have applied the pseudo-spectral method and opath-following techniques to investigate a system of reaction-diffusion equations studied by Catalano et al. (1981) for a model enzyme reaction

$$\begin{aligned} u_t(x,t) &= a - F(u,v) + d_1\Delta u(x,t) \\ v_t(x,t) &= F(u,v) - v(x,t) + d_2\Delta v(x,t) \end{aligned} \qquad (5.1)$$

where

$$F(u,v) = \frac{bu(1+u)(1+v)^2}{[c + (1+u)^2(1+v)^2]}$$

with Neumann boundary conditions on a one-dimensional region $x \in [-L,L]$. By scaling the x variable as for the single model equation considered earlier, a parameter α is introduced in front of the non-diffusive terms in (5.1). Increasing α, corresponding to increasing the size of the system, is equivalent to decreasing the diffusion coefficients (d_1,d_2) along a ray through the origin in the (d_1,d_2) plane. We can use the stability diagrams introduced by Brown and Eilbeck (1982) to analyse the linear stability of the homogeneous steady-state solution (u_o,v_o) of (5.1), with a,b,c and d_1/d_2 fixed, as α varies. This is shown in Figure 5. This figure shows various stability hyperbolae in the d_1,d_2 plane, with the dashed line representing α increasing from top right to bottom left. Starting at the top right of the diagram (small α, large d_1,d_2), only the homogeneous solution is stable. As we increase α we cross the first hyperbola, corresponding to the $n = 1$ $(\cos \pi x)$ eigenfunction. At this point the steady state becomes unstable and an $n = 1$ nonhomogeneous solution appears. Similarly, as the next $(n = 2)$ hyperbola is crossed we expect a solution curve corresponding to $\cos 2\pi x$ to branch off. As α increases still further, the $n = 1$ hyperbola is crossed again marking the termination of an $n = 1$ branch. (There is also an $n = 3$ branch near this point). The bifurcation curves corresponding to the steady-state nonlinear solution following from these bifurcation points on the homogeneous solution curve are shown in Figure 6. (Only $n = 1$ and $n = 2$ branches are plotted, up to $\alpha = 4.2$).

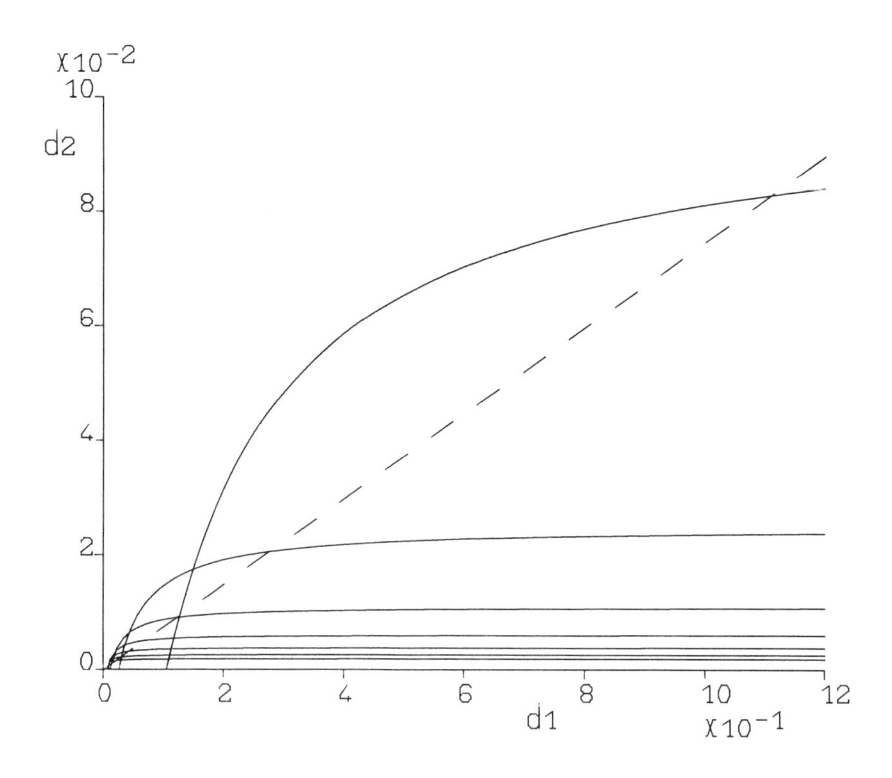

Figure 5: Stability diagram in the d_1, d_2 plane

The symmetry properties of the solutions on these branches are interesting. Near the bifurcation points from the constant solution, the $n = 1$ and $n = 2$ solutions behave like $\cos \pi x$ and $\cos 2\pi x$ respectively. Each branch is double valued: the $n = 1$ branch has two solutions connected by the transformation $x \to 1 - x$ (i.e. $\cos \pi x$ and $-\cos \pi x$ near the constant solution) and the $n = 2$ branch has two solutions connected by the transformation $x \to x + 0.5$ for $x < 0.5$ and $x \to x - 0.5$ for $x \geqslant 0.5$ (corresponding to $\cos 2\pi x$ and $-\cos 2\pi x$ near the constant solution). This latter construction works because the $n = 2$ branch is symmetric about $x = 0.5$ for all α, and because of the Neumann boundary conditions.

The Jacobian of the system of equations is the same for both branches on the $n = 1$ curve but different for the two branches on the $n = 2$ curve. Thus the two bifurcation points away from the constant solution

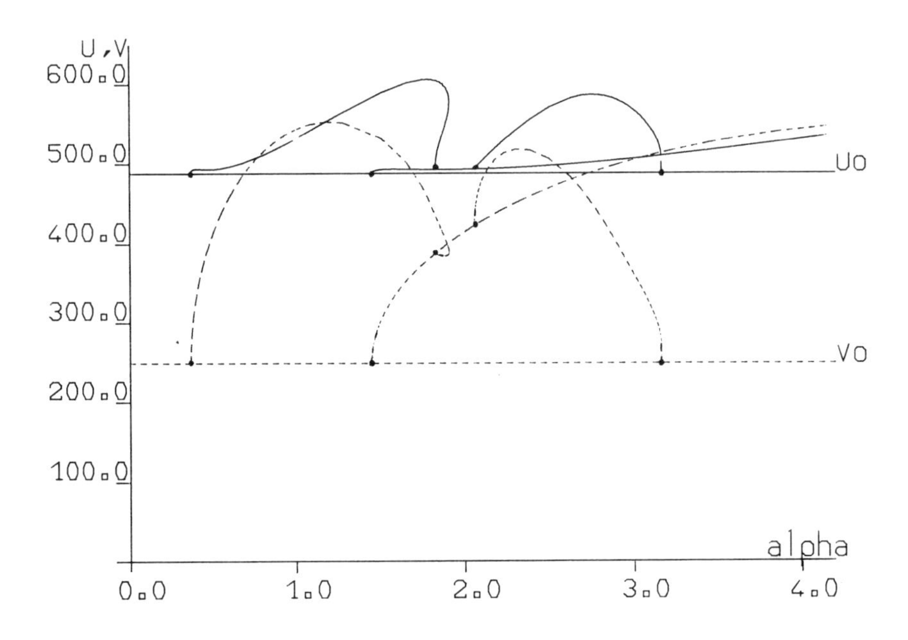

Figure 6: Bifurcation diagram for the coupled system (5.1)

on the $n = 2$ curve correspond to one bifurcation on each branch of this curve. If one proceeds along the $n = 1$ curves from the bifurcation point from the constant solution, the solutions are initially antisymmetric, but lose this symmetry with an increasing symmetric component until at the bifurcation point on the $n = 2$ branch the solution is symmetric. Further details of these branches will be published elsewhere. It is interesting that bifurcation curves of this type have also been noted by Meyer-Spasche (Meyer-Spasche and Keller, 1985; Meyer-Spasche and Wagner, 1986) in studies of Taylor vortex flows with periodic boundary conditions. This pattern is probably generated by the unfolding of the degenerate bifurcation point where the $n = 1$ and $n = 2$ hyperbolae in the (d_1,d_2) plane intersect (Fijii et al. 1982).

6. Conclusions

The result of these and previous investigations shows that the pseudo-spectral method and path-following techniques are useful tools in the study of time-dependent and steady-state reaction-diffusion systems in one or more space dimensions. Although most bifurcation studies have concentrated on single systems in one space dimension, a preliminary extension to coupled systems in two or three space dimensions has already been carried out in the time-dependent case by Eilbeck (1983) using a microcomputer. More detailed studies of reaction-diffusion systems in higher dimensions are now in progress.

Although unconnected with these reaction-diffusion studies, it may be appropriate, in a volume on bifurcation studies, to direct those readers with an interest in bifurcation in Hamiltonian systems to a simple model system with many exact solutions (Eilbeck et al. 1985).

Acknowledgements

It is a pleasure to acknowlege helpful conversations with K.J. Brown, K. Duncan, J. Fürter and R.L. Pego.

7. References

Brown, K.J. and Eilbeck, J.C. (1982) Bifurcation, stability diagrams, and varying diffusion coefficients in reaction-diffusion equation. Bull. Math. Biol. 44, 87-102.

Catalano, G., Eilbeck, J.C., Monroy, A. and Parisi, E. (1981) A mathematical model for pattern formation in biological systems. Physica 3D, 439-456.

De Dier, B. and Roose, D. (1986) Determination of bifurcation points and cusp catastrophes for the Brusselator model with two parameters. (In these proceedings).

Eilbeck, J.C. (1983) A collocation approach to the numerical calculation of simple gradients in reaction-diffusion systems. J. Math. Biol. 16, 233-249.

Eilbeck, J.C., Lomdahl, P.S. and Scott, A.C. (1985) The discrete self-trapping equation. Physica 16D. 318-338.

Eilbeck, J.C. (1986) The pseudo-spectral method and path following in reaction-diffusion bifurcation studies. SIAM J. Sci. Stat. Comput. 7, 599-610.

Eilbeck, J.C. and Manoranjan, V.S. (1986) A comparison of basis functions for the pseudo-spectral method for a model reaction diffusion problem. J. Comp. Appl. Math. 15, 371-378.

Fujii, H., Mimura, M. and Mishiura, Y. (1982) A picture of the global bifurcation diagram in ecological interacting and diffusing systems. Physica 5D, 1-42.

Jepson, A.D. and Spence, A. (1986) Numerical methods for bifurcation problems. To be published in the proceedings of the IMA conference "State of the art in numerical analysis", Birmingham, 1986.

Küpper, T., Mittelmann, H.D. and Weber, H. (1984) Numerical methods for bifurcation problems. Birkhäuser Verlag: Stuttgart.

Manoranjan, V.S. (1984) Bifurcation studies in reaction-diffusion II. J. Comput. Appl. Math 11, 307-314.

Meyer-Spasche, R. and Keller, H.B. (1985) Some bifurcation diagrams for Taylor vortex flows. Phys. Fluids 28, 1248-1252.

Meyer-Spasche, R. and Wagner, M. (1986) Steady axisymmetric Taylor vortex flows and free stagnation point of the poloidal flow. (In these proceedings).

Mullin, T. (1982) Mutations of steady cellular flows in the Taylor experiment. J. Fluid. Mech. 121, 207-218.

J.C. Eilbeck
Department of Mathematics
Heriot-Watt University
Riccarton
Edinburgh EH14 4AS
Scotland, U.K.

International Series of
Numerical Mathematics, Vol. 79

A quick multiparameter test for periodic solutions

Bernold Fiedler, Peter Kunkel
Universität Heidelberg
Sonderforschungsbereich 123
Im Neuenheimer Feld 294
D-6900 Heidelberg

§0. Introduction

We consider ordinary differential equations of the form

$$D(\sigma)\,\dot{x} = f(x,\tau)\,, \qquad x \in \mathbb{R}^n \tag{0.1}$$

with a diagonal matrix $D(\sigma) = \mathrm{diag}[1,\dots,\sigma,\dots,1]$ which differs from the unit matrix by an entry σ in the row (and column) i_c. Here $\tau \in \mathbb{R}$ and $\sigma>0$ are considered as bifurcation parameters. Note that stationary solutions of (0.1) satisfy

$$0 = f(x,\tau) \tag{0.2}$$

and are thus independent of σ.

We ask the following two <u>questions</u>:

(I) Does system (0.1) show periodic solutions?

(II) Which components i_c make the system oscillate?

Below, we present a code BALCON which investigates these questions numerically. It provides suggestions for a reasonable choice of the critical component i_c, and it yields a global result on Hopf bifurcation of periodic solutions in the (τ,σ)-plane. As we shall see below, the costs are determined essentially by the one-parameter pathfollowing routine which solves the steady state equation (0.2). This makes the test BALCON particularly quick, but it also restricts applicability to systems of the form (0.1).

This contribution is organized as follows. In section 1 we sketch the underlying theory on global Hopf bifurcation in two parameter flows. In particular we define the center-pieces: B-point and B-index. In section 2 we describe how BALCON localizes and analyzes such points. Section 3 concludes with remarks and illustrations on a practical example.

§1. Analysis

In this section we give a brief account of global Hopf bifurcation in two parameters; for a complete reference see [4,5]. The theory applies to general two parameter systems. Here we adapt it to the special context of equations of the form

(0.1) $D(\sigma)\,\dot{x} = f(x,\tau)$.

Suppose that $f \in C^2$, and that the solution set of

(0.2) $0 = f(x,\tau)$

has a <u>turn</u> at (x^0,τ^0). I.e. the unique C^2-smooth branch $(x(s), \tau(s))$ through $(x^0,\tau^0) = (x(0),\tau(0))$ of solutions of (0.2) has

(1.1a) "vertical" tangent:
$\tau'(0) = 0,\ x'(0) =: r,\ |r|^2 = 1$

(1.1b) nonzero curvature:
$\tau''(0) \neq 0$

In particular, the linearization of (0.2), given by

(1.2) $D_x f(x^0,\tau^0)$

has an eigenvalue zero with right eigenvector r.

1.1 Definition [5]:

In two parameters (τ,σ) we call (x^0,τ^0,σ^0) a B-point if the linearization of (0.1), given by

$$D(\sigma^0)^{-1}\, D_x f(x^0,\tau^0), \tag{1.3}$$

has an eigenvalue zero with algebraic multiplicity ≥ 2.

With the left eigenvector

$$\ell^T\, D_x f(x^0,\tau^0) = 0 \tag{1.4}$$

this condition is equivalent to

$$\ell^T\, D(\sigma^0)\, r = 0\ , \tag{1.5}$$

which describes a hyperplane of codimension 1 in the space of all diagonal matrices D. Mostly, σ^0 will be determined uniquely by (1.5) because the straight line $\sigma \mapsto D(\sigma)$ intersects the hyperplane transversely.

Under suitable nondegeneracy conditions, our B-point has a precisely double eigenvalue zero and the flow in the two-dimensional center manifold transforms into normal form

$$(1.6)\qquad \begin{aligned} \dot{y}_1 &= y_2 \\ \dot{y}_2 &= \varepsilon_1 + \varepsilon_2 y_1 + y_1^2 + \beta y_1 y_2 \end{aligned}$$

with $\beta = \pm 1$, under a change of coordinates

$$\begin{aligned} \varepsilon &= \varepsilon(\tau,\sigma)\quad , & y &= y(x,\tau,\sigma)\ . \\ 0 &= \varepsilon(\tau^0,\sigma^0)\ , & 0 &= y(x^0,\tau^0,\sigma^0). \end{aligned}$$

This normal form goes back to [1,2,10].

1.2 Definition [5]:

In the above situation, the B-index B of the B-point (x^0,τ^0,σ^0) is defined to be

$$B := (-1)^{E(\sigma^0)}\cdot\beta\cdot\operatorname{sgn}\det \varepsilon'(\tau^0,\sigma^0) \tag{1.6}$$

where $E(\sigma^0)$ denotes the (strict) unstable dimension of the linearization (1.3).

1.3 Theorem [5]:

Consider a closed Jordan curve γ in the (τ,σ)-plane (see e.g. fig. 1.1). Then $\sum_{\mathrm{int}(\gamma)} B \neq 0 \Rightarrow$ global Hopf bifurcation on γ.

Here the sum ranges over all indices of B-points with (τ^0,σ^0) inside γ.

For technically precise assumptions and a discussion of global Hopf bifurcation on γ see again [5]. At any rate, we now know one answer to question (I) on periodic solutions in system (0.1). It is sufficient to find B-points (x^0,τ^0,σ^0) and calculate their index B. Recall, however, that we are restricted to $\sigma^0>0$. In view of

$$(1.5) \qquad \ell_{i_c} r_{i_c} \sigma^0 + \sum_{j\neq i_c} \ell_j r_j = 0$$

this restricts the possible choices for i_c and indicates an answer to question (II).

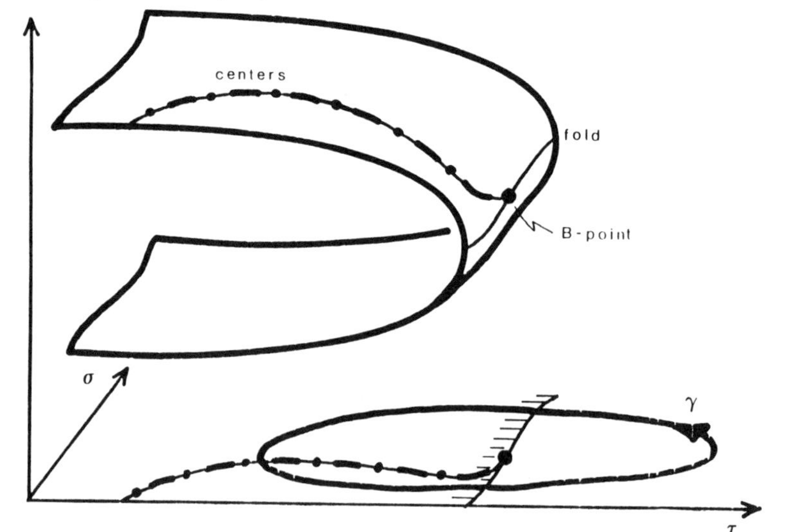

Fig. 1.1: Global Hopf bifurcation in two parameters

§2. Algorithm

In this section we describe how our code BALCON ("B-point ALgebraic CONtinuation") computes B-points and B-indices. Note that our definition 1.2 of the index B is somewhat impractical, since it is based on a normal form reduction in the center manifold. In lemma 2.1 below, we remedy this drawback. We then describe some implementation details. Finally, we put our approach in perspective with the very interesting results of Roose [8,9].

To locate B-points we first solve

$$0 = f(x,\tau) \quad (0.2)$$

We use our code ALCON, for "ALgebraic CONtinuation", which is described in [3]. ALCON is based on tangent approximation with Gauss-Newton iteration and predictor-corrector steplength control. For the Jacobian we use QR-decomposition. Turns are determined by cubic Hermite interpolation, monitoring det $D_x f$ along the branch. Simple bifurcations are treated as well.

Once a turn (x^0,τ^0) is detected, QR decomposition of $D_x f$ yields

$$D_x f = Q\tilde{R}\Pi, \qquad \tilde{R} = \begin{pmatrix} R & S \\ & 0 \end{pmatrix}$$

$$r := \Pi^T \begin{pmatrix} -R^{-1}S \\ 1 \end{pmatrix}, \quad (2.1)$$

$\ell := Q e_n$, e_n the n-th unit vector, and we may also normalize r. In a first run, we print $\ell_j r_j$ to get a quick guide to a reasonable choice of i_c. If all $\ell_j r_j$ have the same sign, no B-point ever occurs - regardless of the choice of i_c. This is immediate from (1.5), $\sigma^0>0$. From now on suppose i_c is given and σ^0 is positive.

2.1 Lemma:

Assume $\ell_{i_c} r_{i_c} \neq 0$ for the B-point (x^0,τ^0,σ^0). Then the B-index satisfies

$$(2.2) \qquad B = (-1)^n \cdot \mathrm{sgn}\ \det\ [D_x f(x^0,\tau^0)]_{\hat{i}_c} \cdot \cdot \mathrm{sgn}\ \tau''(0)$$

where n=dim x, $\hat{i}_c$ indicates that row and column i_c shall be omitted and $\tau''(0)$ is the curvature from definition 2.1.

Proof: The proof is based on [5 , lemma 2.3], but details are omitted - for brevity.

Applying lemma 2.1 we can now determine the index B from one additional QR-decomposition of the (n-1)x(n-1) matrix $[D_x f(x^0,\tau^0)]_{\hat{i}_c}$. Compared to the iteration along f=0, this makes BALCON a very quick and efficient test.

In [8,9], Roose computes B-points for general two parameter problems

$$\dot{x} = f(x,\tau,\sigma).$$

He traces a curve of turning points until he encounters an "origin for Hopf bifurcation", which we call "B-point". Next he follows curves of Hopf bifurcation points, one emanating from each B-point. Instead, we compute the index B and then invoke theorem 1.3 to obtain periodic solutions, analytically. Both approaches should be combined.

§3. Application

For demonstration purposes, we apply BALCON to a system of two identical, diffusively coupled Oregonators. For an example involving the Briggs-Rauscher oscillatory reaction, as well as for a discussion of a scaling technique which puts general chemical reaction systems in the form (0.1), see [6]. Purely analytic examples for B-points and B-indices are contained in [5]. For numerical examples of B-points obtained by different techniques, see e.g. [7,8,9].

Let $x=(x^{(1)},x^{(2)}) \in \mathbb{R}^6$, $x^{(1)},x^{(2)} \in \mathbb{R}^3$, similarly $f=(f^{(1)},f^{(2)})$, and consider the coupled Oregonator cells

$$D(\sigma^0)\,\dot{x} = f(x,\tau)\ ,$$

$$f^{(1)}(x,\tau) := g(x^{(1)}) + \tau\, D'\cdot(x^{(2)}-x^{(1)})$$
$$f^{(2)}(x,\tau) := g(x^{(2)}) + \tau\, D'\cdot(x^{(1)}-x^{(2)})$$

$$g(x^{(1)}) = g(x_1,x_2,x_3) := \begin{pmatrix} s\ (x_2-x_1x_2+x_1-qx_1^2) \\ s^{-1}(-x_2-x_1x_2+\phi x_3) \\ x_1-x_3 \end{pmatrix}$$

$$D' := \mathrm{diag}\,[d_1,d_2,d_3]\ .$$

We choose the constants to be

$$\phi = 1,\ q = 10^{-3},\ s = 50$$
$$d_1 = 0.25,\ d_2 = 10^{-4},\ d_3 = 0.5\ .$$

The symmetric stationary solution $x^{(1)} = x^{(2)}$, $g(x^{(1)}) = 0$ is known explicitly. It is independent of τ. We start with this solution at $\tau=1$, searching the window $0\leq\tau\leq 15$. The resulting plots for the stationary solutions $(x^{(1)},\tau)$ are shown in fig. 3.1. By symmetry, the plots for $(x^{(2)},\tau)$ are the same and are omitted.

At turns and pitchforks we obtain the following sign pattern for $\ell_j r_j$:

τ^0	type	sign $\ell_j r_j$, $j=1,\ldots,6$
0.089819	turn, twice	+ + − + + − − − + − − +
0.10403	pitchfork	− − + − − +
14.311	pitchfork	+ + − + + −

The two turns are related by the symmetry $x^{(1)} \leftrightarrow x^{(2)}$. The table suggests the choice $i_c=3$, cf. (1.5). This provides the following

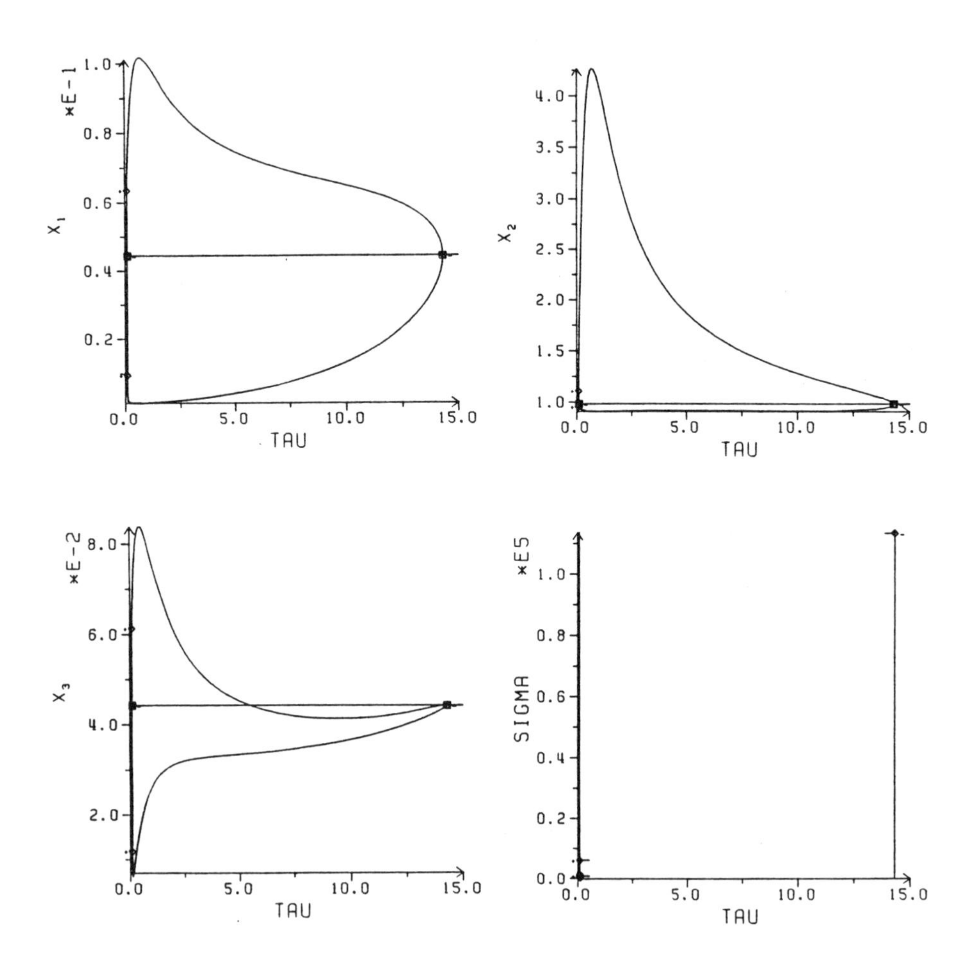

Fig. 3.1: Example: coupled Oregonators

B-points and B-indices.

τ^0	σ^0	B
0.089819	863.05	+1
0.089819	6075.7	+1
0.10403	189.05	-1
14.311	$1.1363 \cdot 10^5$	termination

Termination is due to the prescribed accuracy bound $eps=10^{-5}$; with $eps=10^{-6}$ this result would be accepted with B=-1, cf. fig. 3.1. Note that some of the required σ^0 actually make our example operate on different time scales, dynamically. For the code this problem does not matter, of course.

Note that our choice of $D(\sigma^0)$ breaks the symmetry $x^{(1)} \leftrightarrow x^{(2)}$ of the problem. Indeed $D(\sigma^0) := \text{diag}\,[1,1,\sigma^0,1,1,\sigma^0]$ might appear more appropriate. It is a nice exercise, to determine the corresponding σ^0 and B from those given above. For brevity we leave its solution, as well as other modifications (e.g. nonlinear $\sigma \mapsto D(\sigma)$, general $f(x,\tau,\sigma)$), to the reader.

References:

[1] V.I. Arnold: Lectures on bifurcations and versal systems, Russ. Math. Surveys 27,5 (1972), 54-123.

[2] R.I. Bogdanov: (a) Bifurcations of a limit cycle of a family of vector fields in the plane (russ.), Trudy Sem. I.G. Petrovskogo 2 (1976), 23-36
(b) A versal deformation of a vector field in the plane in the case of zero eigenvalues (russ.), ibid. 37-65.

[3] P. Deuflhard, B. Fiedler, P. Kunkel: Efficient numerical pathfollowing beyond critical points, to appear in SIAM J. Numer. Analysis.

[4] B. Fiedler: An index for global Hopf bifurcation in parabolic systems, Crelles J. reine und angew. Math. 359 (1985), 1-36.

[5] B. Fiedler: Global Hopf bifurcation of two-parameter flows, Arch. Rat. Mech. Anal. 94 (1986), 59-81.

[6] B. Fiedler, P. Kunkel: Multistability, scaling, and oscillations, in "Modelling of Chemical Reaction Systems", Heidelberg 1986, Deuflhard, Ebert, Jäger, Warnatz, Wolfrum (eds.), to appear.

[7] M. Kieser: Mathematische Analyse der Dynamik zweier Populationen mit inverser trophischer Relation, Diplomarbeit, Heidelberg 1986.

[8] D. Roose: Numerical determination of an emanating branch of Hopf bifurcation points in a two-parameter problem, preprint 1986.

[9] D. Roose: Numerical computation of origins for Hopf bifurcation in a two-parameter problem, this volume.

[10] F. Takens: Singularities of vector fields, Publ. IHES 43 (1974), 47-100.

B. Fiedler,
P. Kunkel
Universität Heidelberg
Sonderforschungsbereich 123
Im Neuenheimer Feld 294
D-6900 Heidelberg

International Series of
Numerical Mathematics, Vol. 79

CODIMENSION TWO BIFURCATIONS NEAR AN INVARIANT n-TORUS

Dietrich Flockerzi
Mathematisches Institut
Universität Würzburg
Würzburg, F.R.G.

1. Introduction

We consider a m-parameter C^∞-family of ordinary differential equations possessing an invariant n-dimensional torus. We assume the existence of local coordinates such that the flow near this torus is generated by

$$\begin{aligned} \dot{x} &= Ax + X(x,y,z,\mu) = Ax + \mathcal{O}(|x|^2+|z|^2+|\mu|^2) \\ \dot{y} &= \omega + Y(x,y,z,\mu) = \omega + \mathcal{O}(|x|+|z|+|\mu|) \\ \dot{z} &= B(y)z+Z(x,y,z,\mu) = B(y)z+\mathcal{O}(|x|^2+|z|^2+|\mu|^2) \end{aligned} \tag{1.1}$$

with $X(o,y,o,\mu)\equiv o$ and $Z(o,y,o,\mu)\equiv o$. Here, (x,z,μ) is supposed to vary in a neighborhood of the origin in $\mathbb{R}^k\times\mathbb{R}^\ell\times\mathbb{R}^m$ and y on the standard torus T^n. The invariant n-torus of (1.1) is the trivial one given by $x=o, z=o$. The splitting of the normal coordinates into x and z is motivated by the spectral assumptions

(1.2) The eigenvalues of A are pure imaginary.

(1.3) The fundamental matrix solution $\mathcal{B}(t,s,\eta)$ of $\dot{z} = B(\omega t+\eta)z, \eta\in T^n$, satisfies the estimate
$$\|\mathcal{B}(t,s,\eta)\| \leq \gamma\exp(-b(t-s)) \text{ for } t\geq s$$
with positive η-independent constants γ, b.

The cases of A possessing exactly one and two pairs of simple nonzero pure imaginary eigenvalues have been considered in [3,4, 13] and [5,6] respectively. In this contribution we shall investigate the cases of A being equal to

$$(1.4) \qquad A_1 = \begin{pmatrix} 0 & -1 & 0 \\ 1 & 0 & 0 \\ 0 & 0 & 0 \end{pmatrix} \quad \text{or} \quad A_2 = \begin{pmatrix} 0 & 0 \\ 0 & 0 \end{pmatrix}$$

where we take the dimension of the μ-space to be 2. We shall study the question whether there exist nontrivial invariant tori of dimension $\geq n$ for (1.1) in these interactions of critical modes in the normal spectrum of the trivial n-torus of (1.1)(cf.[11], [13]). In our analysis we confine ourselves to the situation where the quadratic nonlinearities are decisive. The case where, due to an underlying symmetry, the cubic terms determine the bifurcation behavior are covered by [6].

The present paper is based on [5] and the preceeding investigations in [2,7-10,12,14] where special cases of (1.1) have been studied. For further references we refer to these articles. In the sections 2-4 we consider system (1.1) for $A=A_1$, in the final section 5 we remark on the case $A=A_2$ (cf.(1.4)).

2. Approximate Normal Forms for $A=A_1$

For the investigations we have in mind we may restrict our attention to the reduced system of (1.1) on a center manifold represented by a C^N-function $z=\sigma(x,y,\mu)$ for $|x|+|\mu|<\delta, y\in T^n$. By choosing $\delta=\delta(N)>0$ sufficiently small we may assume the reduced system

$$(2.1) \qquad \begin{aligned} \dot{x} &= A_1 x + X(x,y,\sigma(x,y,\mu),\mu) = A_1 x + \mathcal{O}(|x|^2+|\mu|^2) \\ \dot{y} &= \omega + Y(x,y,\sigma(x,y,\mu),\mu) = \omega + \mathcal{O}(|x|+|\mu|), \end{aligned}$$

possessing the trivial n-torus $\{x=0\}$ to belong to the class C^N with an arbitrarily large $N\in\mathbb{N}$ (cf.[5]). Under the nonresonance condition NR(j+1) defined by

NR(j+1) There exist constants $\kappa_o>o,\kappa_1>o$ such that

$$|\ell+(k,\omega)| \geq \kappa_o(|\ell|+|k|)^{-\kappa_1}$$

holds true for "$\ell=o\in Z, k\in Z^n-\{o\}$" and for

"$\ell\in Z, 1\leq|\ell|\leq j+1, k\in Z^n$".

we may without loss of generality start our analysis with the approximate Birkhoff-Normal-Form of order j (cf.[2,5]).Introducing cylindrical coordinates

$$(2.2) \qquad x_1 = r\cos\theta\ , \quad x_2 = r\sin\theta\ , \quad x_3 = w$$

we thus shall study the local C^m-system ($m\geq 2$)

$$(2.3) \qquad \begin{aligned} \dot{r} &= L_1(\mu)r+R_j(r,w,\mu)+\mathcal{O}(j+1) \\ \dot{w} &= L_2(\mu)w+W_j(r,w,\mu)+\mathcal{O}(j+1) \\ \dot{\theta} &= \ 1 \ +\Theta_j(r,w,\mu)+\mathcal{O}(j+1)/r \\ \dot{y} &= \ \omega \ +Y_j(r,w,\mu)+\mathcal{O}(j+1). \end{aligned}$$

Here,the $R_j, W_j=\mathcal{O}(r^2+|w|^2)$ and the Θ_j, Y_j are polynomials in (r,w,μ) that are of order j in (r,w).The perturbing $\mathcal{O}$-terms stand for $\mathcal{O}(|r,w,\mu|^{j+1}), (r,w,\mu)\to(o,o,o)$.We impose the following genericity assumptions on (2.3) under the condition NR(3):

$$(2.4) \qquad L_1(\mu) = \mu_1, \quad L_2(\mu) = \mu_2.$$

$$(2.5) \qquad R_2(r,w,o) = arw, \quad W_2(r,w,o) = bw^2+cr^2, \quad abc\neq o,$$

(cf.[9,12,14]).Then the unperturbed system (2.3) decouples such that the "truncated amplitude equations"

$$(2.6) \qquad \begin{aligned} \dot{r} &= \mu_1 r + arw \\ \dot{w} &= \mu_2 w + bw^2 + cr^2 \end{aligned}$$

indicate where to expect nontrivial invariant tori for the full system (2.3) and hence for (2.1) and (1.1).In the following we shall establish the correspondances between equilibria and periodic solutions of (2.6) and invariant tori of (1.1). Thereby we shall repeatedly employ the results in [5,6] on the persistence of invariant tori.

3. Invariant Tori of Dimension n and n+1

The truncated amplitude equations (2.6) possess the nonnegative equilibria

$$(3.1)\qquad (o,-\mu_2/b)^T \quad \text{for } \mu\in H:=\{\mu:b\mu_2<o\},$$

$$(3.2)\qquad (\gamma_1(\mu),\gamma_2(\mu))^T = (\{\mu_1(b\mu_1-a\mu_2)a^{-2}c^{-1}\}^{1/2}, -\mu_1/a)^T$$
$$\text{for } \mu\in C:=\{\mu:a\mu_1<o, ac(b\mu_1-a\mu_2)>o\}.$$

With scaled variables

$$(3.3)\qquad \mu=\varepsilon\nu\ ,\ r\to\varepsilon r\ ,\ w\to\varepsilon w \quad (\varepsilon>o,\nu\in\mathbb{R}^2)$$

and with the translations

$$(3.4)\qquad w=u+\nu_2/b,\quad v=(r\cos\theta, r\sin\theta)^T \quad \text{for } \nu\in H,$$

$$(3.5)\qquad \rho=r-\gamma(\nu) \quad \text{for } \nu\in C$$

one passes from (2.3) to (perturbed) systems of the form

$$(3.6)\qquad \begin{aligned} \dot u &= \varepsilon\{-\nu_2 u+bu^2+c|v|^2\} && + \mathcal{O}(\varepsilon^2)\\ \dot v &= V(u,v,\varepsilon,\nu)v && + \mathcal{O}(\varepsilon^2)\\ \dot y &= \omega + \varepsilon\Omega(u,\nu) && + \mathcal{O}(\varepsilon^2)\end{aligned}$$

and

$$(3.7)\qquad \begin{aligned} \dot\rho &= \varepsilon P(\rho,\nu) && + \mathcal{O}(\varepsilon^2)\\ \dot\theta &= 1+\varepsilon G(\rho,\nu) && + \mathcal{O}(\varepsilon^2)\\ \dot y &= \omega+\varepsilon H(\rho,\nu) && + \mathcal{O}(\varepsilon^2)\end{aligned}$$

respectively. The relevant properties of the explicitly listed terms in (3.6) and (3.7) are the following:

$$(3.8a) \qquad V(o,o,\varepsilon,\nu) = \begin{pmatrix} \varepsilon(b\nu_1 - a\nu_2)/b & -1+\mathcal{O}(\varepsilon) \\ 1+\mathcal{O}(\varepsilon) & \varepsilon(b\nu_1 - a\nu_2)/b \end{pmatrix}$$

$$(3.8b) \qquad P(\rho,\nu) = C(\nu)\rho + \begin{pmatrix} a\rho_1\rho_2 \\ b\rho_2^2 + c\rho_1^2 \end{pmatrix} \quad \text{with}$$

$$(3.8c) \qquad C(\nu) = \begin{pmatrix} o & a\gamma_1(\nu) \\ 2c\gamma_1(\nu) & \nu_2 + 2b\gamma_1(\nu) \end{pmatrix}$$

$$(3.8d) \qquad \Omega = \mathcal{O}(|u|+|\nu|), \quad G,H = \mathcal{O}(|\rho|+|\nu|).$$

The unperturbed systems (3.6) and (3.7) possess trivial invariant tori M_o^n and M_o^{n+1} corresponding to $(u,v)=(o,o)$ and $\rho=o$ respectively.The following theorem describes the parameter constellations for which these trivial tori persist under the indicated $\mathcal{O}(\varepsilon^2)$-perturbations in (3.6-7).For its formulation we restrict ν to the unit circle via

$$(3.9) \qquad \nu = \nu(\tau) = (\cos\tau, \sin\tau)^T$$

and write the closures of H and C as

$$(3.1o) \qquad \bar{H} = \{(\varepsilon,\tau) : \varepsilon > o, \tau \in [\underline{\tau}, \underline{\tau}+\pi]\},$$

$$(3.11) \qquad \bar{C} = \{(\varepsilon,\tau) : \varepsilon > o, \tau \in [\tau_1, \tau_2]\}$$

where we assume for convenience

$$(3.12) \qquad b\cos\tau_2 - a\sin\tau_2 = o, \quad \cos\tau_1 = o, \quad \tau_1 < \tau_2.$$

We define $\tau^* \in (\tau_1, \tau_2)$ by

$$(3.13) \qquad a\sin\tau^* - 2b\cos\tau^* = o \quad \text{(in case } ab<o,\ ac<o)$$

so that the eigenvalues $\lambda_1(\tau), \lambda_2(\tau)$ of $C(\nu(\tau))$ are pure imaginary iff $\tau=\tau^*$.We note that the eigenvalues of $V(o,o,\varepsilon,\nu(\tau))$ are pure imaginary iff $\tau=\tau_2$.

Theorem 3.1

Under the nonresonance condition NR(3) and the conditions (2.4-5) system (2.3) can be transformed by (3.4) or (3.5) into system (3.6) or (3.7) respectively. In terms of the parametrization by ε,τ (cf.(3.3),(3.9-13)) one has the following:

1) There exist positive constants $\bar{\varepsilon}=\bar{\varepsilon}(\underline{\tau}),\kappa=\kappa(\underline{\tau},m)$ such that for $\varepsilon\in(o,\bar{\varepsilon})$ there is a continuous function $(u,v)=S_o(y,\varepsilon,\tau)$ for all $\tau\in(\underline{\tau},\underline{\tau}+\pi)$ and all such ε with

$$(3.14)\qquad \varepsilon^{1/2} < \kappa\min(|\cos\tau - \tfrac{a}{b}\sin\tau|,|\sin\tau|) =: \kappa\alpha_o(\tau)$$

so that for fixed such ε,τ

$$(3.15)\qquad M^n = \{(S_o(y,\varepsilon,\tau),y,\varepsilon,\tau): y\in T^n\}$$

represents an invariant n-torus for (3.6) at $\nu=\nu(\tau)$.

2) There exist positive constants $\bar{\varepsilon}=\bar{\varepsilon}(\tau_1,\tau_2),\kappa=\kappa(\tau_1,\tau_2,m)$ so that for $\varepsilon\in(o,\bar{\varepsilon})$ there is a continuous function $\rho=S_1(\theta,y,\varepsilon,\tau)$ for all $\tau\in(\tau_1,\tau_2)-\{\tau^*\}$ and all such ε with

$$(3.16a)\qquad \varepsilon < \kappa\gamma_1(\nu(\tau)),$$

$$(3.16b)\qquad \varepsilon^{1/2} < \min_{j=1,2} |\mathrm{Re}\lambda_j(\tau)| =: \kappa\alpha_1(\tau)$$

so that for fixed such ε,τ

$$(3.17)\qquad M^{n+1} = \{(S_1(\theta,y,\varepsilon,\tau),\theta,y,\varepsilon,\tau): (\theta,y)\in T^{n+1}\}$$

represents an invariant (n+1)-torus for (3.7) at $\nu=\nu(\tau)$.

3) The functions S_j belong to C^{m-2} with respect to the angular variables and are of order $\mathcal{O}(\varepsilon/\alpha_j(\tau))$ for $\varepsilon\to o$ (j=1,2).

Remark 3.2

1) The above theorem describes the parameter sets in μ-space for which the persistence of the trivial n-torus M_o^n and (n+1)-torus

M_o^{n+1} of the unperturbed systems (3.6) and (3.7) is guaranteed. We note that for fixed $|\mu|=\varepsilon$ there are gaps around τ_2 and τ^* where the existence of the tori (3.15) and (3.17) has not been established.At τ_2 and τ^* the tori M_o^n and M_o^{n+1} are lacking normal hyperbolicity due to pure imaginary eigenvalues of the matrices $V(o,o,\varepsilon,\nu(\tau_2))$ and $C(\nu(\tau^*))$ respectively.Theses "directions" are associated with Hopf bifurcations in the truncated systems (3.6) and (3.7). Part 2 of the theorem shows that the Hopf bifurcation at τ_2 gives rise to the (n+1)-torus (3.17).In section 4 we shall investigate the implications of the Hopf bifurcation at τ^*.

2)For a detailed proof of Theorem 3.1 we refer to [5] .We just note that (3.14) and (3.16b) guarantee that the coupling between the amplitude and the angular equations in (3.6) and (3.7) is sufficiently weak. In particular these conditions imply that the normal rates $\varepsilon\alpha_j(\tau)$ dominate the tangential rates $\mathcal{O}(\varepsilon^2)$. Condition (3.16a) serves to bound the "critical denominator" $\rho_1+\gamma_1(\nu(\tau))$ in the $\mathcal{O}(\varepsilon^2)$-term of the θ-equation in (3.7) away from zero.

4.Invariant Tori of Dimension n+2

In this section we shall study the case "ab<o,ac<o" where one has the "critical halfline"

$$(4.1) \qquad \ell^*= \{\mu=\varepsilon\nu(\tau^*):\ \varepsilon>o\}.$$

As it is well-known by now system (2.6) is integrable on ℓ^* with integrating factor $(\rho_1+\gamma_1(\nu(\tau^*))^{-2b/a}$. In order to unfold the associated vertical Hopf bifurcation in (2.6) we adjoin higher order y-independent terms to (2.3).In this contribution we confine ourselves to impose the nonresonance condition NR(6) where not only the ε^2-,but also the ε^3-coefficient in the r-equation of (2.3) may be assumed to be y-independent.For weaker nonresonance conditions we refer to [5].

We parametrize a cusp-like neighborhood of (4.1) by ε,η via

$$(4.2)\qquad \mu=\varepsilon\nu,\quad \nu=\nu(\varepsilon,\eta)=(\cos(\tau^*+\varepsilon\eta),\sin(\tau^*+\varepsilon\eta))^T,\ \varepsilon>0,|\eta|<\eta_0,$$

and denote system (2.3) along (4.2) in the scaled variables $\xi=\varepsilon(r,w)^T$ by

$$(4.3)\qquad \begin{aligned}\dot\xi&=\varepsilon\Xi^*(\xi,\varepsilon,\eta)+O(\varepsilon^4) = \varepsilon\Xi_1(\xi)+\varepsilon^2\Xi_2(\xi,\eta)+\varepsilon^3\Xi_3(\xi,\eta)+O(\varepsilon^4)\\ \dot\theta&= 1 + \varepsilon\Theta_1(\xi,\varepsilon,\eta) +O(\varepsilon^3)\\ \dot y&= \omega + \varepsilon Y_1(\xi,\varepsilon,\eta) +O(\varepsilon^3).\end{aligned}$$

By $\Gamma(\varepsilon,\eta)=\gamma(\nu(\tau^*))+O(\varepsilon)$ we denote the positive equilibrium of $\dot\xi=\varepsilon\Xi^*(\xi,\varepsilon,\eta)$. In terms of

$$(4.4)\qquad \rho = \xi - \Gamma(\varepsilon,\eta)$$

system (4.3) takes the form (cf.3.8b))

$$(4.5)\qquad \begin{aligned}\dot\rho &= \varepsilon P(\rho,\nu(\tau^*))+\varepsilon^2P_2(\rho,\eta)+\varepsilon^3P_3(\rho,\eta)+O(\varepsilon^4)\\ \dot\theta &= 1 + \varepsilon G(\rho,\varepsilon,\eta) +O(\varepsilon^3)\\ \dot y &= \omega + \varepsilon H(\rho,\varepsilon,\eta) +O(\varepsilon^3)\end{aligned}$$

with polynomials G,H and $P_2,P_3=O(|\rho|)$. In particular one has

$$(4.6)\qquad \begin{aligned}P_2(\rho,\eta) &= C_2(\eta)\rho + O(|\rho|^2)\quad\text{with}\\ \operatorname{trace} C_2(\eta) &= \lambda^*+ \eta(\cos\tau^*+\tfrac{2b}{a}\sin\tau^*)\end{aligned}$$

with a fixed $\lambda^*\in\mathbb{R}$. Because of (3.13) we can introduce a new parameter δ via

$$(4.7)\qquad \delta = \operatorname{trace} C_2(\eta)$$

so that the eigenvalues of $\varepsilon C(\nu(\tau^*))+\varepsilon^2C_2(\eta)$ cross the imaginary axis transversally at

$$(4.8)\qquad \pm i\{\varepsilon\sqrt{-2ac}\,\gamma_1(\nu(\tau^*))+O(\varepsilon^2)\} =: \pm i\{\varepsilon\beta^*+O(\varepsilon^2)\}$$

for $\delta=o$. Since $\dot{\rho}=\varepsilon P(\rho,\nu(\tau^*))$ represents a polynomial integrable system its Birkhoff-Normal-Form in polar coordinates

$$(4.9) \qquad \rho = (u\cos\psi, u\sin\psi)^T, \quad u<\rho_o,$$

may be assumed to be of the form $\dot{u}=o$, $\dot{\psi}=\varepsilon\Psi(u^2)=\varepsilon(\beta^*+O(u^2))$ (cf.(4.8) and [1,5]). With respect to (4.9) system (4.5) can be written as a perturbation of a system that is analytic in (u,ψ):

$$\begin{aligned}
\dot{u} &= \varepsilon^2 U_2(u,\psi,\delta)+\varepsilon^3 U_3(u,\psi,\delta)+O(\varepsilon^4)\\
\dot{\psi} &= \varepsilon\Psi(u^2)+\varepsilon^2\tilde{\Psi}(u,\psi,\delta) \qquad +O(\varepsilon^3)\\
\dot{\theta} &= 1 + \varepsilon\tilde{G}(u,\psi,\varepsilon,\delta) \qquad +O(\varepsilon^3)\\
\dot{y} &= \omega + \varepsilon\tilde{H}(u,\psi,\varepsilon,\delta) \qquad +O(\varepsilon^3).
\end{aligned}$$

By analytic averaging transformations we can delete the ψ-dependence of the terms listed explicitly (cf.[5]) so that the transformed system is of the form

$$(4.1o) \qquad \begin{aligned}
\dot{u} &= \varepsilon^2 M(u,\delta)+\varepsilon^3 M_3(u,\delta) \ +O(\varepsilon^4)\\
\dot{\psi} &= \varepsilon\Psi(u^2) \ +\varepsilon^2\hat{\Psi}(u,\delta) \ +O(\varepsilon^3)\\
\dot{\theta} &= 1 \quad + \varepsilon\hat{G}(u,\varepsilon,\delta)+O(\varepsilon^3)\\
\dot{y} &= \omega \quad + \varepsilon\hat{H}(u,\varepsilon,\delta)+O(\varepsilon^3).
\end{aligned}$$

With (4.6-7) a lengthy computation shows

$$(4.11) \qquad M(u,\delta) = \frac{1}{2\pi}\int_o^{2\pi} U_2(u,s,\delta)\,ds = u\{\frac{\delta}{2} + Ku^2 + O(u^4)\}.$$

Thus simple positive zeros of (4.11) represent invariant (n+2)-tori M_o^{n+2} for the unperturbed system (4.1o). These persist under the perturbation by the above listed O-terms. We formulate this result for the small simple positive zero

$$(4.12) \qquad \zeta(\delta) = \sqrt{-\delta/2K} + O(|\delta|), \quad \delta K<o,$$

of (4.11) whose existence follows for $K\neq o$. We note that the derivative of (4.11) at (4.12) is of the form $-\delta(1+o(1))$.

Theorem 4.1

Under the nonresonance condition NR(6) and conditions (2.4-5) system (2.3) can be transformed into (4.1o). In terms of the parametrization by ε,δ (cf. (4.2), (4.7)) one has the following for $K\neq o$ in (4.11):

There exist positive constants $\delta_o, \bar{\varepsilon}=\bar{\varepsilon}(\delta_o), \kappa=\kappa(\delta_o,m)$ such that for $\varepsilon\in(o,\bar{\varepsilon})$ there is a continuous function $u=S_2(\psi,\theta,y,\varepsilon,\delta)$ for all δ with $o<|\delta|<\delta_o, \delta K<o$ and all such ε with $\varepsilon<\kappa|\delta|, \varepsilon^{1/2}<\kappa\zeta(\delta)$ so that for fixed such ε,δ

$$M^{n+2} = \{(S_2(\psi,\theta,y,\varepsilon,\delta),\psi,\theta,y,\varepsilon,\delta):(\psi,\theta,y)\in T^{n+2}\}$$

represents an invariant (n+2)-torus for (4.1o). The function S_2 belongs to C^{m-2} with respect to (ψ,θ,y) and is of the form $S_2 = \zeta(\delta)+\mathcal{O}(\varepsilon^2/|\delta|), \varepsilon\to o$. The torus M^{n+2} is stable (unstable) for $\delta>o$ ($\delta<o$).

Remark 4.2

1) The above theorem shows the existence of invariant (n+2)-tori for (4.1o) and hence for (1.1). Again, for fixed $|\mu|=\varepsilon$ there is a gap (o,δ_ε) or $(-\delta_\varepsilon,o)$ on which the persistence of M_o^{n+2} has not been established. An analogous result can be formulated for any simple positive zero of (4.11) in the domain of the above transformations.

2) The proof of Theorem 4.1 follows again from the persistence results in [5,6]. In order to put (4.1o) into the setting required there one just employs a preliminary translation $\underline{u}=u-u^*(\varepsilon,\delta)$ where $u^*(\varepsilon,\delta)=\zeta(\delta)+\mathcal{O}(\varepsilon)$ denotes the simple zero of $M+\varepsilon M_3$.

5. The Case $A=A_2$ (cf. (1.4))

Under the nonresonance condition

"There exist positive constants κ_o,κ_1 so that

$|(k,\omega)| \geq \kappa_o|k|^{-\kappa_1}$ holds for $k\in Z^n-\{o\}$"

the approximate Birkhoff-Normal-Form (corresponding to (2.3)) may be of the form

$$(5.1)\qquad \begin{aligned} \dot{x}_1 &= \mu_1 x_1 + a x_1 x_2 + O(3) \\ \dot{x}_2 &= \mu_2 x_2 + b x_2^2 + c x_1^2 + O(3) \\ \dot{y} &= \omega + Y_2(x,\mu) + O(3) \end{aligned}$$

due to some underlying symmetries.For $abc \neq o$ the results of the previous sections apply.Only the interpretations are to be adjusted to the fact that (5.1) does not have a θ-equation.If the quadratic terms are forced to be zero and the truncated x-equation in (5.1) is

$$\dot{x}_1 = \mu_1 x_1 + a x_1^2 + b x_2^2 , \quad \dot{x}_2 = \mu_2 x_2 + c x_1^2 + d x_2^2 , \quad ad-bc \neq o,$$

then the results of [6] apply if their interpretation is modified similarily.

6.References

1. Bibikov,Y.N.(1979) Local theory of nonlinear analytic ordinary differential equations,Lect.Notes Math.7o2(Springer,Berlin).
2. Chow,S.N.&Hale,J.K.(1982) Methods in bifurcation theory,Grundlehren math.Wiss.251(Springer,New York).
3. Flockerzi,D.(1984) Generalized bifurcation of higher dimensional tori,JDE 55.3,346-367.
4. Flockerzi,D.(1985) On the $T^k \to T^{k+1}$ bifurcation problem,"Singularities&Dynamical Systems",S.N.Pnevmatikos(ed.),North Holland, 35-46.
5. Flockerzi,D.(1986) Invariant manifolds and bifurcation of tori, Habilitationsschrift,Univ.Würzburg.
6. Flockerzi,D.(1986) Persistence and bifurcation of invariant tori,preprint 147,Univ.Würzburg.
7. Guckenheimer,J.&Holmes,P.(1983) Nonlinear oscillations,dynamical systems and bifurcations of vector fields,Appl.Math.Sci.42 (Springer,New York).
8. Guckenheimer,J.(1984) Multiple bifurcation problems of codi-

mension two,SIAM J.math.Anal.15.1,1-49.

9. Iooss,G.&Langford,W.F.(198o) Conjectures on the routes to turbulence via bifurcations,"Nonlinear Dynamics",R.H.G.Helleman(ed.), Annals of the NYAS Vol.357,489-5o5.

1o.Langford,W.F.(1979) Periodic and steady-state interactions lead to tori,SIAM J.appl.Math. 37.1,22-48.

11.Sacker,R.J.&Sell,G.R.(1978) Spectral theory for linear differential systems,JDE 27,32o-358.

12.Scheurle,J.&Marsden,J.(1984) Bifurcation to quasi-periodic tori in the interaction of steady state and Hopf bifurcations, SIAM J.math.Anal.15.6,1o55-1o74.

13.Sell,G,R.(1979) Bifurcation of higher dimensional tori,Arch. Rat.Mech.Anal.69.3,199-23o.

14.Spirig,F.(1983) Sequence of bifurcations in a three-dimensional system near a critical point,ZAMP Vol.34,259-276.

Dietrich Flockerzi, Mathematisches Institut,Am Hubland,
D-8700 Würzburg,West Germany.

International Series of
Numerical Mathematics, Vol. 79

HOPF BIFURCATION IN THE DRILLING PROCESS UNDER REGENERATIVE CUTTING CONDITIONS

Mustapha S.Fofana

Budapest University of Technology, Department of Production Engineering, H-1521 Budapest, Hungary

1. Introduction

Regenerative effects are observed in many naturally occurring nonconservative systems. The purpose of this paper is to describe the regenerative effects; namely, those related to what is called Hopf bifurcation in the drilling process. Here, the bifurcation refers to changes in the qualitative behaviour of the radial drilling machine tool and the solutions of its delay differential equation with a varying parameter.Regenerative effect is a dynamic problem in metal cutting machine tools because of its adverse effects especially on machining accuracy, surface finish and tool life. Vibration may occur as a result of this instability effect.

Experimental tests of the radial drilling machine tool in question have been carried out by the Development Institute of Hungarian Machine Tool Works [1]. In their experiment, pre-drilled holes in mild steel (C 45) are then bored to their required diameter. On the basis of their experimental results summarized in table I, we collect the data for a simple mathematical model. Its mathematical investigation gives a good description of the regenerative effects analyzed in details by Tobias [4].

2. Mathematical model

Consider the following mathematical model of one degree of freedom with a reduced mass m, on a restoring spring of stiffness s and with a damping factor k, as shown in Fig. 1.

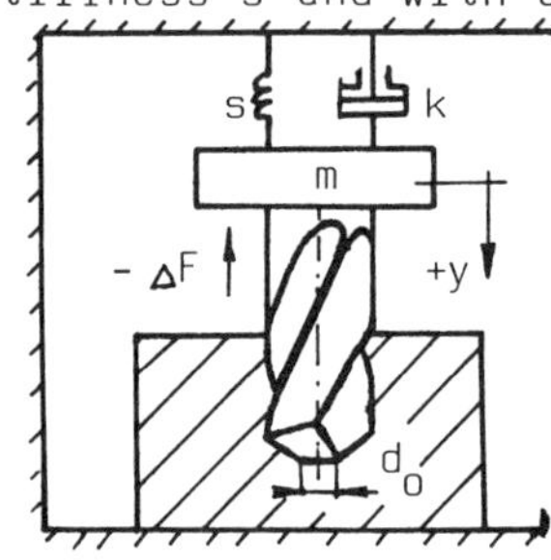

natural frequency, ν = 18 Hz,

angular natural frequency, $\beta = 2\pi\nu$ = 113 1/s

stiffness, s= 3o MN/m,

relative damping, D = o.o8,

reduced mass, $m = \frac{s}{\beta^2}$ = 222o kg,

damping factor, $k = 2D\beta m = 4o \frac{kNs}{m}$

Fig.1.

The delay differential equation of the model is

$$m\ddot{y}(t) + k\dot{y}(t) + sy(t) = -\Delta F\,(y(t)-y(t-T/z),\ \dot{y}(t)), \qquad (1)$$

where t stands for time, z is the number of cutting edges. $T = \frac{6o}{n}$, is the time for one revolution in seconds, where n is the rotational speed. Generally in drilling process, the cutting force acting on the two edges of the drill sets up a force in line with the drill axis (feed force) and a torsional moment. In our investigation, the mathematical model accounts for vibration in the vertical direction only, since the torsional moment has been ignored by assuming very large torsional stiffness in the spindle drive. The cutting force variation, ΔF is a nonlinear function which is a function of the difference between the displacement y(t) and its delayed value y(t-T/z) and the velocity $\dot{y}(t)$. This cutting force variation which develops during regenerative cutting conditions is taken according to Tobias [4].

$$\Delta F(y(t)-y(t-T/z),\dot{y}(t)) = Ck_1 T\dot{y}(t) + zk_1(y(t)-y(t-T/z)) + h.o.t. \qquad (2)$$

The first term in equation (2) describes a damping which is quite large for small cutting speeds (ie for large values of T). As it is explained in [4], the damping effect is caused by the negative rake angle at the chisel edge of the drill. Thus, we expressed the penetration rate, C as a rate of proportionality of this damping effect by the formula

$$C = p \frac{k_1(d) - k_1(d_o)}{k_1(d_o)}, \tag{3}$$

where d_o is the web thickness of the drill and p = o.1. If p is greater than o.1, then the damping effect becomes too strong and there will be no agreement between the experimental observations and the theoretical results. The values of C in table I are computed according to (3). In the case of experiment A.1, d is greater than d_o, which implies that the chisel edge is not engaged during drilling (ie the damping effect does not occur), hence C = 0.

The second term in (2) comes from the linearization of the cutting force with respect to the feed e. Thus, its variation with respect to one edge is

$$\frac{1}{z} \Delta e(t) = y(t) - y(t-\frac{T}{z}), \tag{4}$$

and the coefficient k_1 is

$$k_1 = \frac{dF}{de}\bigg|_{e = e_o}, \tag{5}$$

where e_o is the theoretical feed. F is expressed by the technological parameters as follows (see [2]):

$$F = o{,}45\ k_{1.1}\ e^{x_f} \frac{D_o^2}{D_o+d} (1 - (\frac{d}{D_o})^2)\ [N]\ , \tag{6}$$

where $k_{1.1}$ = 24o and x_f = o.4.

Thus, for the experimental results of [1] summarized in table I, the value of k_1 is computed by the formula

$$k_1 = 43.2\ e^{-o.6} \frac{D_o^2}{D_o+d} (1 - (\frac{d}{D_o})^2) \left[\frac{N}{mm}\right]. \tag{7}$$

With this equation, we determine the experimental points of the stability chart (see Fig 2).

3. Stability Investigation

It is of great importance to examine the presence or absence of vibrations during drilling process by means of the technological and mechanical data of the system. To do this, we investigate the stability of the delayed differential difference equation

$$\ddot{y}(t)+(2D\beta+\frac{1}{m}Ck_1T)\dot{y}(t)+(\beta^2+\frac{1}{m}zk_1)y(t)-\frac{1}{m}zk_1y(t-\frac{T}{z}) = 0, \qquad (8)$$

which comes from (1) and (2) after division by m.
This equation has a trivial solution for $y = 0$. The method described in [3] is applied here to get the following formulae of the border lines of stability for different values of C (see Fig 2).

$$T=\frac{w+\sqrt{w^2+(2\beta yz)^2}}{2\beta^2}, \quad w = \frac{2D\beta zy}{Cy+\sin y}(1-\cos y), \quad n = \frac{60}{T}, \qquad (9)$$

and

$$k_1 = -\frac{m}{T}2D\beta\frac{y}{\sin y}\;\frac{1}{1+C\frac{y}{\sin y}}. \qquad (10)$$

Here, we consider n, the rotational speed to be the bifurcation parameter and its critical value, written $n_{crit.}$ can be read from the stability chart (see Fig 2). By keeping k_1, the feed rate coefficient fixed and increasing the value of n, the characteristic equation of (8) has a pair of complex conjugate eigenvalues

$$\lambda_{1,2}(n) = \alpha(n)\pm i\omega(n), \qquad (11)$$

that moves from the left to the right hand side (see Fig.3). In Fig.2, it can be seen that the number of the complex conjugate eigenvalues increase whenever the curves meet.

TABLE I. Experimental results of [1] during drilling process.

No.	D_0 [mm]	d [mm]	n [rev./min]	e [mm/rev.]	A [µm]	ν [Hz]	k_1 [kN/mm]	C	symbol	stable / unstable
A.1.	40	10	118	0.187	190	17	3.544	0	△	unstable
B.1.	60	10	75	0.25	150	20	4.962	0.005	○	unstable
B.2.			95	0.375	36	-	3.891			stable
B.3.				0.187	175	19	5.907			unstable
B.4.				0.125	195	19	7.522			unstable
C.1.	60	6	118	0.375	20	-	4.202	0.014	□	stable
C.2.				0.25	165	19.5	5.359			unstable
C.3.				0.187	165	19.5	6.379			unstable

D_0 - diameter of the drill,

d - diameter of the pre-drilled hole,

e - feed rate,

A - vibration amplitude in the vertical direction.

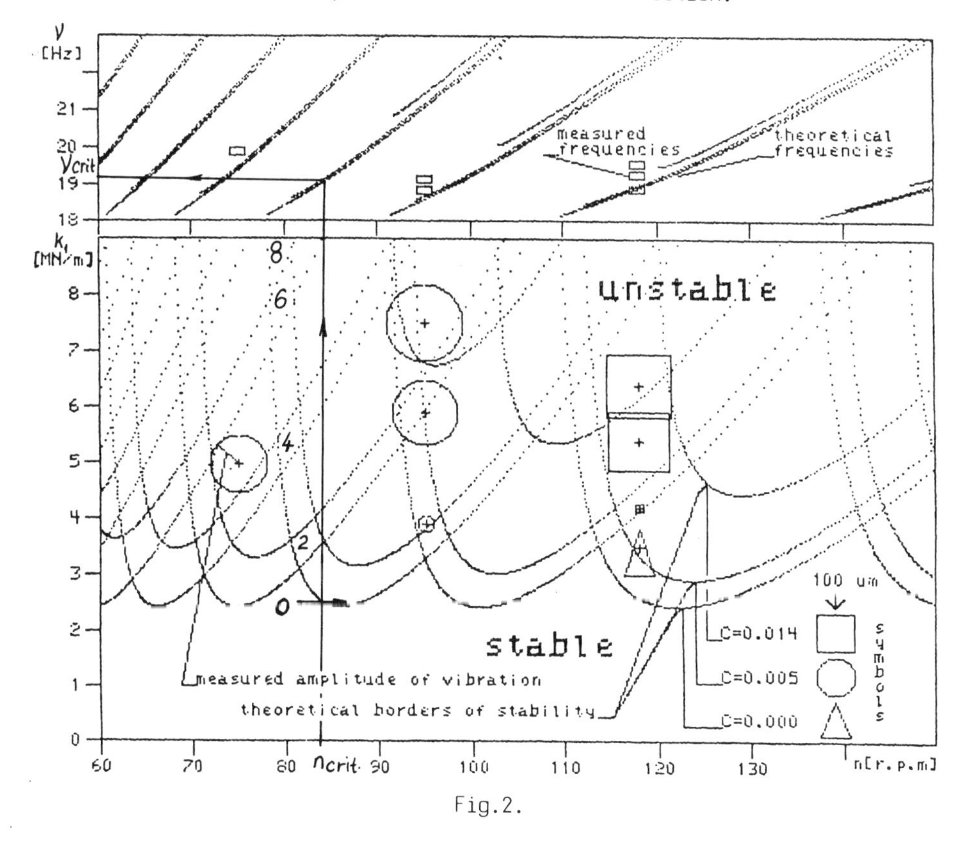

Fig.2.

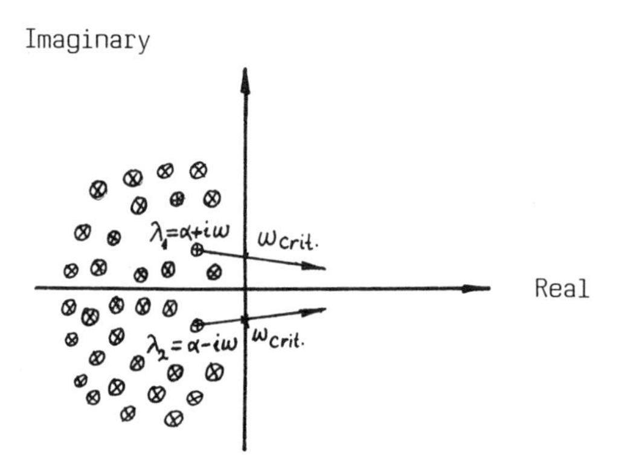

Fig.3.

If $\quad n = n_{crit.}$, then

$$\lambda_{1,2}(n_{crit.}) = \alpha(n_{crit.}) \pm i\omega(n_{crit.}), \tag{12}$$

where $\alpha(n_{crit.}) = 0$, $\omega(n_{crit.}) = 2\pi\nu_{crit.} > 0$.

From the stability investigation, it can be proved that the remaining eigenvalues of (8) have strictly negative real parts (ieRe $\lambda_i < 0$, $i \neq 1,2$). Furthermore, we prove numerically that the negative real parts of the eigenvalues

$$\mathrm{Re}\left.\frac{d\lambda}{dn}\right|_{n=n_{crit.}} > 0. \tag{13}$$

Therefore, one can see from our results that the conditions of the Hopf bifurcation theorem are satisfied and there exists a periodic motion around $n = n_{crit.}$. However, we still cannot determine the supercritical or subcritical bifurcation mathematically because, for the time being, the determination of the penetration rate C contains great uncertainty without dynamical tests of the drilling process. But the experimental results of [1] have shown that the stable periodic motions and the unstable equilibrium position refer to supercritical bifurcation.

REFERENCES

[1] Baráti A: Vibration test of the radial drilling machine tool Rfh-75, research report No.19917, SZIMFI, 1981. (in Hungarian)

[2] Bakondi and Kardos: Machine tool production. Tankönyvkiadó, Budapest, 1979. (in Hungarian).

[3] Stépán G: Stability investigation of retarded differential equations, Acta Technica, 9o (1o9-132), 198o.

[4] Tobias S.A.: Machine tool vibration, Blackie and son LTD, Glasgow, 1965.

M.S.Fofana,M.Sc.(mechanical engineering).
No.48 Pillar street, Freetown,
Sierra Leone.

International Series of
Numerical Mathematics, Vol. 79

CRITICAL POINTS OF MIXED FLUIDS AND THEIR NUMERICAL TREATMENT

A. Griewank and P. Rabier

1. INTRODUCTION

The simulation of oil reservoirs based on compositional models requires efficient numerical methods for computing multiphase equilibria of mixed fluids. The number of coexisting *phases* e.g. liquid, vapour and aqua, depends on the composition of the fluid as well as the temperature and pressure. As these external conditions change, a given mixed fluid may reach a transition state where a new phase emerges or an old one disappears. Most of these so called *saturation points* do not correspond to local bifurcations in the governing equations. Rather we have a global exchange of stability as in the case of a double well potential whose two local minima have exactly the same energy. The proper numerical treatment of such transitions requires techniques that are familiar from the handling of degenerate constraints in optimization algorithms.

More interestingly from the bifurcation point of view are so called critical points where two kinds of saturation states, e.g. bubble and dew points, coalesce. These are also important in practice because enhanced oil recovery by flooding with CO_2 or sulfactants creates critical conditions in a frontal region. Apparently the structure of these critical points and their vicinity has not yet been described in the mathematical literature. Here we will analyze the situation using an analytic energy function. In the chemical engineering literature [9] it is understood that such analytical models are inadequate because they are supposedly unable to reproduce observed *critical exponents*. So far the mathematical arguments for this conclusion appear somewhat unsatisfactory and the whole question warrants a thorough investigation with the tools of bifurcation theory. Since most simulation codes are based on analytical models their proper understanding is of great practical importance.

The first author is endebted to the staff of Mobil Research Corporation for stimulating his interest in the phase equilibrium problem. His work was partially supported by NSF grant DMS-8401023.

2. PROPERTIES OF THE HELMHOLTZ ENERGY FUNCTIONS

Consider a mixture of $n > 1$ fluid components, such as carbon dioxide, decane, water, etc. The composition of this mixed fluid is then defined by the vector of mole numbers

$$m \equiv (\mu_1, \mu_2, \dots \mu_n) \geq 0$$

such that it contains $\mu_i \geq 0$ moles of the i-th fluid component. Even at high pressures and low temperatures the volume v occupied by the mixture is bounded below by

$$v > b(m) > 0 \quad \text{if } m > 0 \quad . \tag{2.1}$$

The so-called *hard-sphere* volume b(n) has norm properties, i.e.

$$b(\bar{m}+\tau * m) \leq b(\bar{m}) + \tau * b(m) \qquad \text{if } \tau \geq 0 \quad .$$

Due to the homogeneity of b the condition (2.1) is equivalent to

$$d \equiv m/v \in \mathcal{D} \equiv \{d \in \mathbb{R}^n : d \geq 0 \quad , \quad b(d) \leq 1\} \quad . \tag{2.2}$$

In other words the *density vector* d must belong to the intersection $\mathcal{D}$ of the positive cone and the unitball with respect to the norm b.

The *Helmholtz energy* of a mixed fluid that occupies a unit volume is given by a lower semi-continuous function

$$E(d) : \mathcal{D} \to \mathbb{R} \cup \{\infty\} \quad .$$

For physical reasons this extended real-valued function satisfies

$$E(0) = 0 \quad , \quad E(d) = \infty \iff b(d) = 1$$

and has no subdifferential on the boundary, i.e. a vector $g \in \mathbb{R}^n$ s.t.

$$E(d) \geq E(d_0) + g^T(d - d_0) \qquad \text{for all } d \in \mathcal{D} \tag{2.3}$$

can only exist if d_0 belongs to the interior $\mathcal{D}^0$ of $\mathcal{D}$. Finally we may assume that E(d) is five times continuously differentiable in $\mathcal{D}$.

For example, all these conditions are met by the semi-empirical Peng-Robinson formula [8]

$$E(d) \equiv R \cdot T \sum_{i=1}^{n} d_i \, \ell n \frac{d_i}{(1 - b(d)} - \frac{d^T A d}{\sqrt{8}\, b(d)} \, \ell n \frac{1 + (1 + \sqrt{2}) b(d)}{1 + (1 - \sqrt{2}) b(d)}$$

where A is a symmetric positive n x n matrix and b(d) is usually a weighted ℓ_1-norm. Not only the product R·T but also the matrix A and the function b(d) may depend on the temperature T which will be considered constant throughout.

3. CHARACTERIZATION OF MULTIPHASE EQUILIBRIA

Theoretically a *feed* fluid with mole vector $m \in \mathcal{D}$ in a container of unit volume can always split into $p > 0$ separate phases (i.e. fluid parcels), whose density vectors $d_j \in \mathcal{D}$ and *phase volumes* $v_j \geq 0$ must satisfy the mass balance equation

$$(3.1) \qquad \sum_{j=1}^{p} v_j \cdot d_j = m \quad \text{and} \quad v \equiv \sum_{j=1}^{p} v_j \leq 1 \quad .$$

Subject to these bilinear constraints one has to solve the task

$$(3.2) \qquad \text{Min} \sum_{j=1}^{p} v_j \cdot E(d_j) \qquad \text{over all } p > 0$$

in order to find the physically stable equilibrium configuration. Under our assumptions on $E(d)$ this *semi-infinite programming problem* [2] has always a global minimizer with $p \leq n+1$, in agreement with *Gibbs phase rule*. Moreover all phase densities d_j involved will belong to the interior of the domain $\mathcal{D}$ so that its boundary can not become an active constraint. The *Kuhn-Tucker-Karusch* conditions for such a local minimum consist of the constraints (3.1) and the additional equations

$$(3.3) \qquad \begin{aligned} t_j &\equiv E(d_j) + P - g^T d_j \geq 0 = v_j \cdot t_j \\ \nabla E(d_j) &= g \text{ if } t_j = 0 \end{aligned}$$

where the $t_j \in \mathbb{R}$ as well as $g \in \mathbb{R}^n$ and $P > \mathbb{R}_+$ are Lagrange multipliers. P represents the physical pressure that results from the bound 1 on the volume v. Under reservoir conditions the equilibrium is attained by minimizing the Gibbs free energy for fixed pressure P but with arbitrary volume v. Here we discuss the Helmholtz problem (3.1), (3.2) because it is mathematically more convenient, yet equivalent to the Gibbs problem with respect to the bifurcation aspect. Provided $v_j > 0$ we have by (3.3)

$$(3.4) \qquad P = \pi(d_j) \qquad \text{with} \quad \pi(d) \equiv \nabla E(d) \cdot d - E(d)$$

While the conditions (3.3) are easily checked at a finite number of vectors d_j, they also imply the global inequality

$$(3.5) \qquad t(d) \equiv E(d) - g^T d + P \geq 0 \quad \text{for all } d \in \mathcal{D} \quad .$$

Geometrically this *tangent plane criterion* [1] requires that

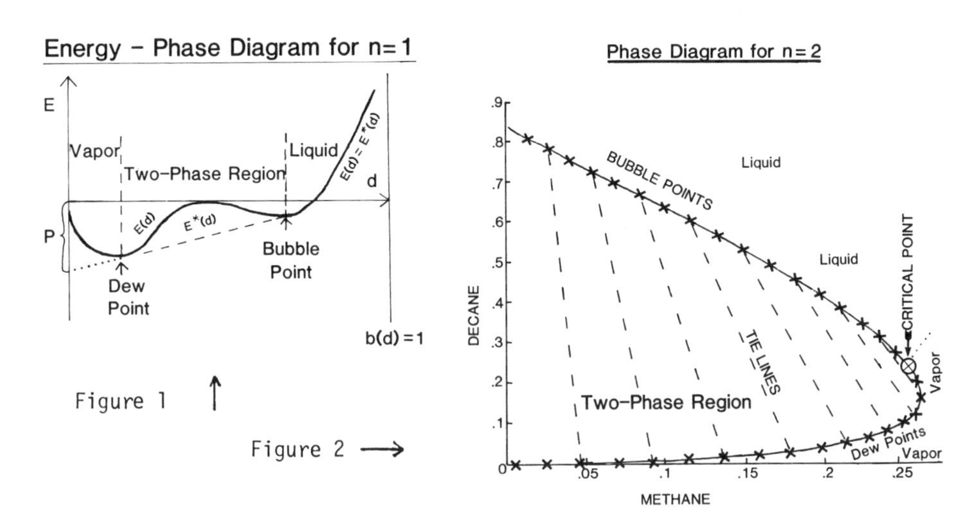

Figure 1 ↑

Figure 2 ⟶

$$g^T m - P = E^*(m) \equiv \text{convhull}\,(E)(m) \tag{3.6}$$

as depicted in Fig. 1 for the case $n+1 = 2 \geq p$. Hence, the determination of the stable phase split for given feed $m \in \mathcal{D}$ and $v \leq 1$ is equivalent to the evaluation of the convex hull $E^*(d)$ of $E(d)$ at $d = m$.

The computational determination of the equilibrium configuration has at least three difficult aspects. Firstly the verification of the inequality (3.5) requires the global minimization of $t(d)$ over the domain $\mathcal{D}$. Even though $\mathcal{D}$ is compact this cannot be done with certainty unless E is evaluated on a rather dense grid in $\mathcal{D}$. Secondly the KKT-conditions (3.3) loose the property of *strict complementarity* [6] at saturation points where $v_j = 0 = t_j$ for some j. While suitably designed constrained optimization codes can cope with this difficulty their speed of performance is severely impaired whenever the *reduced Hessian* is singular. This third difficulty must occur near *critical points* where several phases coalesce so that the corresponding phase volumes are arbitrary. Baring degeneracies of much higher codimension, critical points are a two phase phenomenon, and we will therefore assume from now on that $p \leq 2$.

A typical situation is depicted in Fig. 2 for a binary mixture of methane and decane at 381°K in a unit volume. The tick marks measure the fractions μ_i/β_i between μ_i, the number of moles, and β_i, the hard sphere volume, of one mole of component i. Unless its density is quite high or extremely low, the comparatively heavy hydrocarbon decane splits into a vapor and a liquid phase. Consequently the bubble point and the dew point curves intersect the vertical axis ($\mu_1 = 0$, i.e. no decane) rather high and very close to the origin respectively. As methane is added it desolves into the vapour and to a lesser extent into the liquid phase. If the feed composition m is varied along one of the *tie-lines* in the *two-phase region*, the liquid and vapour density vectors d_j for $j = 1,2$ stay the same and merely the phase volumes v_j change. The pressure $P = \pi(d_j)$ for $j = 1,2$ is also constant. For increasing methane density μ_1/β_1 either the vapour or the liquid phase will eventually disappear depending upon whether the fixed decane density μ_2/β_2 lies above or below a critical value of about .235. The corresponding value of μ_1/β_1 is about .252. Near the critical point (.252, .235) the labeling as vapour or liquid phases becomes meaningless because the physical properties are almost indistinguishable, as indicated by the dotted line.

As one can see quite clearly the bubble and dew point curves merge smoothly at the critical point. This is not only interesting from an analytical point of view but also important for the design of tests [7] to decide whether a given feed composition is stable or lies in the two phase region. In the remainder of the paper we characterize the critical points as singular minima of the tangent function t(d) and sketch the argument that establishes the smoothness of the saturation boundary across critical points.

4. NONDEGENERATE CRITICAL POINTS

Accoridng to (3.3) and (3.5) for $p = 1$, a feed m will be stable as a single phase exactly when

(4.1) $$t(d) = E(d) - \nabla E(m) \cdot d + \pi(m) \geq 0 = t(m) \quad .$$

Then m has to be at least a local minimizer of t(d), which requires that the *smallest eigenvalue* $\alpha(d)$ of the Hessian $\nabla^2 E(d) = \nabla^2 t(d)$ is nonnegative at $d = m$. Moreover if $\alpha(m) = 0$ but all other eigenvalues are positive, then the following higher order optimality conditions developed in [3] must be satisfied. With u(d) a differentiable unit eigenvector corresponding to $\alpha(d)$ the quantities

(4.2) $$\beta(d) \equiv \nabla\alpha(d) \cdot u(d) \quad \text{and} \quad \nu(d) \equiv \nabla\beta(d) \cdot d$$

must be zero and nonnegative at $d = m$ respectively, i.e.

(4.3) $$\beta(m) = 0 \leq \nu(m) \quad \text{if} \quad \alpha(m) = 0 \quad .$$

Conversely if this condition holds with

(4.4) $$\nu(m) > 0 \quad \text{and} \quad \nabla\alpha(m) \neq 0$$

then m is indeed an isolated local minimizer of $t(d)$. Such points will be called *(nondegenerate) critical points*, provided that they are also the only global minimizer of $t(d)$. The latter condition as well as the inequalities (4.4) are stable with respect to small perturbations. Therefore the critical points form locally the solution set

$$(4.5) \qquad CP \equiv \{d \in \mathcal{D}: \alpha(d) = 0 = \beta(d)\}$$

which is a smooth manifold of codimension 2 because of (4.4).

5. ANALYSIS OF THE SATURATION BOUNDARY

Any two density vectors d_1 and d_2 involved in a multiphase equilibrium must satisfy by (3.3) the equations

$$(5.1) \qquad 0 = G(d_1,d_2) \equiv \nabla E(d_1) - \nabla E(d_2) \quad : \quad \mathcal{D} \times \mathcal{D} \to \mathbb{R}^n$$

and

$$(5.2) \qquad 0 = H(d_1,d_2) \equiv \pi(d_1) - \pi(d_2) \quad : \quad \mathcal{D} \times \mathcal{D} \to \mathbb{R} \ .$$

with π as defined in (3.4). Obviously both equations are anti-symmetric and have therefore the trivial solution space

$$Diag \equiv \{(d,d) : d \in \mathcal{D}\} \ .$$

All other solution pairs (d_1,d_2) form a smooth $n-1$ dimensional manifold as long as

$$(5.3) \qquad \alpha(d_1) > 0 \quad \text{and} \quad \alpha(d_2) > 0$$

and provided $\nabla\pi \neq 0$ in the region of interest.

It follows from fairly standard arguments that the solution set $G^{-1}(0)$ of (5.1) undergoes a pitchfork-like bifurcation along the singular set $\alpha^{-1}(0) \cap Diag$. The nontrivial branch M forms a smooth n dimensional manifold that is symmetric about $Diag$ and has at $(d,d) \in \alpha^{-1}(0)$ the nontrivial tangent $(u(d),-u(d))$. Here $u(d)$ is again the null vector of $\nabla^2E(d)$. Now one can define a local chart

$$f(\xi,x) : \mathbb{R} \times \mathbb{R}^{n-1} \to M$$

such that changing the sign of ξ corresponds to a reflection of $f(\xi,x)$ about $Diag$ and

$$f_\xi(0,x) = (u(d),-u(d)) \quad \text{at} \quad (d,d) = f(0,x) \in \alpha^{-1}(0) \ .$$

Then one can show that the restriction

$$h(\xi,x) \equiv H(f(\xi,x)) : \mathbb{R} \times \mathbb{R}^{n-1} \to \mathbb{R}$$

of H as defined in (5.2) to M has the following properties

$$(5.4) \qquad h(-\xi,x) = -h(\xi,x), \quad h_{\xi\xi\xi}(0,x) = \beta(d), \quad h_{\xi\xi\xi}\,(0,x) = \nabla\beta(d)$$

where $(d,d) = f(0,x)$ and $\beta(d)$ is as defined in (4.2). Pictorially the derivatives of h at some point $(0,x)$ can be represented as follows.

$\partial^i/\partial\xi^i \downarrow$ \ $\partial^j/\partial x^j \rightarrow$					
	0	0	0	0	0
	0	0	0	0	0
	0	0	0	0	0
	β	$\nabla\beta$	?	?	?
	0	0	0	0	0

Thus we conclude that $G^{-1}(0) \cap H^{-1}(0)$ can only bifurcate from *Diag* at those singular points $(d,d) = f(0,x) \in \alpha^{-1}(0)$ that are also critical, i.e. we need $\beta(d) = 0$. Moreover if $\nabla\beta(d) \neq 0$ at such a point $d \in CP$, then it follows from the theory of nondegenerate functions that $G^{-1}(0) \cap H^{-1}(0)$ contains a smooth nontrivial branch of codimension 1 near (d,d). Finally if in fact $0 \neq \nu(d)$ as assumed in (4.4), then the projection of this branch into $\mathcal{D}$ is also a smooth manifold of codimension 1 as shown in Fig. 2. The proof that this *phase boundary* does locally separate $\mathcal{D}$ into a single and two phase region as well as all details of the argument above are contained in the forthcoming paper [5]. The algorithmic consequences of these observations are discussed in the technical report [4] and numerical test are currently under way.

6. SUMMARY AND DISCUSSION

The multiphase equilibrium problem for mixed fluids was formulated as a semi-infinite programming problem in terms of the Helmholtz energy density E(d). Under the assumptions on E(d) listed in Section 2 there exists always an optimal split of a given feed m into $p \leq n+1$ phases. Mathematically the computation of this configuration is equivalent to the evaluation of the convex hull of E(d) at m. Apart from a global optimization aspect and the possible loss of strict complementarity one has to deal with (near-) singularity of the governing equations in the critical region. The treatment of the last contingency requires the analysis of saturation points in the vicinity of critical points. The former were found to form a smooth manifold of codimension 1 that contains the latter as the submanifold CP.

Even though bifurcation problems of the form (5.1), (5.2) occur in some other applications [2], their analysis as described above is apparently new. A more standard approach through the algebraic bifurcation equations and nondegeneracy conditions is not applicable because the second condition (5.2) has a higher degree of singularity than the first (5.1). Further analysis should reveal the structure of *tricritical points* at which several two-phase regions merge. Another worthwhile topic for further research is the relation between the *critical exponents* and the differentiability of the energy model.

REFERENCES

[1] Baker, L.E., Pierce, A.C.: Gibbs energy analysis of phase equilibria. *Proceedings of SPE/DOE Second Joint Symposium on Enhanced Oil Recovery* (1981). pp. 471-480.

[2] Grasshof, K., Gustafson, S.: Linear Optimization and Approximation. *Applied Mathematical Sciences 45*, (1983), Springer-Verlag, Heidelberg.

[3] Griewank, A., Osborne, M.R.: Analysis of Newton's method of irregular singularities. *SIAM Journ. of Num. Anal.*, Vol. 20 (1984), pp. 747-773.

[4] Griewank, A.: The phase behaviour of mixed fluids; a difficult semi-infinite programming problem. Report CMA-R42-85, ANU Canberra, Austr.

[5] Griewank A., Rabier, P.: Bifurcation analysis for phase equilibria.

[6] Jittorntrum, K.: Solution point differentiability without strict complementarity in nonlinear programming. *Mathematical Programming Study*, Vol. 21 (1984), pp. 127-138.

[7] Michelsen, M.: The isothermal flash problem, Part I Stability. *Fluid Phase Equilibria*, Vol. 9 (1982), pp. 1-19.

[8] Peng, D.Y., Robinson, D.B.: A new two constant equation of state. *Ind. Eng. Chem. Fundam.*, Vol. 15 (1976), pp. 59-64.

[9] Stanley, H.E.: Introduction to phase transitions and critical phenomena. *Oxford University Press.* (1971) New York.

Department of Mathematics
Southern Methodist University
Dallas, Texas 75275

International Series of
Numerical Mathematics, Vol. 79

BIFURCATION THEORY IN PHYSICS: RECENT TRENDS AND PROBLEMS

Werner Güttinger
Universität Tübingen, FRG

Abstract

The application of bifurcation theory to nonlinear physical problems is reviewed and an outlook is given on future developments. It is shown, in terms of representative examples, that, at the macroscopic and microscopic levels, the topological singularities and bifurcation processes deriving from the principle of structural stability determine the dominating phenomena and features observed in both structure formation and structure recognition. Because of their universality and classifiability, these bifurcation processes also provide a unifying topological framework for our understanding of the analogies that have been discovered in the critical behavior of nonlinear systems of quite different genesis. After a survey on the basic concepts of bifurcation theory, some new developments are outlined. These include bifurcation phenomena in pattern recognition and remote sensing, in fluid dynamics and nonlinear optics, in condensed matter physics and materials sciences, and in high energy physics and cosmology. We conclude by briefly exploring the frontiers and limits of present bifurcation concepts in dealing with instabilities producing fractal patterns and phase transitions in complex networks.

1. Introduction

The increasing diversification of the physical sciences during recent years has made it more and more imperative to search for unifying principles. These reveal themselves in a wealth of fascinating analogies discovered in the critical behavior of systems of various genesis, which, when passing through instabilities, suddenly exhibit new spatio-temporal patterns or modes of behavior. Bifurcation theory aims at an understanding of the mechanisms by which forms are generated in nature and at a classification and unifying description of generic pattern-forming processes independent of system details.

The phenomenological picture looks indeed seductively general. There is a striking similarity among the instabilities that lead to convection patterns in fluids [1], cellular flame and solidification fronts in crystal growth [2], geophysical textures [3], phase transitions in condensed matter physics [4], chemical patterns [5], and so forth. Their common characteristic is that one or more significant behavior variables or order parameters undergo spontaneous, large and discontinuous changes or cascades of these if slow, competing but continuously driving control parameters or forces pass through critical values so that an equilibrium existing between system-immanent effects breaks down. As a consequence, an initially quiescent system becomes unstable and, in a sequence of bifurcations, restabilizes into ever more complex space- or time-dependent configurations. Primary bifurcations induce limit cycles, spatial patterns and spatio-temporal patterns in the form of standing waves when the bifurcation branches remain disjoint. If other controls cause these branches to interact, multiple degenerate bifurcation points produce higher instabilities. Then the system undergoes additional transitions into more complex states, giving rise to travelling waves, hysteresis, resonance and entrainment effects. These ultimately lead to states which are intrinsically chaotic. In the vicinity of degenerate bifurcation points a system becomes extremely sensitive to small ambient factors like imperfections, external fields or fluctuations that lead to symmetry breaking. This in turn enhances the system's ability to perceive its external environment and, adapting to it by capturing its asymmetry, to form preferred patterns or modes of behavior. Most prominent among the theoretical programs venturing into the area of general principles are Prigogine's concept of dissipative structures [5], Haken's synergetics [6] and Thom's catastrophe theory [7]. Among these, Thom's program on structural stability and morphogenesis has both the potential to provide a geometrical explanation for the variety of analogies encountered in the critical behavior of systems of different genesis and, as we shall see, also the power to predict new phenomena.

The basic role physics plays in the sciences may be traced to the fact that most systems and structures in nature enjoy an inherent stability property: They preserve their quality under slight perturbations, i.e., they are structurally stable. Otherwise we could hardly think about or describe them, and today's experiment would not reproduce yesterday's results. We do not know how it got that way. But accepting structural stability as a fundamental principle that complements the known physical laws, universal critical physical phenomena have a common topological origin. They are describable and classifiable by stable unfoldings of singularities, i.e., in terms of topological normal forms, that organize the bifurcation processes exhibited by dynamical systems. We use the term bifurcation to refer to changes in the qualitative structure of solutions to differential equations. A phenomenon is said to be structurally stable if it persists under all allowed perturbations in the system.

Bifurcation is primarily a geometrical phenomenon, independent of the details of the underlying physical mechanism and largely determined by geometrical constraints, e.g., by the dimension and shape of the system, inherent symmetries and boundary conditions. This fact explains the qualitative similarity among the instabilities that lead to phase transitions of quite different genesis. Philosophically speaking, bifurcation geometry is at the origin of a theory of analogy. The main result of bifurcation theory is that the bifurcations producing stable patterns are not arbitrary but occur in certain definite and classifiable ways because a bifurcation, i.e., an unstable event, appears as a singular element in a stable process. To understand this, consider an equilibrium point of the nonlinear dynamical system $\dot{x} = X(x,y,\lambda)$, $\dot{y} = Y(x,y,\lambda)$ in the plane as the locus of intersection of the two solution curves C_1 and C_2 of the equations $X = 0$ and $Y = 0$, respectively. For a given λ, C_1 and C_2 will (i) either intersect transversely, i.e., without tangential contact (with Jacobian $J \neq 0$), or (ii) nontransversely, i.e., with tangential contact (when $J = 0$). A slight variation of λ results in a slight deformation of the two curves, $C_1 \to C_1'$, $C_2 \to C_2'$. In case (i), transversal intersection persists under such deformation and nothing new happens when the curves are deformed: transverse crossings are structurally stable. In case (ii), however, the deformed curves in general intersect at several points representing new equilibria: Nontransverse crossings are structurally unstable. Then, the curves' deformation causes the original intersection point to bifurcate into several intersections. The latter, however, remain stable against further deformation of the intersecting curves C_1', C_2'. The generalization of this fact is known as Thom's transversality lemma: Two manifolds meet almost always transversely and stably. Consequently, in a world made up of events generated by transverse intersections nothing new could happen upon perturbation. On the other hand, an event generated by contact intersection carries an inherent potential to evolve spontaneously, through bifurcation, into new events by a slight perturbation or unfolding. The latter may, e.g., be caused by external forces, whereas the degree of contact, i.e., the codimension, is a measure of the richness of the phenomena that may spring up.

While pattern formation by bifurcation and, as we shall see, also pattern recognition, may thus be understood locally in terms of topological principles, we are only just beginning to comprehend the enormity of the task of understanding how global structures are formed in complex, e.g., living, systems. Bifurcation and singularity theory can only explain the local geometry and not the forces that are shaping it. On the other hand, looking at local forces alone cannot explain the geometry unless structure formation can be tied to such physical principles as minimum entropy production. This presents one of the most challenging problems at this time, viz., to unite bifurcation geometry with physical principles. It is, indeed, surprising that the concept of structural stability -- nature's central dogma -- has so far been ignored in the formulation of physical theories. The fundamental laws of physics derive from geometrical invariance principles (space-time symmetries) which provide us with conservation laws and equations of motion. Besides, there are material invariance principles, such as gauge symmetries, which provide relationships between interactions and the forces that are producing them. It is the third, qualitative, i.e., topological invariance of a structurally stable geometry that may be expected to provide the missing link between the former and to enable us to answer the basic questions of structure formation, viz., how structures emerge from a structureless environment and under what circumstances does nature make just these and select specific patterns by adapting to a changing environment. Today we cannot. But it may well be that the next great era of awakening of human intellect may produce a method for understanding the qualitative content of complex systems and phenomena.

After a survey of the principles of bifurcation theory in Section 2 we discuss in Section 3 some recent applications to the problem of pattern recognition. In Section 4, basic concepts of structure formation are considered while Section 5 is devoted to an outline of recent problems.

2. Bifurcation Theory

Bifurcation occurs in a nonlinear equation when a variation of a parameter through a critical value causes a qualitative change in the behavior of a solution, e.g., when an equilibrium splits into two.

A general bifurcation problem [8] g consists of finding the solutions $x = x(\lambda)$ of a system of equations

$$g(x,\lambda) = 0 \qquad (2.1)$$

with $g(0,0) = g_x(0,0) = 0$, where $g = (g_1,\ldots,g_n)$ is a smooth function, $x = (x_1,\ldots,x_n)$ are state variables and λ is a distinguished bifurcation parameter representing the control variable in a physical experiment. It may be assumed that (2.1) is the equation system for the amplitudes x of the solution of a given nonlinear evolution equation, obtained, e.g., by a Lyapunov-Schmidt reduction (Sec. 4). The point $(x,\lambda) = (0,0)$ is a bifurcation point or a singularity of g, and the solutions (x,λ) of (2.1) constitute the bifurcation diagram. Two functions $g(x,\lambda)$ and $g'(x,\lambda)$ are called equivalent or "qualitatively similar" if there exist smooth local coordinate changes $x \to X(x,\lambda)$ and $\lambda \to \Lambda(\lambda)$ with $\det X_x(0,0) \neq 0$ and $\Lambda_\lambda(0) > 0$ so that

$$g(x,\lambda) = T(x,\lambda)\cdot g'(X(x,\lambda),\Lambda(\lambda)) \qquad (2.2)$$

where T is a square matrix with $\det T(0,0) > 0$. Since Λ may not depend on x, the control parameter λ influences x, but not conversely. Since T is invertible, two equivalent bifurcation problems possess qualitatively the same solution set. The effects of perturbations in a system are incorporated into the bifurcation problem (2.1) by the unfolding of g. A k-parameter function $G(x,\lambda,\alpha)$, with $G(x,\lambda,0) = g(x,\lambda)$ is called an unfolding of g with unfolding or imperfection parameters $\alpha = (\alpha_1,\ldots,\alpha_k)$. If for any sufficiently small perturbation $\varepsilon p(x,\lambda,\varepsilon)$ there exists an α so that the perturbed bifurcation problem $g + \varepsilon p$ is equivalent (in the sense of (2.2)) to $G(x,\lambda,\alpha)$ and if k is the minimum number of unfolding parameters needed to describe the perturbation, then $G(x,\lambda,\alpha)$ is called a stable or universal unfolding of g, and k is called the codimension of g. The codimension is a measure for the degree of complexity of the bifurcation problem. We have the following theorem [8]: If g has finite codimension then there exists a polynomial $g' = N(x,\lambda)$, called a normal form, which is equivalent to g, and a universal unfolding $G' = F(x,\lambda,\alpha)$ of N which is also a polynomial. Here the new variables X and Λ are denoted again by x and λ. By varying α in $F(x,\lambda,\alpha) = 0$ one obtains a finite number of qualitatively different, perturbed or unfolded bifurcation diagrams. The α-parameter space can be divided into a finite number of regions such that for any two parameter values lying within the same region the corresponding bifurcation problems are equivalent and structurally stable. Crossing the boundaries separating these regions produces new qualitatively different diagrams.

Changes in the stability of a solution x of $F=0$ follow by considering x as equilibrium solution of the system $\dot{x}=F(x,\lambda,\alpha)$. The signs of the real parts of the eigenvalues of the Jacobian of F evaluated at x determine the stability properties of the bifurcating solution branches of $F=0$. In order that bifurcation occurs in (2.1) and g is equivalent to a normal form, g must satisfy certain defining (or degeneracy) conditions which express the vanishing of some of its first derivatives and certain nondegeneracy conditions which relate to the nonvanishing of some next higher derivatives. With these techniques it is possible to classify bifurcation problems by their codimension, i.e., to give a list of all possible bifurcation diagrams. The imperfect bifurcation theory outlined above can be generalized to problems with symmetry [8]. g is called to be a bifurcation problem with symmetry group Γ if g is Γ-covariant, i.e., if $g(\gamma x,\lambda)=\gamma g(x,\lambda)$ for all $\gamma\in\Gamma$. g and g' are Γ-equivalent if (2.2) holds with the symmetry conditions $X(\gamma x,\lambda)=\gamma X(x,\lambda)$ and $T(\gamma x,\lambda)=\gamma T(x,\lambda)\gamma^{-1}$. The notions of Γ-covariant unfolding, Γ-codimension etc. are defined in an analogous way. If $g=0$ is a Γ-equivariant bifurcation problem, a bifurcating solution will have an isotropy subgroup which is not Γ, i.e., the new solution will have less symmetry than the old: The symmetry has broken spontaneously. The following example is instructive. Let $x\in\mathbb{R}$ and suppose that g satisfies the defining conditions $g=g_x=g_{xx}=g_\lambda=0$, and the nondegeneracy condition $g_{xxx}g_{x\lambda}<0$ at $(x,\lambda)=(0,0)$. Then the bifurcation problem $g(x,\lambda)=0$ is equivalent to the codimension-2 pitchfork $g'=N=x^3-\lambda x=0$ with universal unfolding $F=x^3-\lambda x+\alpha_2\lambda+\alpha_1$. The solution set of $F=0$ consists of the bifurcation diagrams shown in Fig. 1. Setting $\lambda=-u$ and $\alpha_2\lambda+\alpha_1=v$, the pitchfork's $F=0$ is just Thom's overhanging cliff S of Fig. 2. The bifurcation diagrams of the perturbed pitchfork are the intersection curves of S with vertical planes having straight lines through the cusp as basis (Fig. 3). Fig. 4

Fig. 1. Perturbed pitchfork for various α_1,α_2

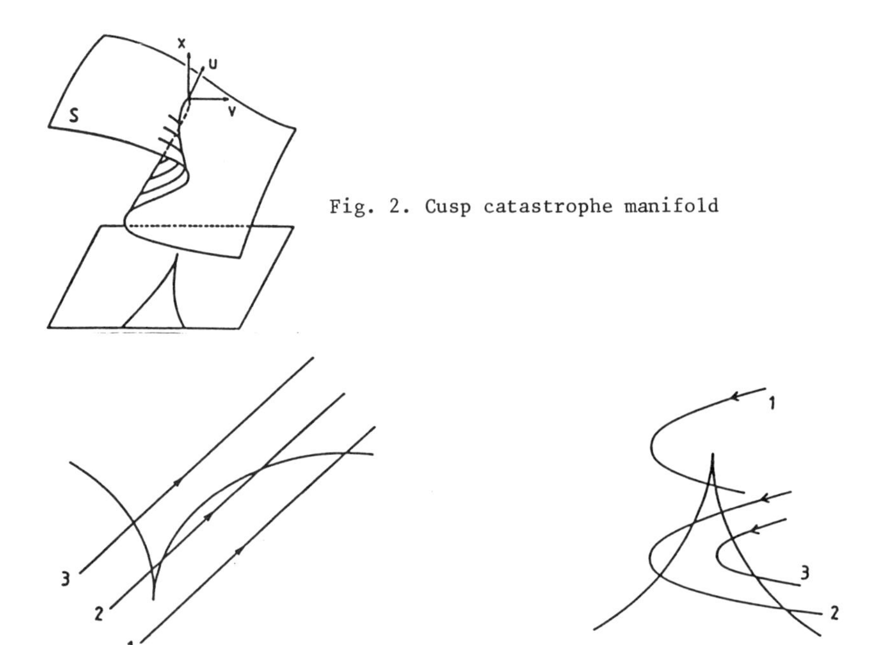

Fig. 2. Cusp catastrophe manifold

Fig. 3. Straight paths through the cusp producing the perturbed pitchfork

Fig. 4. Parabolic paths through the cusp producing Fig. 5

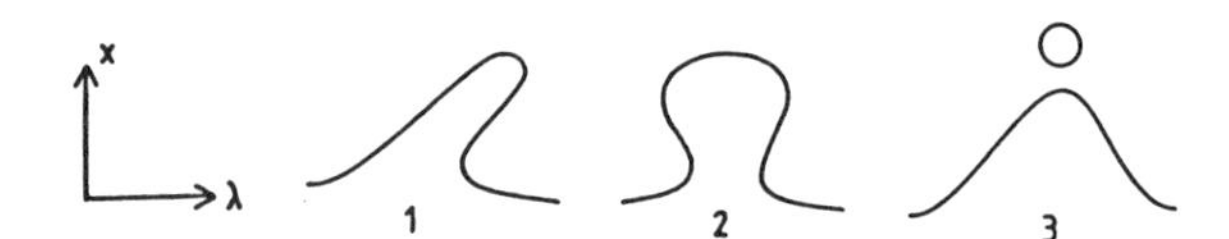

Fig. 5. Bifurcation diagrams of the perturbed winged cusp

shows paths generating the bifurcation diagrams of Fig. 5 of the winged cusp. Since the higher Thom-Arnol'd singularities can be generated by successive Legendre transformations, starting from the normal forms with lowest codimension, we conjecture that the Golubitsky-Schaeffer theorems of imperfect bifurcation theory [8] can be proved along these lines. To make the formalism readily accessible for physical applications, the next step ought to consist of formulating a theory for concatenating bifurcation diagrams in a way similar in spirit to Feynman diagram techniques.

3. Pattern Recognition as a Bifurcation Problem

In all fields of physics, wave systems are used as a tool for investigating unknown structures. Such a structure impresses geometrical singularities upon a smooth incident wavefield and the question arises what information about that unknown structure can be inferred from these singularities. This is the "inverse scattering problem" of pattern recognition and imaging. We confine ourselves here (1) to the problem of reconstructing a scattering surface from backscattered waves [9] and (2) to the problem of imaging by ballistic phonons [10].

3.1 Inverse Scattering

Suppose a point source at x_o in $\mathbb{R}^3$-space emits a spherical wave that is reflected back to x_o by an unknown two-dimensional surface S. The rays are orthogonal to the spherical wavefront and, since source and receiver are at the same place, only those points x_s on S (the specular points) whose distance vectors $\underline{R}_s = x_o - x_s$ are normal to S contribute to the echo received at x_o: $\nabla_{x_o} R_s = n$ where $R_s = |\underline{R}_s|$ and n is the surface unit normal. Consequently, every point $x_s = x_o - R_s \nabla R_s$ of S can be reconstructed by measuring the two-way traveltime $2R_s/C$, C being the speed of the wave. Let $x = x(u)$ be any point of S, parametrized by surface coordinates $u = (u_1, u_2)$ in principal directions. The distance $R = |x_o - x|$ obviously has an extremum $R = R_s$ at $x = x_s : \nabla_u R = 0$ whose solution $x = x_s(x_o)$ is a singularity of R, with Hessian $H = (1 - \kappa_1 R_s)(1 - \kappa_2 R_s)$ and principal curvatures κ_i. If $H \neq 0$, R is Morse. If $H = 0$, singularity theory tells us that R is a Thom polynomial provided that the scattering process is structurally stable. Two consequences are immediate: (i) The singularities in echo recordings can be classified by means of singularity theory and (ii) elimination of u from $\nabla_u R = 0$ and $H = 0$ gives the equation for the evolute E of the surface S whence Thom's theory provides a classification of surface evolutes. The technique can be extended to include diffraction effects and a classification of the bifurcations of level curves of the unknown surface seen from an observer at x_o [9]. Fig. 6 shows the caustics in a geophysical seismogram and Fig. 7 the layered medium that produces them.

3.2 Phonon imaging

Suppose a burst of heat with frequency ω is created at a point of a crystal surface with a laser beam. The heat produces phonons (sound quanta associated with elastic waves in the crystal) radiating in all directions. In

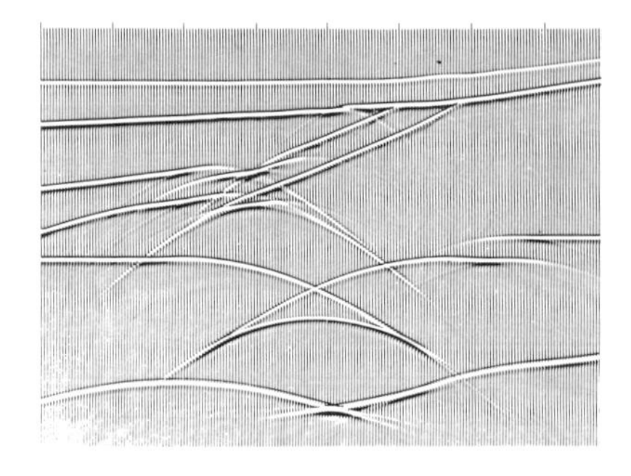

Fig. 6. Caustics in an echo recording

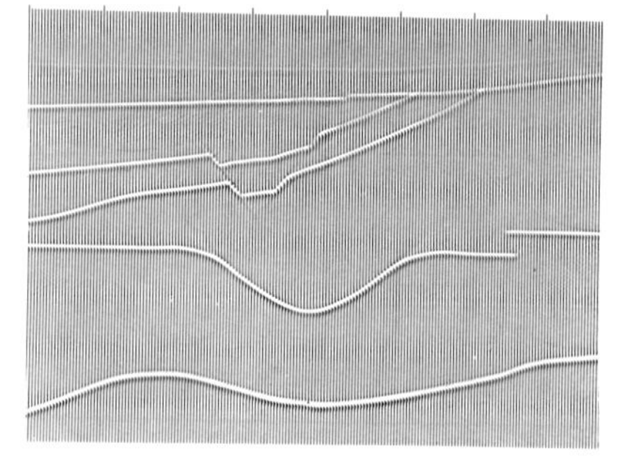

Fig. 7. Layered medium which produces the caustics in Fig. 6

a crystal cooled to low temperatures (2K) the phonons travel without scattering. Since the crystal's elastic properties are anisotropic, dispersion takes place and the energy flux propagates in the direction of the group velocity $v = \nabla_k f(k)$ where k is the phonon wave vector and $S: \omega = f(k)$ the energy surface in wave-vector (k-) space. The surface S is in general not spherical, and since v is normal to S, the phonon flux is channeled into intense beams along preferred crystal directions. Bolometric recordings show then high intensity images exhibiting caustic effects as, e.g., in Fig. 8. The latter can be classified by determining the topological singularities of the Gaussian map of the

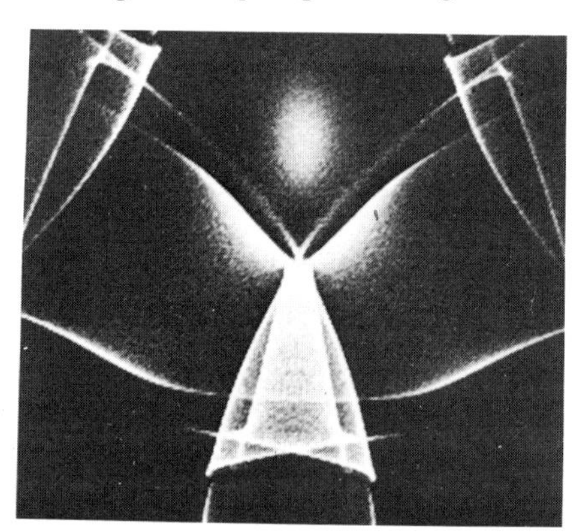

Fig. 8. Phonon caustics in a crystal.

surface S in k-space onto the sphere S^2 of unit normals $v/|v|$: Focusing directions v come from the inflection points of S along principal curvature lines with zero Gaussian curvature which plays the role of the Hessian of the map [10].

More generally, bifurcating wave modes in dispersive media can be classified in terms of imperfect bifurcation theory by representing the waves in the form

$$\psi(x,\lambda) := \int_{R^n} dk\,\delta(g(k,\lambda))e^{ikx} = \int_{S:g=0} dS \frac{e^{ikx}}{|\nabla g(k,\lambda)|} \tag{3.1}$$

where δ is the Dirac distribution. Eq. (3.1) establishes a relation between the topological singularities of $S:g=0$ and the analytical singularities which ψ possesses at those values of λ where both g and ∇g vanish [11]. This theory still awaits further elaboration for vector waves, surface scattering etc.. It is obvious from these examples that pattern recognition is a field inviting further applications of bifurcation theory.

4. Bifurcation Theory of Structure Formation

This section addresses topological problems of structure formation whose general physical setting has been discussed in the Introduction. The objective here is to classify transitions between spatio-temporal patterns described by systems of evolution equations $F(u,\lambda)=0$ with $F(0,0)=0$ in terms of normal forms. Here, F is a nonlinear operator, u an element of a Banach space and λ a bifurcation parameter. Assume that $\dim(\ker F_u(0,0)) = n$, $N = \ker F_u(0,0) = \{\phi_1,\ldots,\phi_n\}$ and denote by P the projection onto N. Then, $F=0$ yields the Lyapunov-Schmidt reduced bifurcation equations (2.1) with $g=PF(v+w(\lambda,v),\lambda)=0$, where $v = Pu = \Sigma x_i\phi_i$ and $w=(I-P)u$, and x_i are the amplitudes of the bifurcating solutions. Writing (2.1) symbolically as

$$g_i(x,\lambda) = \int F(v+w,\lambda)\phi_i d\xi = 0 \tag{4.1}$$

allows a simple physical interpretation: For $\lambda=0$ the system exists in an eigenmode ϕ_i. When λ varies away from zero, selfinteraction takes place in virtue of the nonlinearity of F and the system blows up unless the work g_i done by the force F on the i-th mode vanishes. This implies that, in order to avoid instability, the system transfers energy to other modes thereby changing its course through bifurcation. The bifurcation branches are the curves in (x,λ)-space in which the solution surfaces of (4.1) intersect. When an intersection occurs with tangential contact, then, by varying the imperfection parameters in F, i.e., by unfolding, the surfaces can be deformed in such a way that they intersect transversely, thereby ensuring structurally stable bifurcation diagrams. Symmetry constraints on F, e.g., through boundary conditions, carry over to g and reduce the number of diagrams. Applications of this formalism include degenerate Hopf bifurcations (eq. (2.1) with $n = 1$ and Z(2)-symmetry) [8] and coupled Hopf and steady-state bifurcations (eq. (2.1) with $n = 2$) [12] whose typical bifurcation diagrams include mushroom hysteresis and island formation similar to that in Fig. 5. One observes such effects in nonlinear optics, nerve impulse propagation and in fluid dynamics [13]. In what follows, we discuss two new applications.

4.1 Cellular Solidification

Consider a long, thin sample of a dilute binary alloy which is drawn, at velocity v, through a fixed temperature gradient established by a hot contact, where the sample is molten, and a cold contact where it is solid, with the liquid-solid interface visible in between. At low pulling velocity, the interface is flat. As v increases, the solidifying front bulges into the melt and becomes unstable. This effect is counterbalanced by the stabilizing force of the surface tension. Then, at a critical velocity $v = v_c$, the interface breaks spontaneously into spatially periodic cells. As v increases further, cells of different wavelengths appear, the interface develops dendrites and, ultimately, fractal structures (Fig. 9, Fig. 12).

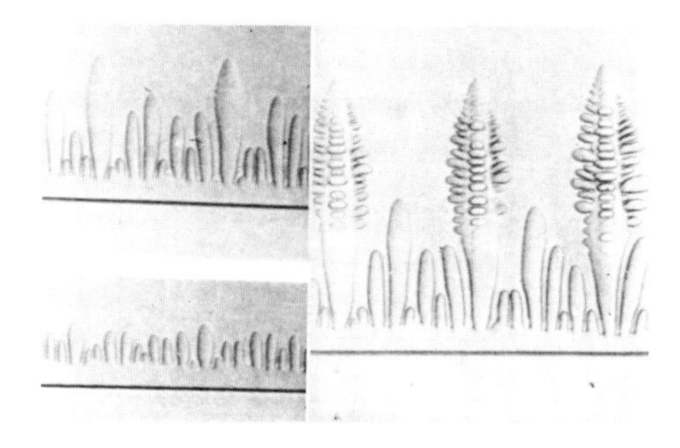

Fig. 9. Cellular solidification and dendrite formation

Let the sample lie in the (ξ,η)-plane with $0 \leq \xi \leq b$ and v parallel to η, let $W(\xi,\eta,t)$ be the solute concentration and $\eta = s(\xi,t)$ the interface I, t being the time. The equations of motion are $W_{\xi\xi} + W_{\eta\eta} + vW_{\eta} = 0$ and, on I,

$W = -s + H(s)$, $s_t + v + W_\eta - s_\xi W_\xi = 0$ where $H = -s_{\xi\xi}/(1 + s_\xi^2)^{3/2}$ is the surface tension. We impose non-flux (Neumann) boundary conditions $s_\xi = 0$ at the side walls $\xi = 0, \xi = b$. Expanding $s = \Sigma\varepsilon_m \cos(mk_o\xi)$ where k_o is a constant wavenumber, one can show in the stationary case [14] that at $v = v_c$: Either a single cellular mode with amplitude $\varepsilon_n = x$ bifurcates from the flat interface ($s \equiv 0$) and satisfies a Z(2)-symmetric codim-2 bifurcation equation $g \equiv xh(x^2,\lambda) = 0$, where $\lambda = v - v_c$, whose bifurcation diagrams are identical with those known from degenerate Hopf bifurcations. Or else two neighboring cellular modes with amplitudes $\varepsilon_n = x, \varepsilon_{n+1} = y$ bifurcate from the flat interface and interact through the bifurcation equations

$$
\begin{aligned}
g_1(x,y,\lambda) &= xa + x^n y^n b &= 0 \\
g_2(x,y,\lambda) &= yc + x^{n+1} y^{n-1} d &= 0
\end{aligned}
\tag{4.2}
$$

where $a = a(x^2,y^2,\lambda)$, b, c and d are smooth functions of x^2, y^2 and λ. The form of eqs. (4.2) is typical for systems on a line of finite width with non-flux boundary conditions since they possess an inherent O(2)-symmetry. An unfolding theory was developed in [15]. Besides pure mode solutions S_x, S_y, there are mixed mode solutions S_m connecting the former (Fig. 10). They produce spatial beats and mode jumping. In addition, tertiary bifurcations occur, i.e., Hopf bifurcations to standing waves.

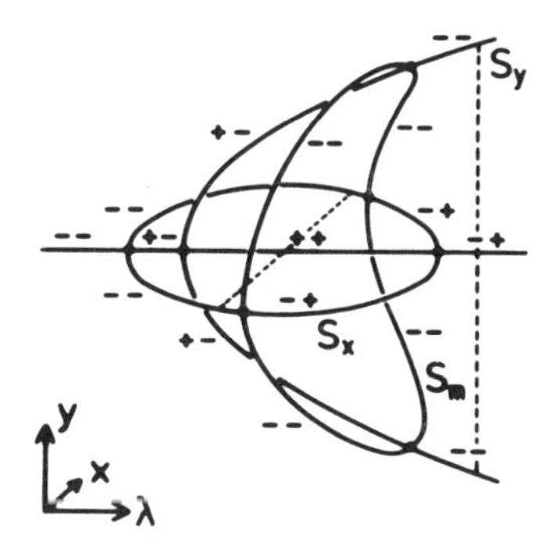

Fig. 10. Bifurcation diagram for coupled cellular states

4.2 Bifurcations in Particle Physics

Grand unified theories (GUTs) attempt to unify all elementary particles and their interactions into one single gauge theory. The idea is to start from a highly symmetric universe with symmetry group, e.g., $\Gamma = SU(5)$. As the universe cools down, with temperature playing the role of the bifurcation parameter λ, a sequence of spontaneous symmetry-breaking bifurcations occurs, $SU(5) \to \ldots \to SU(3)SU(2)SU(1) \to SU(2)U(1)$, whose branches describe the interactions observed today. This leads to a bifurcation problem (2.1), in which the x_i are traceless Hermitean matrices, and to a classification of the bifurcating solutions $x(\lambda)$ through their broken symmetries, cf. Fig. 11. This theory, although in its infancy [14], may well lead the way towards an understanding of some of the most fundamental problems of physics.

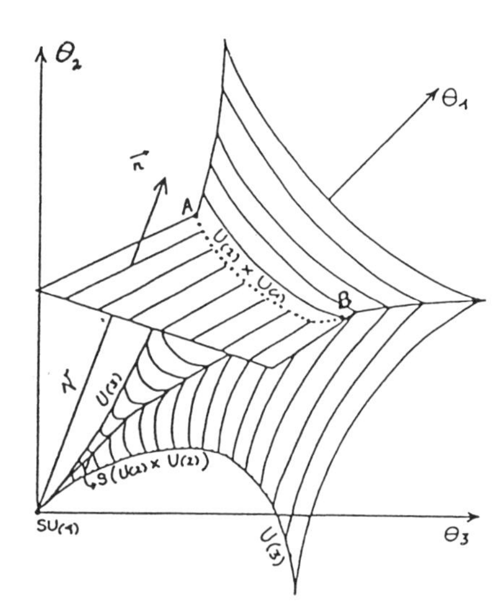

Fig. 11. U(2)×U(1) isotropy subgroup solutions on the two-dimensional stratum of SU(4)

Fig. 12. Fractal clusters

5. Perspectives on Future Developments

In many respects present-day bifurcation theory is ahead of physics. For example, the complex cellular bifurcation phenomena predicted to take place in a solidifying material (Sec. 4.1) have not yet been submitted to an experimental test. On the other hand, in such major fields of research as dynamical bifurcation theory, partial amplitude equations and bifurcations with symmetry we have hardly scratched the surface (cf., [15]). The predictive power of bifurcation geometry is incontestable in the case of problems with low corank, i.e., when the number of significant state variables x_i remains small. In recent years, however, pattern formation problems in the physical sciences have received increasing attention whose complexity transcends the conceptual framework of present bifurcation theory. These include interfacial growth of dendrites in solidifying systems and the evolution of fingers in viscous fluid mixtures [16], diffusion limited aggregation (DLA) and the formation of self-similar fractal structures [17], and the formation and retrieval of patterns in organized disordered systems with a very large number of degrees of freedom [18], ranging from cellular automata to spin glasses and neural networks.

Dendritic growth in solidifying systems can be modeled by singular nonlinear integro-differential equations for the interface y(x) of the type

$$\alpha + \frac{\kappa y''(x)}{(1+y'(x)^2)^{3/2}} = e^{-y(x)} \int_{-\infty}^{\infty} K_0([(x-x')^2+(y(x)-y(x'))^2]^{1/2})e^{y(x')}dx'$$

where K_0 is a Hankel function. Finite surface tension ($\kappa \neq 0$) introduces a "solvability condition" at zero anisotropy which is essentially singular at $\kappa = 0$ and, therefore, invisible in perturbation theory, and it is this condition which determines the velocity of dendritic tips. A similar role is played by surface tension in the formation of fingers which arise in a Hele-Shaw experiment, where a more viscous fluid is displaced by a less viscous one whose dynamics are dominated by tip bifurcation leading to branched ramified structures [16]. The instabilities and pattern forming processes in these nonlinear systems, and the fractal properties [17] emerging from them, are presently the subjects of beautiful experiments and intensive computer simulations. But with general organizing principles still missing, the physics and mathematics of these structures is, in many ways, a subject waiting to be born.

Modeling complex networks consisting of very many identical nonlinearly interacting and locally connected elements constitutes another class of problems which calls for new bifurcation theoretical concepts. Although random-looking, one of the most striking properties of such disordered systems is their capability of selforganization involving multiple possible attractors that create and store patterns. We refer here, in particular, to spin glass models and to equivalent formal neural networks [18] which are structurally stable ("robust") under random changes of elements and whose dynamics are described by trajectories in a space of "bit word" states. In many instances, a central problem is that of finding an absolute minimum of an energy function which has many local minima. Information retrieval is then identified as the persistence of a pattern under the dynamic. Although the various models considered in this field of research are still in a preliminary phase, the instabilities involved in the pattern forming and recognizing processes described by them make the subject particularly amenable to new bifurcation theoretical ideas.

A systematic understanding and classification of complex, selfgenerated patterns must ultimately involve the amount of information required to specify pattern forming or recognizing events [19]. Furthermore, in biological systems, the pattern-forming media themselves change their properties in response to complex sequences of external stimuli. These problems raise difficult but fascinating questions that promise a great challenge for future research in bifurcation theory.

Acknowledgments

It is a pleasure to acknowledge helpful discussions with D. Armbruster, G. Dangelmayr, Ch. Geiger, D. Lang and P. Haug, and the support of the Stiftung Volkswagenwerk.

References

[1] S. Lundquist (ed.), The Physics of Chaos and Related Problems, Phys. Scripta T9 (1985).

[2] A.R. Bishop, L.J. Campbell & P.J. Channel (eds.), Physica 12D (1984).

[3] U. Bayer, Pattern Recognition Problems in Geology and Paleontology (Springer, Berlin 1985).

[4] F. Claro (ed.), Nonlinear Phenomena in Physics (Springer, Berlin 1985).

[5] G. Nicolis & I. Prigogine, Selforganization in Nonequilibrium Systems (Wiley, New York 1977).

[6] H. Haken, Synergetics and Advanced Synergetics (Springer, Berlin 1983).

[7] R. Thom, Structural Stability and Morphogenesis (Benjamin, Reading 1975).

[8] M. Golubitsky & D. Schaeffer, Singularities and Groups in Bifurcation Theory (Springer, New York 1985).

[9] G. Dangelmayr & W. Güttinger, Geophys. J. R. Astro. Soc. 71 (1982) 79.

[10] D. Armbruster, G. Dangelmayr & W. Güttinger in Phonon Scattering in Condensed Matter, ed. by W. Eisenmenger (Springer, Berlin 1984).

[11] G. Dangelmayr, Proc. R. Soc. Edinburgh 95A (1983) 301.

[12] D. Armbruster, G. Dangelmayr & W. Güttinger, Physica 16D (1985) 99.

[13] W. Güttinger in Frontiers in Nonequilibrium Stastical Physics, ed. by G.T. Moore & M. Scully (Plenum, New York 1986).

[14] Ch. Geiger, W. Güttinger & P. Haug in Complex Systems -- Operational Approaches, ed. by H. Haken (Springer, Berlin 1985).

[15] G. Dangelmayr & D. Armbruster in Contemporary Mathematics 56, ed. by J. Guckenheimer & M. Golubitsky (Amer. Math. Soc., Providence 1986).

[16] R. Dashen, D. Kessler, H. Levine & R. Savit, Physica 21D (1986) 371; D. Kessler & H. Levine, Phys. Rev. B33 (1986) 7867.

[17] H.E. Stanley & N. Ostrowsky (eds.), On Growth of Form (M. Nijhoff, Dordrecht 1986).

[18] E. Bienenstock, F. Fogelman Soulié & G. Weisbuch, Disordered Systems and Biological Organization (Springer, Berlin 1986).

[19] E.T. Jaynes in refs. [13] and [14].

Institut für Informationsverarbeitung
Köstlinstr. 6
D-7400 Tübingen
Fed. Rep. of Germany

International Series of
Numerical Mathematics, Vol. 79

AN APPLICATION OF COMPLEX BIFURCATION TO A PROBLEM IN COMPUTER GRAPHICS

Michael E. Henderson
IBM Watson Research Center 81-Z06
P.O. Box 218
Yorktown Heights, NY 10598 USA

Abstract

The ray-tracing algorithm for rendering three dimensional objects in two dimensions has created some of the most realistic computer generated images ever made. However, it requires a large amount of computation time, as much as seven CPU hours per image.

The algorithm can be formulated as finding the solutions of a system of equations whose coefficients depend on two real parameters. For a particular set of objects we compute all the roots of this system using pseudo-arclength continuation, and complex bifurcation. This reduces the time required to produce an image by exploiting the spatial coherence of the image, and introduces a parallel structure which can be used to gain a factor of 18 in the computation time. It also allows objects to be ray-traced which cannot be rendered by existing implementations of the algorithm.

The Ray-Tracing Algorithm

The fundamental problem of computer graphics is to produce a two-dimensional image of a set of three dimensional objects. Of all the algorithms for doing this the ray-tracing algorithm produces the most realistic images, including shadows, reflections, and refraction in transparent objects.

A set of three dimensional objects is given, along with a set of point light sources, the position of the observer, and the position and orientation of an imaging plane located between the observer and the objects. The objects are projected onto the imaging plane by coloring each point in the plane according to the amount of light that reaches the observer from the sources by passing through that point. To do this the ray-tracing algorithm works backwards from the observer to the light sources, tracing the paths of light from the observer back to the sources. It is assumed that light travels along straight rays, which are reflected and refracted when they strike an object.

A primary ray is cast from the observer through the point in the imaging plane. The first object that this ray intersects may direct light along the ray toward the observer. The intensity and color are determined by the surface reflectivity, usually given by one of several models (e.g. Phong or Torrance-Sparrow), and the refractive index of the object. The reflectivity consists of two parts, diffuse reflection, which is isotropic, and specular reflection, which has a sharp peak in one direction.

To compute the amount of light that is diffusely reflected toward the observer, primary illumination rays are cast from the primary intersection to each of the light sources. If the ray does not intersect an object, light from that source reaches the primary intersection and is diffusely reflected to the observer. The intensity is the intensity of the source times the diffuse reflection coefficient, which depends on the direction to the observer, the direction to the source, and the surface normal.

To find the amount of light specularly reflected and refracted toward the observer, secondary rays are cast in the directions of maximum specular reflection and refraction. The first object that these rays intersect may direct light toward the primary intersection, where it is specularly reflected or refracted toward the observer. This requires that secondary illumination rays be cast from the secondary intersections to the sources, so that the amount of light diffusely reflected toward the primary intersection may be computed. In general, secondary reflection and refraction rays would also be cast.

For a complete description of the ray-tracing algorithm see GOLDSTEIN and NAGLE, KAJIYA (1), or WHITTED. For a description of the models used for surface reflectivity see BLINN or PHONG.

Formulation of the Problem

Most existing ray-tracing codes allow only very simple objects to be displayed, such as spheres, cylinders, and other quadratic surfaces. These have the advantage that the ray-object intersections can be computed explicitly, but require that smooth surfaces be approximated. This produces such undesirable effects as piecewise smooth shadows on smooth surfaces, and the incorrect rendering of patterns painted on the surface. It is therefore necessary to find methods of ray-tracing more complicated objects. We will consider the problem of ray-tracing bicubic surfaces, or Coons patches, which are commonly used in computer aided design.

A bicubic surface is a parametric surface of the form

$$x(u,v) = \sum_{i=0}^{3}\sum_{j=0}^{3} a_{ij}u^i v^j.$$

Here $x \in \mathbb{R}^3$ is a point on the surface, and the coefficients $a_{ij} \in \mathbb{R}^3$ determine the surface position and derivatives at the corners of a unit square in the (u,v) space. They are (COONS)

$$A = \begin{bmatrix} 2 & -2 & 1 & 1 \\ -3 & 3 & -2 & -1 \\ 0 & 0 & 1 & 0 \\ 1 & 0 & 0 & 0 \end{bmatrix} \begin{bmatrix} x(0,0) & x(0,1) & x_v(0,0) & x_v(0,1) \\ x(1,0) & x(1,1) & x_v(1,0) & x_v(1,1) \\ x_u(0,0) & x_u(0,1) & x_{uv}(0,0) & x_{uv}(0,1) \\ x_u(1,0) & x_u(1,1) & x_{uv}(1,0) & x_{uv}(1,1) \end{bmatrix} \begin{bmatrix} 0 & 0 & 0 & 1 \\ 1 & 1 & 1 & 1 \\ 0 & 0 & 1 & 0 \\ 3 & 2 & 1 & 0 \end{bmatrix}.$$

Several algorithms have been proposed to ray-trace bicubic and other algebraic surfaces (CATMULL, HANRAHAN, KAJIYA (2), SIDERBERG and ANDERSON, TOTH, and LANE and CARPENTER.) Catmull's algorithm subdivides the surface, then determines which of the subdivisions contain an intersection. By repeating the process the real intersections can be located to any accuracy. The algorithms of Hanrahan, Kajiya, Siderberg and Anderson, and Toth rely on reducing the intersection problem to a polynomial equation, whose real roots are found using an iterative method.

The algorithm proposed by Lane and Carpenter is based on the idea of a scan line algorithm, in which the intensities are computed in scan line order. The scan starts at the top left of the imaging plane and proceeds left to right, top to bottom, similar to the scan in a television set. As the primary ray travels across the imaging plane its real intersections with the surface are computed, using the previous point in the scan as an initial guess for an iterative solver. At silhouette edges the number of real intersections may change, so they are computed before the scan, and intersections are added or deleted as required. Only the real primary intersections are found.

This is a type of continuation, and takes advantage of the spatial coherence of the objects. We propose two improvements. The first is to use pseudo-arclength continuation, which allows longer steps to be taken, and avoids problems near folds. The second is to compute all of the intersections, both real and complex, on both the primary and secondary rays. The number of these intersections is independent of the position in the scan, so all of the real intersections may be found without any special procedures near silhouettes.

We first formulate an algebraic system whose solutions are the set of intersections of the rays with the bicubic surface. The intersection of any ray $x(t) = o + tn$ with the surface is determined by the three algebraic equations

$$x(t) - \sum_{i=0}^{3}\sum_{j=0}^{3} a_{ij}u^i v^j = 0. \tag{1}$$

Using some basic results from algebraic geometry KAJIYA (2) has shown that for $(t,u,v) \in \mathbb{C}^3$ there are always 18 solutions of (1).

The primary ray $x_p(t)$ is determined by p, a point in the imaging plane and e, the position of the observer. It is

$$x_p(t) = e + t(p - e)/\|p - e\|. \tag{2}$$

If the primary intersection is $x_p(t_p)$, and the light sources are at the points $\{l_m\}$, the illumination rays $x_l^m(t)$ are given by

$$x_l^m(t) = x_p(t_p) + t(l_m - x_p^0)/\|l_m - x_p^0\|. \tag{3}$$

The reflected ray $x_r(t)$, is

$$x_r(t) = x_p(t_p) + t(n_p - 2(n_p, \sigma)\sigma(x_p^0))/\|n_p - 2(n_p, \sigma)\sigma(x_p^0)\|, \tag{4}$$

where $n_p = (p - e)/\|p - e\|$ is the direction of the primary ray, and $\sigma(x_p^0)$ is the normal to the surface at the primary intersection. The refracted ray, determined by Snell's law, is

$$x_t(t) = x_p(t_p) + t(-\sigma + \alpha(n_p - (n_p, \sigma)\sigma))/\| - \sigma + \alpha(n_p - (n_p, \sigma)\sigma)\|, \tag{5}$$

where $\alpha = 1/\sqrt{\eta^2 + (n_p, \sigma)^2 - 1}$, and η is the ratio of indices of refraction on either side of the surface.

Since we do not know a priori which primary intersection is closest to the observer, we write a system for all 18, and their associated secondary intersections. There are 17*L intersections along each illumination ray (L is the number of light sources, and one intersection is the primary intersection), and 17*2 intersections along the reflected and refracted rays, each of which has 17*L intersections on its illumination rays. This is a total of 18*(1+17*L+2*17*(1+17*L))= 630 + 565*L intersections, each determined by three equations. The algebraic system which determines the intersections therefore consists of 18 copies of (1) for the primary ray (2), 17 copies of (1) for each illumination ray (3) and each primary intersection, and so on.

There is nothing in the system which guarantees that the intersections are distinct. However, if we start with distinct intersections and use small enough steps in a continuation method they will remain distinct, provided that the branch switching is done carefully.

Although this system is large, and depends on two real parameters, it is not as difficult to solve as it might be. Each primary intersection depends only on the point in the imaging plane, and the illumination intersections and secondary intersections depend only on one of the primary intersections. The Jacobian is therefore block 3 by 3 lower triangular. In addition we will use a scanning pattern on the imaging plane, which reduces the number of parameters to one.

Complex Bifurcation and the Continuation Method

We use Pseudo-arclength continuation (KELLER) to compute the ray-patch intersections as a function of the distance along the scan. For complex solutions this involves solving complex equations with a real pseudo-arclength constraint. Keller's block elimination algorithm for Newton's method must therefore be modified slightly. Newton's method is

$$\begin{aligned} g_u^0 \Delta u + g_\lambda^0 \Delta \lambda &= - g \\ Re(\dot{u}_0 \Delta u + \dot{\lambda}_0 \Delta \lambda) &= - Re(N). \end{aligned}$$

Here g is the system to be solved for the solution u, and N is the pseudo-arclength constraint. If we let $g_u^0 v = - g_\lambda^0$ and $g_u^0 z = - g$, and if g_u^0 is nonsingular, then

$$\begin{aligned} \Delta u &= z + \Delta \lambda v \\ \Delta \lambda &= - Re(N + \dot{u}_0 v) / Re(\dot{\lambda}_0 + \dot{u}_0 z). \end{aligned}$$

We compute all of the intersections, so the behavior of the system at singular points is a bit unusual. At a simple fold for example, where two intersections meet, the nullspace of the system will be two dimensional not one. Here complex bifurcation (HENDERSON and KELLER) is useful. Suppose that near the singular point an incoming branch has the form

$$u(s) = u_0 + s u_0^{(1)} + O(s^2)$$

$$\lambda(s) = \lambda_0 + \frac{1}{K!} s^K \lambda_0^{(K)} + O(s^{K+1}),$$

and that $s = 0$ is the singular point. An examination of the bifurcation equations shows that if the incoming branch is isolated, additional complex branches of the form

$$u(s) = u_0 + (\pm 1)^{1/K} s u_0^{(1)} + O(s^2)$$

$$\lambda(s) = \lambda_0 \pm \frac{1}{K!} s^K \lambda_0^{(K)} + O(s^{K+1}),$$

must also exist near the singular point. (This applies both to simple and multiple bifurcation points.)

We wish to switch branches so that the position in the scan is always increasing. If K is even we may switch branches by multiplying the incoming tangent by $(-1)^{1/K}$. This yields a branch that opens to the opposite side from the incoming branch. If K is odd no switching is necessary, as the incoming branch continues smoothly through the singular point.

Results

We have computed images of three bicubic surfaces, which are shown in Figures 1, 2, and 3. In Figures 1 and 2, the unit square in the parameter space is painted dark gray, as are the lines of constant u and v. In Figure 3 the unit square is colored white. Only the primary intersections were computed, so there are no shadows or reflections. The path of each primary intersection was computed independently, using a different step size on each. At singular points branches were switched by multiplying the incoming tangent by the imaginary unit i. (All of the singular points were folds, for which $K = 2$.)

The images were displayed on an IBM 5080 graphics terminal and photographed. They are 200 pixels high and 400 pixels wide. Each required about 4000 continuation steps per primary intersection, and a total of about an hour of CPU time on an IBM 3083. Each step requires 2-3 Newton iterations and one tangent computation, for a total of 5-7 solves of a block 3x3 diagonal system. This compares favorably with other algorithms, which require 2-5 iterations of a root finder for each real intersection at each point in the imaging plane.

There are several ways to further decrease the running times. The scan used had a one pixel vertical spacing, so used very small steps in the vertical direction. Using a continuation in two parameters would reduce the number of scans required from 200 to approximately 10, (the number of horizontal steps needed.) This would reduce the number of steps required to 200. We are also considering better predictors. These images were made using an Euler predictor.

A factor of 18 can be gained in computation time by computing the intersections associated with each primary intersection in parallel. We are currently implementing such an algorithm on the IBM EPEX RP3 simulator. No communication is necessary between the processors apart from a compare and store to place the computed intensities into a raster. Including secondary rays would not destroy this parallelism.

Lastly, most of the intersections computed are not needed. If the folds on the solution set are computed as functions of the point in the imaging plane, the plane can be divided into regions in which the order of the intersections along their respective rays does not change. This means that in each region only the intersections necessary to the intensity calculation would have to be computed. We have used the system of JEPSON and SPENCE to compute the real folds in the images shown in Figures 1 and 3. These are shown in Figure 4 The folds satisfy an algebraic system, so may be computed using the algorithm described above. For the image shown in Figure 1 we show a few of the paths of complex folds (dotted.)

With these improvements, Pseudo-Arclength Continuation together with Complex Bifurcation will provide a fast means of computing images of bi-cubic surfaces. In addition, it does not depend on the special form of the surfaces, so may be applied to a large class of parametrically defined surfaces.

Figures

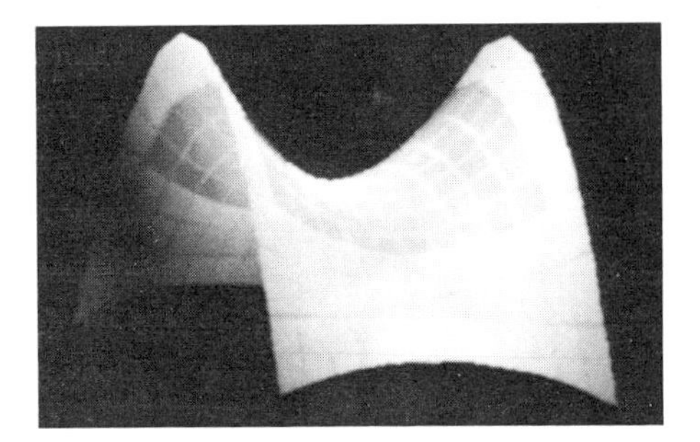

Figure 1. A hyperbolic surface

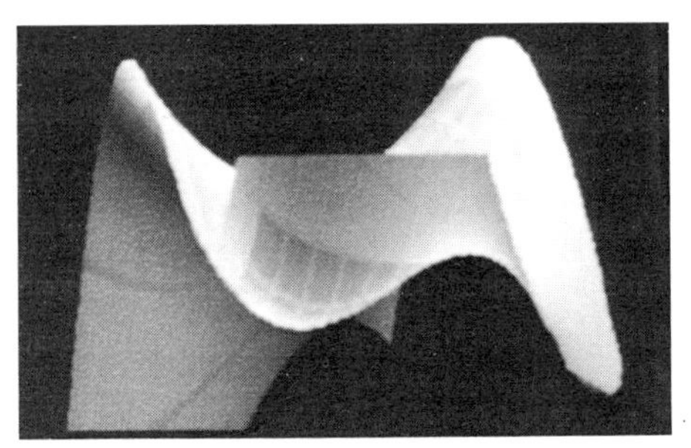

Figure 2. A cubic surface

Figure 3. A surface with many folds

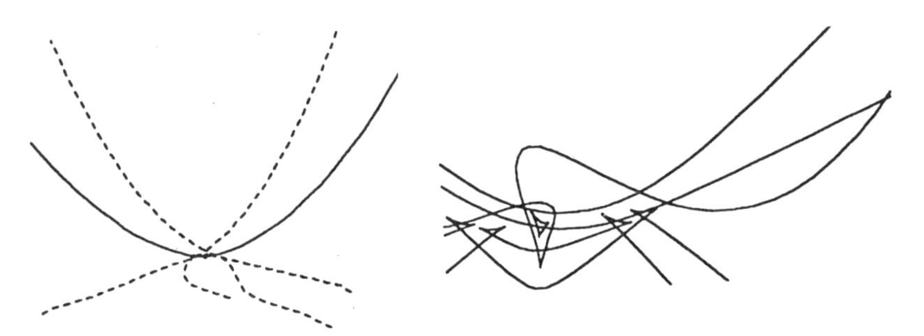

Figure 4. Silhouette edges for Figures 1 and 3. (Complex silhouettes are dashed)

References

- Blinn, J.F. *Computer Display of Curved Surfaces*, Ph.D. Thesis, Department of Computer Science, University of Utah, December 1978.
- Coons, S.A. *Surfaces for Computer-Aided Design of Space Forms*, Technical Report, Mechanical Engineering Department, Massachusetts Institute of Technology, June 1967.
- Goldstein E., and Nagle R. *3d Visual Simulation*, Simulation, 16(1) 1971, pp. 25-31.
- Hanrahan, P. *Ray Tracing Algebraic Surfaces*, Computer Graphics v 17 #3 July 1983, pp. 83-90.
- Henderson, M.E. and Keller, H.B. *Complex Bifurcation*, Submitted to the SIAM Journal on Applied Mathematics.
- Jepson and Spence *Folds in Solutions of Two Parameter Systems and their Calculation Part I*, SIAM Journal on Numerical Analysis, 22(2) April 1985, pp. 347-368.
- Kajiya, J.T., *(1) SIGGRAPH83 Tutorial on Ray-Tracing*, ACM SIGGRAPH 1983.
- Kajiya, J.T., *(2) Ray-tracing Parametric Patches*, Computer Graphics, 16 #3 (July 1983).
- Lane J.M., Carpenter L.C., Whitted T., Blinn J.F. *Scan Line Methods for Displaying Parametrically Defined Surfaces*, Communications of the ACM, 23 #1 (January 1980).
- Phong B-T., *Illumination for Computer Display of Computer Generated Pictures*, Communications of the ACM, 18 #6 (June 1975), p 311.
- Sederberg T.W., and Anderson D.C. *Ray Tracing of Steiner Patches*, Computer Graphics v 18 # 3 July 1984, pp. 159-164.
- Toth, D.L. *On RayTracing Parametric Surfaces*, ACM SIGGRAPH 85, v 19 # 3 1985.
- Whitted, T. *An Improved Illumination Model for Shaded Display*, Communications of the ACM, 23 (June 1980) pp. 343-349.

Michael E. Henderson, IBM Watson Research Center 81-Z06, P.O. Box 218, Yorktown Heights, NY 10598 USA

International Series of
Numerical Mathematics, Vol. 79

CONTINUATION OF PERIODIC SOLUTIONS IN PARABOLIC PARTIAL DIFFERENTIAL EQUATIONS

M. Holodniok, P. Knedlík and M. Kubíček

Prague Institute of Chemical Technology, Prague, Czechoslovakia

1. Introduction

Numerical methods for construction of solution diagram of steady state and periodic solutions in ODE are well developed, e.g. [1,4]. Similarly, methods for study of parametric dependences of time independent (steady-state) solutions in PDE were also described, see e.g. [4]. However, no numerical algorithm for construction of a continuous dependence of periodic solutions of a system of PDEs of parabolic type on a parameter has appeared in literature until now. In this paper we shall describe two such methods and discuss the results of their application to a relatively simple system of two coupled nonlinear PDEs describing a common type of reaction-diffusion problem.

2. The reaction-diffusion problem

The reaction and diffusion of two characteristic components X and Y in spatially one-dimensional medium with constant diffusion coefficients D_x, D_y may be described by a set of two parabolic PDEs:

$$\frac{\partial x}{\partial t} = \frac{D_x}{L^2} \frac{\partial^2 x}{\partial z^2} + f(x,y), \tag{1}$$

$$\frac{\partial y}{\partial t} = \frac{D_y}{L^2} \frac{\partial^2 y}{\partial z^2} + g(x,y).$$

Here L is a characteristic dimension of the system, $z \in [0,1]$ is dimensionless spatial coordinate, x,y concentrations of components X,Y, and f and g describe nonlinear kinetic relations. For simplicity choose boundary conditions of the Dirichlet type, which define constant concentrations of components x and y on the boundaries

$$x(t,0) = x(t,1) = \overline{x}\ , \quad y(t,0) = y(t,1) = \overline{y}\ . \tag{2}$$

Here $\overline{x}$ and $\overline{y}$ satisfy $f(\overline{x},\overline{y}) = 0$, $g(\overline{x},\overline{y}) = 0$. Continuation algorithms developed below will be applied to the above system (1) with the Brusselator reaction schema [5], i.e.

$$f(x,y) = x^2y-(B+1)x+A, \quad g(x,y) = Bx-x^2y\ . \tag{3}$$

Here A and B are parameters, $x=A$, $y=B/A$.

3. Periodic solutions of PDEs

Time-periodic solutions of (1) with a period T satisfy the condition of periodicity

$$x(t,z) = x(t+T,z), \quad y(t,z) = y(t+T,z)\ . \tag{4}$$

Stable periodic solutions may be determined as a result of dynamic simulations of Eqs (1), where after sufficiently long time interval the solution reaches periodically repeating concentration profiles $x(t,z)$ and $y(t,z)$.

Usually, difference methods are used for numerical solution of Eqs (1). The Crank-Nicolson difference approximation with an adaptive control of the grid, together with an approximation of the nonlinear terms by a quasilinearization technique may provide an effective numerical method [6].

To continue a stable periodic solution we can consider a slow variation of the characteristic parameter (e.g., L) value in time; for example we may consider

$$L(t) = L_0 + at \tag{5}$$

where $|a| \ll 1$. An evolution diagram [4,6] is thus obtained.

4. Continuation of periodic solutions

To determine stable and unstable periodic solutions of (1) we have to solve, in principle, a mixed boundary value problem (1)(4) together with the boundary conditions (2). The first method is based on the finite difference approximation.

We divide the interval [0,1] of the spatial variable z into N parts by N+1 equidistant node points. The time coordinate is then divided into M parts in the interval [0,T]. The partial derivatives are approximated by the difference formulas ($u \sim x$ or y), $h=1/N$, $k=T/M$:

$$\left.\partial^2 u/\partial z^2\right|_i^{j+1/2} \sim (u_{i-1}^j - 2u_i^j + u_{i+1}^j + u_{i-1}^{j+1} - 2u_i^{j+1} + u_{i+1}^{j+1})/2h^2$$
$$\left.\partial u/\partial t\right|_i^{j+1/2} \sim (u_i^{j+1} - u_i^j)/k \qquad i=1,\dots,N-1,\ j=0,\dots,M-1. \tag{6}$$

The nonlinearity is approximated: $\left.f(u)\right|_i^{j+1/2} \sim 1/2(f(u_i^j)+f(u_i^{j+1}))$. The periodicity conditions (4) are included in the following system:

$$x_i^0 = x_i^M, \quad y_i^0 = y_i^M \quad . \qquad i=1,\dots,N-1. \tag{7}$$

Then we have to solve a set of 2(N-1)(M+1) nonlinear equations for 2(N-1)(M+1) unknowns

$$x_i^j,\ y_i^j\ ;\quad i=1,\dots,N-1,\quad j=0,\dots,M$$

and for the period T. The values x_0^j, y_0^j, x_N^j, y_N^j are given by the boundary conditions (2).

Assign a value to one of the variables x_i^j, y_i^j. The period T cannot be assigned because the solution exists for discrete (and apriori unknown) values of T only. Thus we have obtained a square system. This system can be solved by means of the Newton method. The Jacobi matrix of the above set of nonlinear equations has, by proper ordering of variables and equations, almost a band structure; there is (2N+3) diagonals with generally nonzero elements, and additional nonzero elements are in 2(N-1) columns originating from (7). A modified standard continuation algorithm DERPAR [4] is adapted to obtain the dependence of periodic solutions on a parameter. The memory requirements can be then approximatelly expressed as 2(N-1)(M+1)(6N+1) words for arrays.

Table I: Internal/external memory requirements for the finite difference method in dependence on N and M (WL=8 Byte).

N \ M	50	100
20	70 kB/1.9 MB	90 kB/3.7 MB
40	240 kB/7.7 MB	270 kB/15 MB
80	900 kB/30 MB	950 kB/61 MB

We used a sparse solver specially developed for such linear systems (called DIAKUB [3]). This solver uses external memory (discs). Total memory requirements are presented in Table I in dependence on N and M. Requirements either on the computer memory or on the computer time are usually very high. The above described algorithm is therefore used mostly to obtain new (unknown) periodic solutions only. The algorithm described below is then used for the continuation.

The second algorithm is based on transformation of the system of PDEs into a large set of ODEs by means of the method of lines, cf., e.g. [7]. The standard algorithm DERPER [1] is then used for continuation of periodic solutions of this system. The algorithm DERPER is a predictor-corrector algorithm based on continuation along arc-length of solution locus, and on the shooting method. Now we shall briefly describe this second approach. Let us denote ($z_i=ih$)

$$x_i(t) \sim x(t,z_i), \quad y_i(t) \sim y(t,z_i), \quad i=1,\dots,N-1. \tag{8}$$

The spatial derivative $\partial^2/\partial z^2$ is replaced by a finite difference approximation and the following system of $2(N-1)$ ODEs results for $i=1,2,\dots,N-1$:

$$\begin{aligned} dx_i/dt &= \frac{D_x}{L^2h^2}(x_{i-1}-2x_i+x_{i+1})+f(x_i,y_i), \\ dy_i/dt &= \frac{D_y}{L^2h^2}(y_{i-1}-2y_i+y_{i+1})+g(x_i,y_i). \end{aligned} \tag{9}$$

The boundary conditions (2) give

$$x_0=x_N=\bar{x}, \quad y_0=y_N=\bar{y}. \tag{10}$$

The periodicity conditions (4) provide

$$x_i(T)-x_i(0) = 0, \quad y_i(T)-y_i(0) = 0, \quad i=1,\dots,N-1 \tag{11}$$

Eqs. (9) with the mixed boundary conditions (11) form a nonlinear boundary value problem. The shooting method is used for its solution. At $t=0$ we choose altogether $2(N-1)$ initial conditions

$$x_i(0) = \eta_i, \quad y_i(0) = \eta_{N-1+i}, \quad i=1,\dots,N-1. \tag{12}$$

After integrating Eqs (9) from $t=0$ to $t=T$ we obtain a system of $2(N-1)$ nonlinear equations from (11):

$$\begin{aligned} F_i(\eta_1,\dots,\eta_{2N-2},T,L) &= x_i(T)-\eta_i=0, \\ F_{N-1+i}(\eta_1,\dots,\eta_{2N-2},T,L) &= y_i(T)-\eta_{N-1+i} = 0 \end{aligned} \tag{13}$$

for $2(N-1)$ unknowns $\eta_1,\dots,\eta_{2N-2}$, the unknown value of the period T and the parameter L (we might choose any other parameter of the problem as a continuation parameter). One of the initial conditions has to be fixed (say η_k) to obtain a set of $2(N-1)$ nonlinear (algebraic) equations for $2(N-1)$ unknowns and one parameter L. We have thus obtained a standard continuation problem in a finite-dimensional space, and the standard software, e.g. DERPAR [4] may be used for its solution. The detailed description of the derived algorithm DERPER is given in [1].

We need to compute the functions F_i in (13), the Jacobi matrix $J = \{\partial F_i/\partial \eta_j\}$ and the partial derivatives $\{\partial F_i/\partial T\}$ and $\{\partial F_i/\partial L\}$. These values are obtained by means of integration of variational equations for variational variables, e.g. $p_{ij}(t) = \partial x_i(t)/\partial \eta_j$. The monodromy matrix $B=J+I$ is computed on basis of these variables, too. The eigenvalues of the matrix B determine stability of the considered periodic solution. Dimension of the monodromy matrix B is usually high and, therefore, the computation of eigenvalues is not a simple task.

5. Application to the Brusselator model

Let us return to the reaction-diffusion system with the Brusselator kinetics for the following values of parameters: $A=2$, $B=5.45$, $D_x=0.008$, $D_y=0.004$. The trivial stationary solution independent of L is $\overline{x}=A=2.0$,

$\overline{y}$=B/A=2.725. On the basis of linear stability analysis (e.g. [5]) we determine the value of L_1^{**}=0.5130 at the Hopf bifurcation point. We can observe from the numerical results that at this length a branch of spatially symmetric periodic solutions bifurcates with a zero amplitude and with a period $T^{**} \cong 2.937$. There bifurcate additional branches of periodic solutions at every multiple of L_1^{**}, i.e. at $L_m^{**}=m.L_1^{**}$, m=1,2,3,... . For m odd the bifurcated solutions are spatially symmetric. The results of continuation are summarized in Fig. 1 where the period of the solution is presented in dependence on L. The detailed analysis has shown that N=20 is necessary to obtain sufficient accuracy in approximation of the continued periodic solutions for $L < 2.5$, cf. Table II. This spatial discretization was then used for constructing the solution diagram presented in Fig.1.

We may observe a symmetry breaking bifurcation on the branch originating at $L_1^{**}(L \sim 1.2)$. Three characteristic periodic solutions for L=1.5 are presented in Fig. 2.

Table II: Accuracy vs N for the method of lines.
L=1.5 branch with m=1. Results of the Poincarè map with x(t,0.05)=2.

N	T	x(0.5)	y(0.5)
20	3.46189	1.14199	3.75756
40	3.46325	1.13551	3.77638
80	3.46360	1.13391	3.78107

6. Conclusions

The above given algorithm for continuation of periodic solutions in the system of parabolic PDEs, based on the method of lines, the shooting method and on the continuation algorithm DERPER, has proved its usefulness in constructing both stable and unstable periodic solutions of relatively low wavenumbers. At the same time it becomes clear that the procedure requires a large computer and that consumption of the computer time can be high. It seems that vector computers can be used advantageously for continuation of periodic solutions in PDEs.

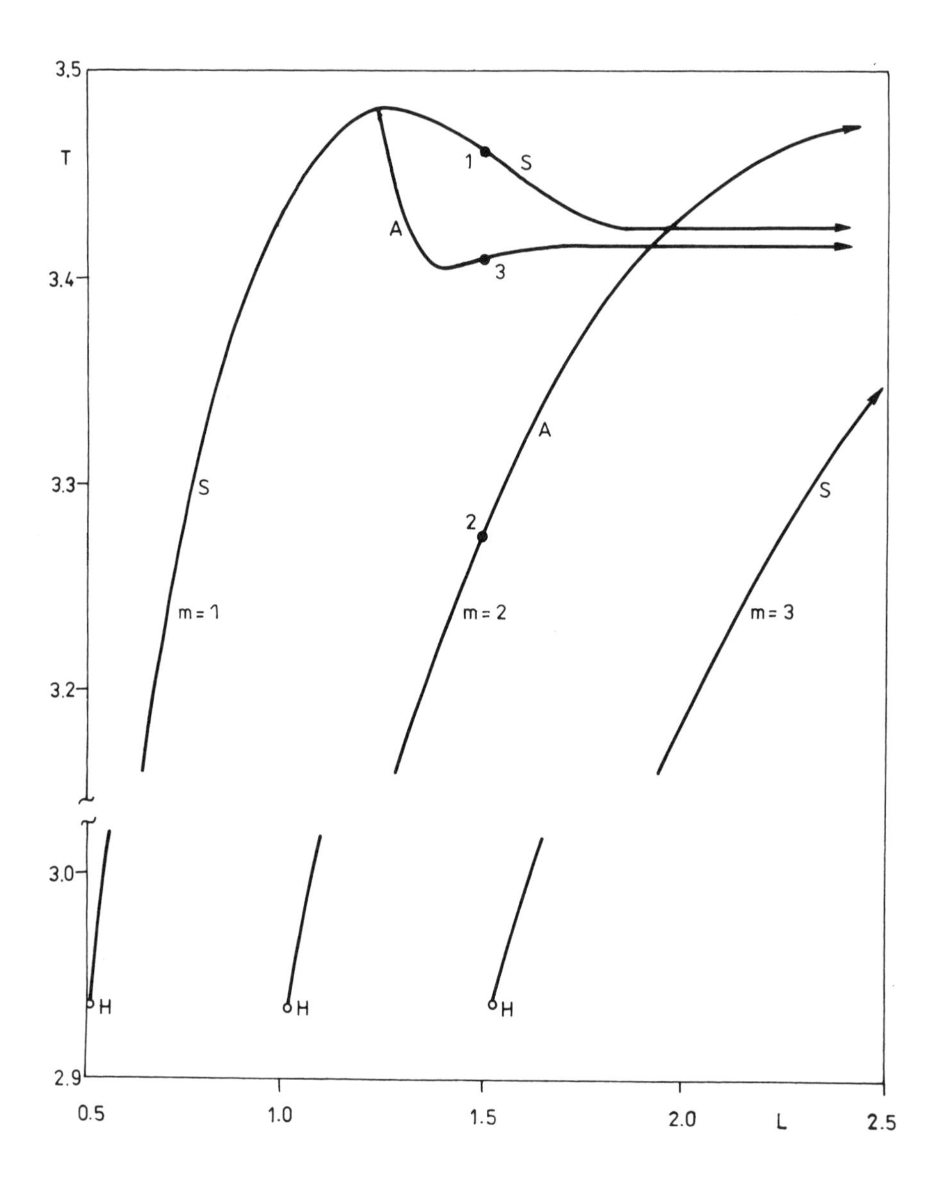

Fig.1: Solution diagram of periodic solutions of the Brusselator model. Dependence of the period T on L. S - spatially symmetric solutions, A - spatially asymmetric solutions. H - points of Hopf bifurcation from the trivial steady state solution. The number denote periodic solutions presented in Fig. 2.

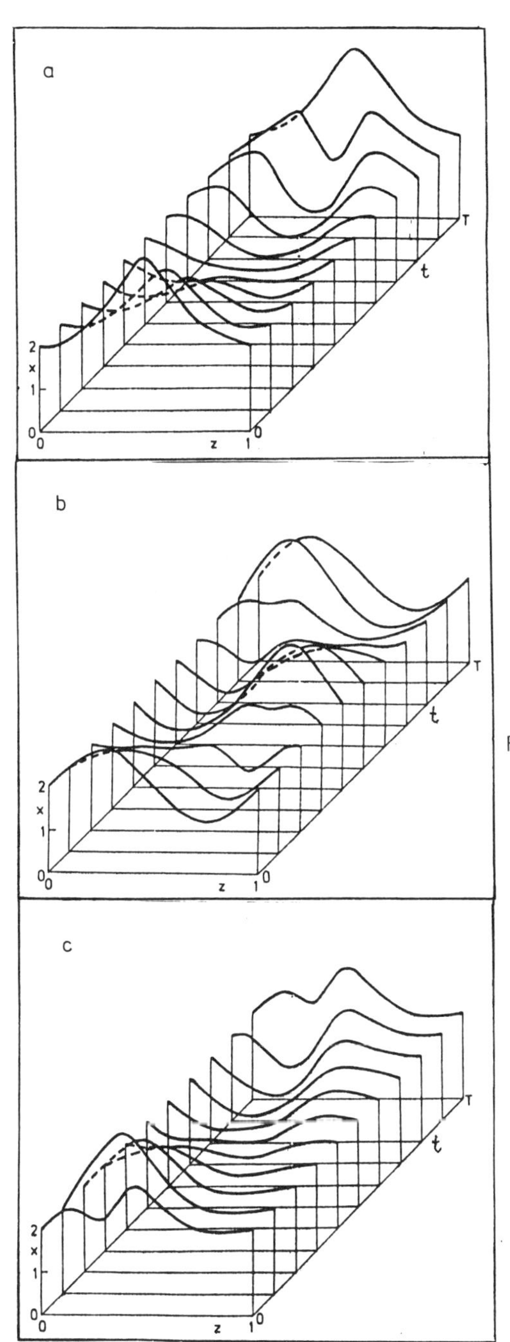

Fig.2: Periodic solutions of the Brusselator model, L=1.5. A_x = amplitude of x at $z = 0.5$.

a) spatially symmetric solution, branch m=1, $T \doteq 3.5$, $A_x \doteq 2.9$, cf. point 1 in Fig. 1

b) branch m=2, $T \doteq 3.3$ $A_x \doteq 0.08$, cf. point 2 in Fig. 1. The solution is symmetric in the following sense: $x(t,z)=x(t+T/2,1-z)$, $y(t,z)=y(t+T/2,1-z)$

c) asymmetric branch, $T \doteq 3.4$, $A_x \doteq 1.4$, cf. point 3 in Fig. 1.

Acknowledgment: The authors thank Prof. Miloš Marek for discussion during preparation of the manuscript.

7. References

1 Holodniok, M., Kubíček, M. (1984) DERPER - An algorithm for continuation of periodic solutions in ordinary differential equations. J.Comput.Physics 55, 254-267.
2 Knedlík,P., Holodniok, M., Kubíček, M., Marek, M. (1984) Periodic solutions in reaction-diffusion problems. 7. International Congress CHISA, Prague, September 1984.
3 Kubíček, M., (1977) Linear systems with almost band matrix. Sci. papers of the Inst. of Chem. Technol. Prague, K12, 5-9.
4 Kubíček, M., Marek, M. (1983) Computational methods in bifurcation theory theory and dissipative structures. (Springer, New York).
5 Nicolis, G., Prigogine, I. (1977) Self-organization in nonequilibrium systems (Wiley, New York).
6 Raschman, P., Kubíček, M., Marek, M. (1982) Concentration waves in reaction-diffusion systems. Sci. papers of the Inst. of Chem. Technol. Prague, K17, 151-175.
7 Seydel, R. (1985) Calculating the loss of stability by transient methods, with application to parabolic PDEs. Progress in Scientific Computing, Vol. 5, (Birkhäuser, Boston), 261-270.

Dr. Martin Holodniok
Department of Mathematics and Computer Centre
Prague Institute of Chemical Technology
166 28 Praha 6, Czechoslovakia

International Series of
Numerical Mathematics, Vol. 79

BIFURCATION IN DEGENERATE DIRECTIONS

Edgar Jäger
Universität Konstanz, FRG

1. Introduction

A typical task in the practical treatment of bifurcation problems is the determination of the set of zeros of a mapping $G: \mathbb{R}^{n+1} \to \mathbb{R}^n$ near a given singular solution z_o of $G(z)=0$. The approach of DESCLOUX and RAPPAZ [1] unifies preceding constructive methods as in [4], [5]. Following [1], one determines the so-called "characteristic rays" by computing the zeros of the multilinear mapping associated with the first nonvanishing derivative at zero of the Lyapunov-Schmidt reduction. Characteristic rays are the candidates for tangents to solution curves passing through z_o. Each characteristic ray σ satisfying a certain nondegeneracy condition corresponds to a branch of solutions of $G(z)=0$ passing through z_o and tangent to σ at z_o. If all characteristic rays are nondegenerate, a complete description of the zero set of G near z_o is possible. However, there are applications where the above mentioned procedure gives no information about the local structure of $G^{-1}\{0\}$.

1.1 Example

Consider the following discrete version of a morphogenetic model due to GIERER and MEINHARDT [3]:

$$\text{(1)} \qquad \begin{aligned} \dot{x}_i &= cx_i^2 y_i^{-1} - \mu x_i - D_A(Ax)_i \\ \dot{y}_i &= cx_i^2 - \nu y_i - D_I(Ay)_i \end{aligned} \qquad (i=1,\dots,N).$$

The model describes a reaction-diffusion process in an assembly of N cells. In any cell there are two chemicals, an "activator" and an "inhibitor". The activator's (inhibitor's) concentration in the i'th cell is denoted by x_i (y_i), and we write $x=(x_1,\dots,x_N)$, $y=(y_1,\dots,y_N)$. The individual cells are coupled by diffusion, and the boundaries of the cell system are assumed to be impermeable. As such, the $(N\times N)$ diffusion matrix has the entries

$$A_{11} = A_{NN} = 1, \quad A_{12} = A_{N,N-1} = -1,$$

$$A_{i,i-1} = A_{i,i+1} = -1, \quad A_{ii} = 2 \quad (i=2,\dots,N-1),$$

$$A_{ij} = 0 \text{ elsewhere.}$$

We are interested in steady-state solutions of (1) showing non-uniform concentration patterns. Obviously, there always exists a stationary solution of (1) with constant concentrations of both substances. This "trivial solution" $\bar{u}=(\bar{x},\bar{y})\in \mathbb{R}^N\times\mathbb{R}^N$ is given by $\bar{x}_i=\nu\mu^{-1}$, $\bar{y}_i=c\nu\mu^{-2}$ $(i=1,\dots,N)$. Adopting the activator diffusion coefficient D_A as a bifurcation parameter, we denote by

$$\text{(2)} \qquad H(u,D_A) = 0 \quad (u = (x_1,\dots,x_N; y_1,\dots,y_N))$$

the steady-state equation associated with (1). The aim is to find zeros of H near the trivial branch $\{(\bar{u},D_A):D_A>0\}$. To determine the points $(\bar{u},D_A)$ at which the Jacobian H_u of H with respect to u becomes singular, let $g(s)=\mu(D_I s-\nu)[s(D_I s+\nu)]^{-1}$. With the eigenvalues $\mu_j=2(1-\cos\frac{j\pi}{N})$, $j=0,\dots,N-1$ of the matrix A, one verifies that $H_u(\bar{u},D_A)$ is singular iff $D_A\in S:=\{g(\mu_1),\dots,g(\mu_{N-1})\}$. Further, for $D_A\in S$, the dimension of the null space $N(H_u(\bar{u},D_A))$ equals the number of elements of $\{j\in\{1,\dots,N-1\}: g(\mu_j)=D_A\}$, which is obviously ≤ 2. If $D_A=g(\mu_j)$ for exactly one $j\in\{1,\dots,N-1\}$, then $(\bar{u},D_A)$ is a simple bifurcation point. Let us therefore consider the case where zero is a double eigenvalue of $H_u(\bar{u},D_A)$. Given $k,l\in\{1,\dots,N-1\}$, $k<l$, we can choose the parameter ν such that $g(\mu_k)=g(\mu_l)=:D_A^{k,l}$. Denote by H' the total derivative of H, and let $\bar{z}=(\bar{u},D_A^{k,l})$. Then $\dim N(H'(\bar{z}))=3$, and the range $R(H'(\bar{z}))$ has codimension 2. Through the Lyapunov-Schmidt procedure, problem (2) is reduced locally to the "bifurcation equation" $S(\sigma)=0$, where S is defined on an open neighborhood of the origin in $N(H'(\bar{z}))$, and it maps into some complement of $R(H'(\bar{z}))$. At the origin, S and its first derivative S' vanish.

Following [1], we determine the characteristic rays, i.e., the elements $0\neq\sigma\in N(H'(\bar{z}))$ satisfying $S''(0)\sigma^2=0$. It is a property of the diffusion matrix A that, in most cases, (precisely whenever $k,l\neq 2N/3$, $k\neq 2(N-l)$, $l\neq 2k$, $l\neq 2(N-k)$) the following situation occurs: the set of characteristic rays is composed of $T:=\{(0,\dots,0,\alpha)\in\mathbb{R}^{2N+1}:\alpha\neq 0\}$ and of all nonzero elements in a certain two-dimensional subspace V of $N(H'(\bar{z}))$. The only non-degenerate characteristic rays (these are characterized by $N(S''(0)\sigma)=\mathrm{span}(\sigma)$) are the elements of T, corresponding to the trivial branch. For $0\neq v\in V$, we have $N(S''(0)v)=V$. Thus, our approach tells nothing about nontrivial zeros of H near $\bar{z}$. Nevertheless, we may obtain information using more general conditions that ensure bifurcation. To derive such conditions, we study the following situation, which can be considered as a result of a Lyapunov-Schmidt procedure [1].

2. Conditions ensuring solution branches.

Let $1\leq k,m$ and $2\leq q$ be integers, and assume B_1 and B_2 are real normed spaces with $\dim B_1=m+1$, $\dim B_2=m$. Let $r>0$ and suppose a C^{q+k+2} mapping $f:U_r:=\{x\in B_1:\ \|x\|<r\}\to B_2$ is given such that

$$(3)\qquad f(0)=0,\quad f'(0)=0,\dots,f^{(q-1)}(0)=0,\quad f^{(q)}(0)\neq 0.$$

First, we derive necessary conditions for branches in $f^{-1}\{0\}$ passing through 0. Assume therefore that $\varphi \in C^{\infty}(I, U_r)$ ($I \subset \mathbb{R}$ an open interval containing 0) is given with $\varphi(0)=0$. Taylor's formula followed by some rearranging gives

$$
(4) \quad \begin{aligned} f(\varphi(t)) &= \frac{t^q}{q!} D^0(\varphi'(0)) + \frac{t^{q+1}}{(q+1)!} D^1(\varphi'(0), \varphi''(0)) \\ &+ \frac{t^{q+2}}{(q+2)!} D^2(\varphi'(0), \varphi''(0), \varphi'''(0)) + \frac{t^{q+2}}{(q+2)!} \widetilde{R}(t) \end{aligned} \qquad (t \in I),
$$

where $\widetilde{R}$ is of class C^k, $\widetilde{R}(0)=0$, and the D^j are defined by

$$
\begin{aligned}
D^0(\xi_1) &= f^{(q)}(0)\xi_1^q, \\
D^1(\xi_1, \xi_2) &= \tfrac{1}{2}q(q+1)f^{(q)}(0)\xi_1^{q-1}\xi_2 + f^{(q+1)}(0)\xi_1^{q+1}, \\
D^2(\xi_1, \xi_2, \xi_3) &= (q+1)(q+2)\big(\tfrac{1}{6}qf^{(q)}(0)\xi_1^{q-1}\xi_3 \\
&\quad + \tfrac{1}{4}\tbinom{q}{2}f^{(q)}(0)\xi_1^{q-2}\xi_2^2 + \tfrac{1}{2}f^{(q+1)}(0)\xi_1^q\xi_2\big) \\
&\quad + f^{(q+2)}(0)\xi_1^{q+2} \qquad (\xi_1, \xi_2, \xi_3 \in B_1).
\end{aligned}
$$

From (4) we conclude that, if φ parametrizes a curve of zeros of f, then $D^0(\varphi'(0))=D^1(\varphi'(0),\varphi''(0))=D^2(\varphi'(0),\varphi''(0),\varphi'''(0))=0$. We also obtain sufficient conditions for solution branches provided a certain nondegeneracy condition is fulfilled. To formulate this last condition, define for fixed $\xi_1, \xi_2 \in B_1$ the linear mapping $M(\xi_1, \xi_2): B_1 \to B_2$ by

$$
h \to \tfrac{1}{2}(q+1)(q+2)\big(\tbinom{q}{2}f^{(q)}(0)\xi_1^{q-2}\xi_2 h + f^{(q+1)}(0)\xi_1^q h\big).
$$

Note that $M(\xi_1, \xi_2) = \frac{\partial}{\partial \xi_2} D^2(\xi_1, \xi_2, \xi_3)$ for any $(\xi_1, \xi_2, \xi_3) \in B_1^3$.

<u>Theorem 2.1.</u> Under the above assumptions on f, suppose that $(\widetilde{\varphi}_1, \widetilde{\varphi}_2, \widetilde{\varphi}_3) \in B_1^3$ satisfies conditions (5):

(5a) $$0 = D^{0}(\tilde{\varphi}_1) = D^{1}(\tilde{\varphi}_1,\tilde{\varphi}_2) = D^{2}(\tilde{\varphi}_1,\tilde{\varphi}_2,\tilde{\varphi}_3),\ \tilde{\varphi}_1 \neq 0,$$

(5b) $$M(\tilde{\varphi}_1,\tilde{\varphi}_2)v\in R(f^{(q)}(0)\tilde{\varphi}_1^{q-1}),\ v\in N(f^{(q)}(0)\tilde{\varphi}_1^{q-1}) \Rightarrow v\in \operatorname{span}(\tilde{\varphi}_1).$$

Then there exist $\delta>0$ and $\varphi \in C^{k}((-\delta,\delta),B_1)$ such that

$$\varphi_2(0)=\tilde{\varphi}_2,\ \varphi(t):=t\tilde{\varphi}_1+\frac{1}{2}t^2\varphi_2(t)\in U_r,\ f(\varphi(t))=0\ \ (|t|<\delta).$$

Proof: Let $V=N(f^{(q)}(0)\tilde{\varphi}_1^{q-1})$. Choose W such that $V\oplus W=B_1$, and denote by $P:B_1\to V$ the projection along W. Define on a suitable domain $B\subset \mathbb{R}\times V\times W$ containing $(0,P\tilde{\varphi}_2,(I-P)\tilde{\varphi}_3)$ the mapping $T:B\to B_2$ by

$$T(t,v,w)=\begin{cases} t^{-(q+2)}f(t\tilde{\varphi}_1+\frac{1}{2}t^2[v+(I-P)\tilde{\varphi}_2]+\frac{1}{6}t^3w) & (t\neq 0)\\ \frac{1}{(q+2)!}D^2(\tilde{\varphi}_1,v+(I-P)\tilde{\varphi}_2,w) & (t=0).\end{cases}$$

Formula (4) with $\varphi(t)=t\tilde{\varphi}_1+\frac{1}{2}t^2[v+(I-P)\tilde{\varphi}_2]+\frac{1}{6}t^3w$ shows that T is a C^k mapping (note that $D^1(\tilde{\varphi}_1,v+(I-P)\tilde{\varphi}_2)=0$ $(v\in V)$). By (5a), $T(0,P\tilde{\varphi}_2,(I-P)\tilde{\varphi}_3)=0$. Now, choose $\chi^*\in B_1^*$ (= dual space of B_1) such that $\langle\chi^*,\tilde{\varphi}_1\rangle\neq 0$ and put

$$G(t,v,w):=(T(t,v,w),\langle\chi^*,v-P\tilde{\varphi}_2\rangle\ \ ((t,v,w)\in B).$$

Obviously, $G(0,P\tilde{\varphi}_2,(I-P)\tilde{\varphi}_3)=0$. Using (5b), one verifies that $G_{(v,w)}(0,P\tilde{\varphi}_2,(I-P)\tilde{\varphi}_3):V\times W\to B_2\times\mathbb{R}$ is an isomorphism. The assertion now follows from the implicit function theorem.

3. Application

We consider the situation discussed in connection with Example 1.1. In particular, we assume for $f=S$ that there exists a subspace V of $B_1=N(H'(\bar{z}))$ with $\dim V=2$ and

(6) $$N(f''(0)v) = V \quad (0\neq v\in V).$$

In the notation of Section 2, we have $m=q=2$. Our goal is to select elements in V that lead to solution curves. Towards that end we seek solutions of (5a). Suppose $0\neq\tilde{\varphi}_1\in V$, $\varphi_2\in B_1$ are such that $D^1(\tilde{\varphi}_1,\varphi_2)=0$. Let $\psi\in B_1-V$. Then φ_2 has a unique representation $\varphi_2=\alpha\psi+v$, $\alpha\in\mathbb{R}$, $v\in V$, and obviously $D^1(\tilde{\varphi}_1,\alpha\psi)=0$. Due to (6), for $v\in V$ the equation $D^2(\tilde{\varphi}_1,\alpha\psi+v,0)=0$ reduces to a linear equation in v that can be solved uniquely for $\hat{v}\in V$ if we require

(7) $$M:=[\alpha f^{(2)}(0)\psi+f^{(3)}(0)\tilde{\varphi}_1^2]_{/V} \quad \text{invertible.}$$

Thus, $(\tilde{\varphi}_1,\tilde{\varphi}_2,\tilde{\varphi}_3)=(\tilde{\varphi}_1,\alpha\psi+\hat{v},0)$ satisfies (5). By Theorem 2.1, there exists a curve in $H^{-1}\{0\}$ passing through $\bar{z}$, and tangent to $\tilde{\varphi}_1$ at $\bar{z}$.

To compute the pairs $(\tilde{\varphi}_1,\alpha)\in(V-\{0\})\times\mathbb{R}$ that satisfy $D^1(\tilde{\varphi}_1,\alpha\psi)=0$, let (b_1,b_2) be a basis for V, and observe that $D^1(xb_1+yb_2,\alpha\psi)=0$ is equivalent to a system of two equations in $x,y,\alpha\in\mathbb{R}$. This can be reduced to a single homogeneous equation of degree 4 in x and y, which can be solved explicitly. Thus we may expect 0, 2 or 4 nontrivial solution branches. For (2), we only observed cases with 2 or 4 "new" solution curves.

4. A Singularity

For $N=6$, $\mu=c=0.01$, $D_I=0.4$, $\nu=[-3+\sqrt{17}]/5$, we obtained three solution curves passing through $(\bar{u},D_A^{2,3})$. To get a clue about the singularity underlying this situation, we studied numerically the effect of some perturbations of H. It turned out that the resulting bifurcation diagrams can be explained by the 2-parameter-unfolding of the singularity (y^3-xy,z^3+xz) given in the legend to Figure 1. In Figure 1, bifurcation points are marked by "•". Perturbing H, we found numerically all the pictures corresponding to values of α and β with $D(\alpha,\beta)\neq 0$. The diagrams associated to the α-axes are also observed in [2].

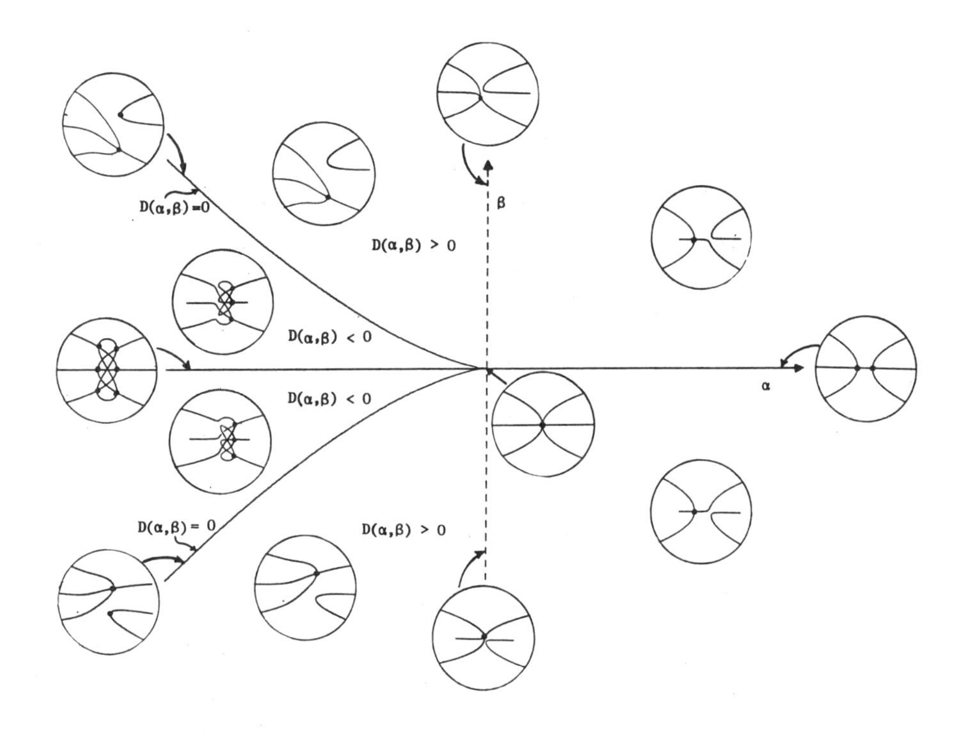

Figure 1: Solution diagrams for $F_{\alpha,\beta}(x,y,z) = \begin{pmatrix} y^3 - xy + \alpha y + \beta \\ z^3 + xz \end{pmatrix} = 0$;

$$D(\alpha,\beta) = \left(\frac{\alpha}{3}\right)^3 + \left(\frac{\beta}{2}\right)^2 .$$

References

[1] Descloux, J., Rappaz, J. (1982) Approximation of solution branches of nonlinear equations. R.A.I.R.O. Anal. Numér. 16, 319-349

[2] Fujii, H., Mimura, M., Nishiura, Y. (1982) A picture of the global bifurcation diagram in ecological interacting and diffusing systems. Physica 5D, 1-42

[3] Gierer, A., Meinhardt, H. (1972) A theory of biological pattern formation. Kybernetik 12, 30-39

[4] Keller, H.B., Langford, W.F. (1972) Iterations, perturbations and multiplicities for nonlinear bifurcation problems. Arch. Rat. Mech. Anal. 48, 83-108

[5] Keller, H.B. (1977) Numerical solution of bifurcation and nonlinear eigenvalue problems. In: Applications of Bifurcation Theory (ed. P.H. Rabinowitz), Academic Press, New York, 359-384

Edgar Jäger, Fakultät für Mathematik der Universität Konstanz,
Postfach 5560, D-7750 Konstanz.

International Series of
Numerical Mathematics, Vol. 79

ON CORNU SPIRALS. DISORDER, SELFSIMILARITY, AND JACOBI'S $\theta_3(v,\tau)$

by Nicholas D. Kazarinoff and Evangelos A. Coutsias, S.U.N.Y. at Buffalo, Department of Mathematics, Buffalo, N.Y., U.S.A. and U. New Mexico, Department of Mathematics and Statistics, Albuquerque, N. M. , U.S.A.

1. Introduction.

We discuss some phenomena exhibited by sums

$$S_N = \Sigma_0{}^N \exp[i\pi(\alpha n^p + \beta n)] \tag{1.1}$$

of unit vectors in $\mathbb{C}$ as $N \to \infty$. The phenomena exhibited by (1.1) for various choices of α, β and p have implications for numerical computations and disorder in dynamical systems and are beautiful as well. The sums (1.1) arise in several contexts, for example, as Riemann sums approximating oscillatory integrals typified by

$$I_p(z) = \int_0^z \exp(i\pi s^p/2)ds. \tag{1.2}$$

If $p = 2$ $(\beta = 0)$, (2) is the Fresnel integral, and $I_2(z) = e^{\pi i/4}$. The graph in $\mathbb{C}$ of (2) is then a Cornu spiral (C-spiral) (Fig.1). For a discussion of the

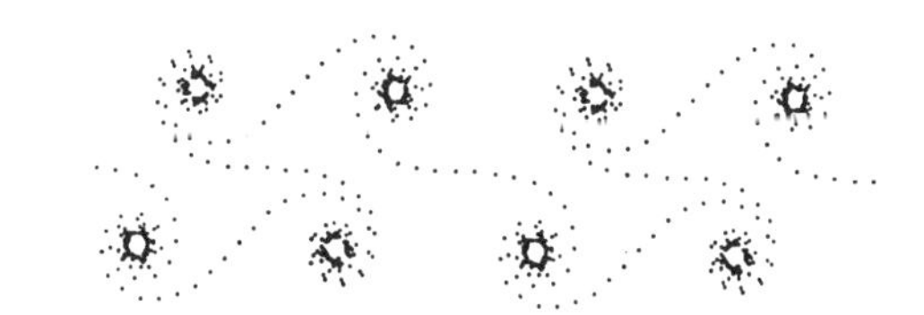

Fig 1. Part of a periodic lattice of simple C-spirals: $\Sigma_1{}^{834}\exp(4i\pi n^2/417)$. One point is plotted for each partial sum.

C-spiral in relation to geometrical optics, see *The Feynman Lectures* [4; vol. 1, Sect. 30.6]. The approximation of (2) by Riemann sums is reliable provided the stepsize is smaller than the period of the integrand. Since for $p > 1$ this period decreases as u increases, the approximation of (2) by Riemann sums of a given stepsize breaks down for z large. There is a $z = z_{max}$ for which there is optimum agreement between (2) and its approximating Riemann sum, and the sum forms a spiral very close to the true C-spiral which is the graph of (2). But, for $z \to \infty$ and fixed stepsize the Riemann sums diverge. At first glance, it is surprising that, depending on the stepsize, the Riemann sums form new spirals, infinitely many spirals of spirals ... for $z = \infty$.

If $p = 2$ and $\beta = 0$, the elementary spirals are all of the same size as the first one, while for $p \neq 2$ their size varies with n, decreasing

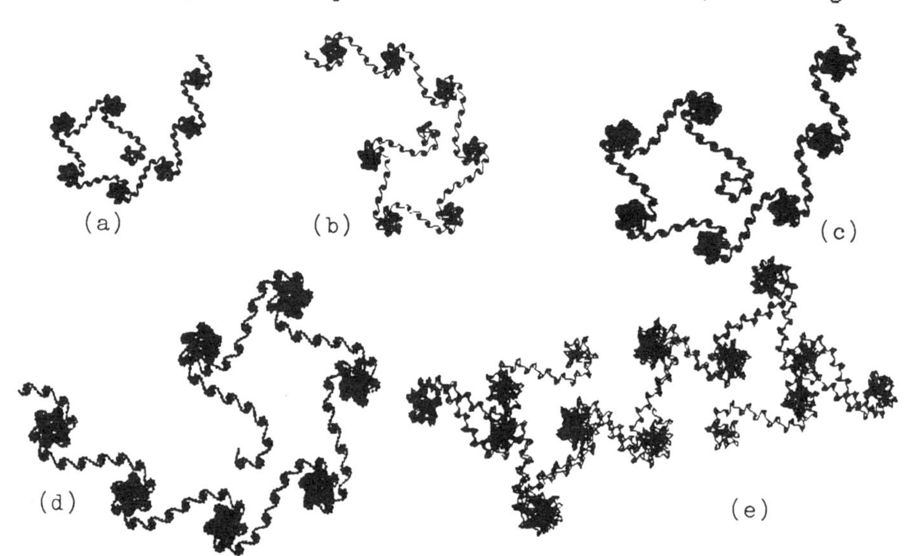

Fig. 2. Parts of $\Sigma_0^\infty \exp(i\alpha\pi n^2)$, computed in TML Pascal (extended arithmetic). (a) Full or part of kth level selfsimilar spirals (sss's): $\alpha = 13-\sqrt{168}$ for $k = 1, 2, 3$; $n = 1, 5{,}000$. (c) continues (a) for $n = 1, 127{,}500$ including part of the 4th level. (b) and (d) are anti-sss's ($\alpha = \sqrt{170}-13$) for the same n. (e) Part of the "lattice" found by replacing α by 4α in Figs. (b),(d) ($\alpha = -13+\sqrt{170}$). In all but Fig. 1 and Figs. 4a,b, successive partial sums are joined by straight line segments of unit length.

(increasing) for $p > 2$ ($1 < p < 2$). The selfsimilar shapes of the spirals of spirals, ... (Fig. 2) we found for special values of α suggested to us that the series (1) for $p = 2$ and $\beta = 0$ might be summable in blocks. For special α's we show this to be nearly so in Section 3 below. The varying size of spirals for $p \neq 2$ makes our results for $p \neq 2$ weaker than those for $p = 2$. Nevertheless, for rational values of $p \neq 2$ and special α's ($\beta = 0$), we produce intervals of n where spirals of spirals are found; see Fig. 3.

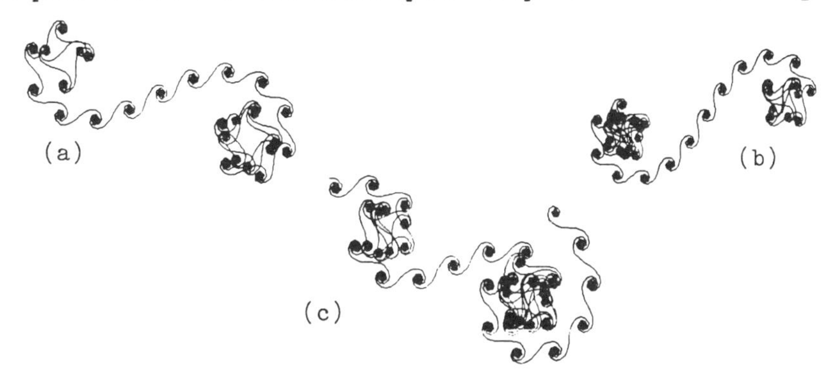

Figure 3. A portion of the graph of $\{z_n\}$ showing spirals of spirals of points, all with $n_c = 10^6$ for: (a) n = 997,750 to n = 1,000,350, with p = 1.5 and $\alpha = 8(30 - \sqrt{899})/3$; (b) 999,250 to 1,000,800, 2.5, $8(20 - \sqrt{399})/15000$; (c) 999,500 to 1,000,550, 3 , $24 \cdot 10^6 (20 - \sqrt{399})$.

The basis of all our understanding of (1.1) is the concept of discrete curvature (defined below). It led us to discovery of a renormalization transformation, both for $p = 2$ and $p \neq 2$, which, in turn, led us to the cases in which we found selfsimilarity in the graph of (1.1). For $p = 2$ and $\beta = 0$ we use a delicate result of Hardy and Littlewood (H. and L. in the sequel), Theorem 2.128 in [5], which is an approximate functional equation for Jacobi's $\theta_3(v,\tau)$, to prove the renormalization is correct up to an error that is at most a constant multiple of $\alpha^{-1/2}$. We thus predict that for $\alpha = m - (m^2 - 1)^{1/2}$ ($m \in \mathcal{J}^+$), $p = 2$ and $\beta = 0$, the spirals are arranged in a selfsimilar sequence of ascending scales, and we establish relevant scaling laws. The selfsimilar pattern is easiest to observe numerically for $p = 2$, but for $p \neq 2$ we establish generalizations

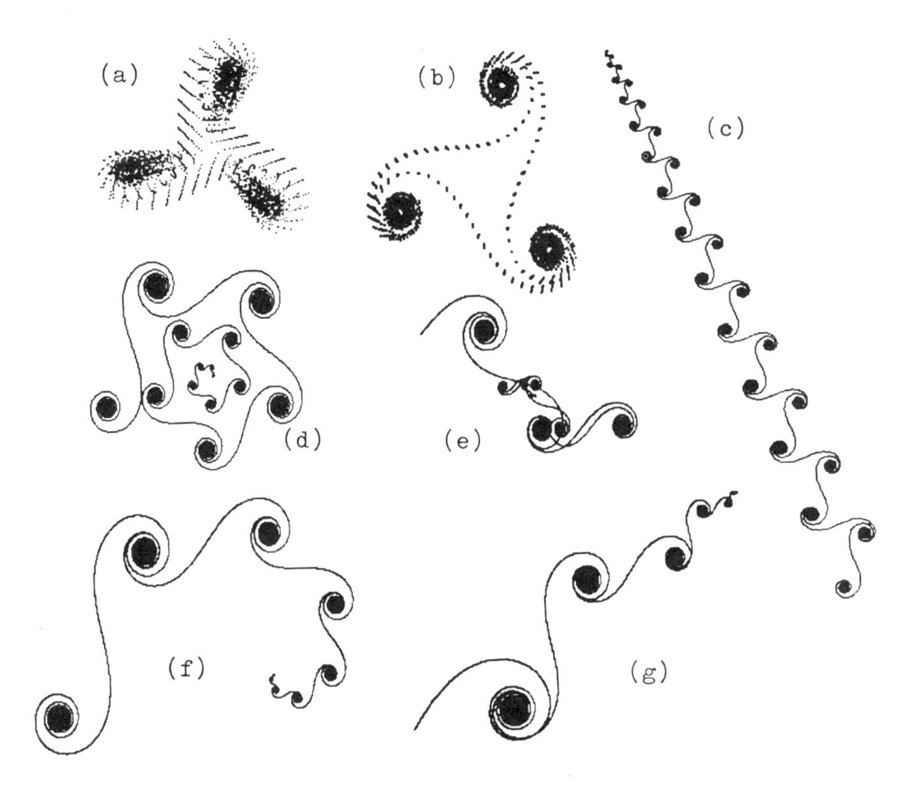

Fig. 4. The p, α, and N and p', α' of the renomalized sum are in (a)-(g), : p = 3/2, 3/2, 4/3, 5/4, 6/5, 7/6, 8/7; α = 4/3,4/3, 6/4, 8/5, 10/6, 12/7, 14/8; N = 3,000, $1.0 \cdot 10^5$ to $1.03 \cdot 10^5$, $\approx 2.5 \cdot 10^4$, $5.0 \cdot 10^4$, $1.0 \cdot 10^5$, $9.99 \cdot 10^5$, $2.227 \cdot 10^6$; p' = 3, 3, 4, 5, 6, 7, 8; α' = -2/3,-2/3,-2/4, -2/5, -2/6, -2/7, -2/8.

and present representative graphs; see Figs. 3 and 4. If $\alpha = (m^2 + 1)^{1/2} - m$, $p = 2$, and $\beta = 0$, then (1) also yields spirals of spirals of ... points, but the spirals at successive levels are oppositely oriented (α is replaced by $-\alpha$). Further, the renormalization of (1) for p = 2 can be extended to a multi-dimensional generalization of (1); see Fig.5. Use of discrete curvature also enables us to define the number of points n_ℓ (s_ℓ) per spiral. It It then follows for p = 2 that the sequence $\{s_\ell\}$ is a Beatty sequence [1, 9, 10] for each α that yields a selfsimilar, or anti-selfsimilar, pattern; namely, if $\alpha = m-(m^2-1)^{1/2}$ (or $(m^2+1)^{1/2}- m$), the sequence of integers $\{s_\ell\}$

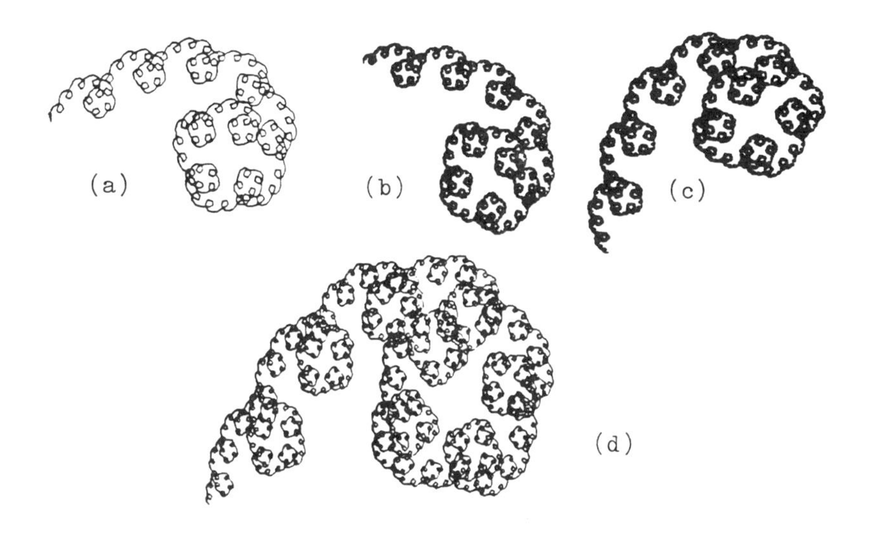

Fig. 5. (a)-(c) The graph of $\sum^{8}_{n_5=0} \sum^{8}_{n_4=0} \sum^{8}_{n_3=0} \sum^{8}_{n_2=0} \sum^{8}_{n_1=0} \exp(i\pi\alpha n^2)$ with $\alpha = 15 - (224)^{1/2}$: five levels of sss's. The largest spiral in (a) is of order 3, in (b) order 4, and in (c) order 5. (d) With $\alpha = 13-(168)^{1/2}$ the graph of

$$\sum^{7}_{n_4=0} \sum^{7}_{n_3=0} \sum^{7}_{n_2=0} \sum^{7}_{n_1=-1} \exp(i\pi\alpha n^2).$$

is a Beatty sequence, of the integers $2m$ and $2m-1$ (or $2m$ and $2m+1$). We supress further discussion of this fact here; for details see [3]. The arrangement of the sequel is: In Section 2 the geometry that led to our prediction of selfsimilarity and scaling laws is introduced. In Section 3 we give similar results for $p = 2$ and some results special to this case. Lastly we present results for $p \neq 2$. We omit details of our derivations to save space; they can be found in [3].

2. Discrete Spirals and Their Curvature.

We consider vectorial addition in $\mathbb{C}$ of points S_N $(N \in \mathfrak{I})$ with

$$S_N = \sum_0^N e^{i\pi\alpha n^p} \equiv \sum_0^N z_n, \tag{2.1}$$

where $z_n = \exp(i\pi\alpha n^p)$ is a unit vector in $\mathbb{C}$, $p > 1$, and $0 < \alpha \ll 1$ (later we shall generalize to $0 < |\alpha| \bmod 2 \ll 1$). We let

$$\phi_n = \pi\alpha n^p \quad \text{and} \quad \Delta\phi_n = \phi_{n+1} - \phi_n \quad (n = 0, 1, 2, \ldots). \tag{2.2}$$

Definitions. (i) The *local discrete radius of curvature* R_n of the pattern generated by the points S_n is the radius of the circle passing through the three consecutive points S_{n-1}, S_n, S_{n+1}; namely,

$$R_n = \tfrac{1}{2}|\csc(\Delta\phi_n/2)|.$$

(ii) For $n = n_{2\ell}$, z_n is the *mid-vector (point of inflection)* of the ℓth C-spiral of the pattern S_∞ if

$$\Delta\phi_{n_{2\ell}-1} \le 2\ell\pi < \Delta\phi_{n_{2\ell}}. \tag{2.4}$$

(iii) Similarly, if $n = n_{2\ell+1}$, z is the *end-vector (cusp)* of the ℓth C-spiral and if $n = n_{2\ell+1} + 1$, z_n is the *initial-vector (cusp)* of the $(\ell+1)^{st}$ C-spiral if

$$\Delta\phi_{-1+n_{2\ell+1}} \le (2\ell+1)\pi < \Delta\phi_{n_{2\ell+1}}; \tag{2.5}$$

Fig. 6. (a) The number of points per spiral and their size decreases: $\Sigma_{25}^{1250} \exp(i\alpha\pi n^4)$ with $\alpha = 10^{-8}\sqrt{2}$. (b) The number of points per spiral and their size increases: $\Sigma_0^{5730}\exp(i\alpha\pi n^{1.01})$ with $\alpha = 750$. The last 1200 points are in the largest spiral.

We combine (2.4) and (2.5) and use the binomial theorem to find that approximately, for ℓ large,

$$bn_\ell \approx [(\ell/p\alpha)^{1/(p-1)}], \tag{2.6}$$

where the square brackets denote the greatest integer function. From the behavior of R_n as n increases we obtain the

Theorem. For $0 < \alpha \ll 1 \bmod 2$ *the pattern formed by the points*

$$\{S_{n_{2\ell-1}}, \ldots, S_{n_{2\ell}}, \ldots, S_{n_{2\ell+1}}\}$$

is a double spiral, resembling a Cornu spiral from diffraction theory [4]. *The number of points* η_ℓ *in the* ℓ^{th} *C-spiral is*

$$\eta_\ell = n_{2\ell+1} - n_{2\ell-1}. \tag{2.7}$$

For ℓ *large and* $p > 1$ $(p \neq 2)$,

$$\eta_\ell = 1 + (p-1)^{-1}(2\ell^{2-p}/p\alpha)^{1/(p-1)}\{1 + \mathcal{O}(1/\ell)\} \tag{2.8}$$

For $p = 2$,

$$n_k = [½ + k/(2\alpha)] \text{ and } \eta_\ell = [½+(\ell+½)/\alpha] - [½+(\ell-½)/\alpha] \approx 1/\alpha. \tag{2.9}$$

It follows from this theorem that for $1 < p < 2$ $(p > 2)$ the number of points per C-spiral decreases (increases) like $\mathcal{O}(\ell^e)$, where $e = (2-p)/(p-1)$. This can be observed in Figs. 4 and 6.

3. The Quadratic Case.

For $p = 2$ the sum (2.1) is related to the Jacobi's $\theta_3(v,\tau)$ defined by

$$\theta_3(v,\tau) = \sum_{-\infty}^{\infty} e^{i\pi\tau n^2 + 2\pi i n v} \qquad (\mathcal{Im}\,(\tau) > 0). \tag{3.1}$$

In our problem, $\mathcal{Im}\ \tau = 0$, so that the infinite series expression for θ_3 diverges. H. & L. [5] investigated the behavior of $\theta_3(v,\tau)$ on the real τ-axis. They derived an approximate functional equation for partial sums of $\theta_3(v,\tau)$. This equation is an extension, to real τ, of the functional equation, known as Jacobi's imaginary transformation [2; pp.72-80], It relates $\theta_3(v,\tau)$ to $\theta_3(v/\tau,-1/\tau)$ for $\mathcal{Im}\,(\tau) > 0$. For $v = 0$. Jacobi's transformation becomes $(\tau/i)^{½}\theta_3(0,\tau) = \theta_3(0,-1/\tau)$ or

$$\sum_{-\infty}^{\infty} e^{i\pi\tau n^2} = (i/\tau)^{½} \sum_{-\infty}^{\infty} e^{-i\pi n^2/\tau}. \tag{3.2}$$

Obviously, (3.2) diverges if $\mathcal{Im}\ \tau = 0$; one has instead the approximate formula [5; Thm. 2.128, p. 209], adapted to our case

$$(\tau/i)^{½}\sum_{0}^{n} e^{i\pi\tau k^2} - \sum_{0}^{[n\tau]} e^{-i\pi k^2/\tau} = \mathcal{O}(1). \tag{3.3}$$

H. & L. used this remarkable result to get several estimates on the rate of divergence of the sum (2.1) for $p = 2$. For example, if τ is irrational, then the sum grows with n like $n^{1/2}$. More precise estimates depend upon the growth rate of the coefficients in the continued fraction expansion of τ. H. & L. also studied (1) for $p = 2$ in relation to the continued fraction expansion of τ. Thus it appears strange that, although they studied these sums for τ a quadratic irrationality, they missed their renormalizability. A short proof of (3.3) was given by MORDELL [8].

Here we sketch a derivation of the approximate functional equation based upon the geometric properties of C-spirals. We begin by summing the terms in the block $\mathcal{B}_\ell$ of terms that correspond to the ℓth spiral: It is convenient to express each term within the ℓth spiral in terms of the mid-vector $z_{n_{2\ell}}$:

$$\mathcal{B}_\ell = \sum_{k=1}^{n_{2\ell+1}-n_{2\ell-1}} z_{k+n_{2\ell}-(n_{2\ell}-n_{2\ell-1})} = \sum_{k=1}^{n_{2\ell+1}-n_{2\ell}} \exp i\pi\alpha(n_{2\ell}+k)^2. \quad (3.4)$$

If we define δ_ℓ by $\delta_\ell = (\ell/\alpha) - [½ + (\ell/\alpha)]$, then

$$-½ \le \delta_\ell \equiv (\ell/\alpha) - [(-½ + \ell)/\alpha] < ½ \quad (3.5)$$

If we recall (2.8) ($n_{2\ell} = [½+(\ell/\alpha)]$), we may write the exponent in (3.6) as:

$$\upsilon\alpha(n_{2\ell}+k)^2 = \alpha(k + ½ + [\ell/\alpha] - \ell/\alpha)^2 - \ell^2/\alpha + 2\ell(k + n_{2\ell}), \quad (3.6)$$

so that, finally,

$$\mathcal{B}_\ell = e^{-i\pi\ell^2/\alpha} \sum_{1+n_{2\ell-1}-n_{2\ell}}^{n_{2\ell+1}-n_{2\ell}} \exp[i\pi\alpha(k-\delta_\ell)^2. \quad (3.7)$$

To estimate this sum, we approximate (1.2) by a Riemann sum

$$I_2(z) = ½\int_{-z}^{z} \exp(i\pi s^2/2)ds = ½\Delta s\sum_{1-M}^{M} \exp[i\pi(k\Delta s - \xi)^2 + o(\Delta s), \quad (3.8)$$

where Δs is the stepsize, $M = z/\Delta s$, and $-½\Delta s \le \xi < ½\Delta s$. By letting $½\Delta s = \alpha^{1/2}$ ($\ll 1$), $\xi = \delta_\ell/\alpha$ and $M = n_{2\ell+1} - n_{2\ell} \approx n_{2\ell} - n_{2\ell-1} \approx ½\alpha$, we obtain

$$\sum_{1+n_{2\ell-1}-n_{2\ell}}^{n_{2\ell+1}-n_{2\ell}} \exp[i\pi\alpha(k-\delta_\ell)^2 \approx \alpha^{-1/2}\int_{-½\alpha^{-1/2}}^{½\alpha^{-1/2}} \exp(i\pi s^2)ds. \quad (3.9)$$

The integral converges to $e^{i\pi/4}$ as $\alpha \to 0$. Thus, by (3.7-9), we conclude

$$\alpha^{1/2}e^{-\pi i/4}\sum_{n=-1}^{n_{2\ell+1}} \exp(i\pi\alpha n^2) \approx \sum_{n=0}^{\ell} \exp(-i\pi n^2/\alpha) \quad (3.10)$$

The result of H. & L. [5] guarantees that the error in (3.10) is $\mathcal{O}(1)$.

The block summation in (3.10) embodies the intuitive idea that the elementary C-spirals of (2.1) may be thought of as "points" of a new sum of vectors joining these "points". These vectors have length about $\alpha^{-1/2}$ (the "size" of a C-spiral), and are rotated counterclockwise by $\pi/4$ from the mid-vector of their corresponding C-spiral. The new sum is of the same type as the original one except that α is replaced by $-1/\alpha$.; Summation by blocks left the form of the sum invariant but "renormalised" the parameter α and scaled the new sum by the factor $\alpha^{-1/2}e^{\pi i/4}$.

The renormalization map

$$(-1, 1)\backslash 0 \circlearrowleft, \ \alpha \mapsto -1/\alpha \mod 2, \tag{3.11}$$

associates to every number $\alpha = \alpha_0$ in $(-1, 1)$ the sequence $\{\alpha_i\}$ of iterates of the map. If α is rational. the sequence terminates with a 0 or with infinitely many 1's. If α is irrational, so are all the α_i.

Indeed, if

$$\alpha \equiv \alpha_0 = 2n_0 + (1/2n_1 + (1/2n_2 + (1/n_3 + \ldots \equiv (2n_0, 2n_1, 2n_2, \ldots)$$
$$(-1 < \alpha < 1;\ \text{each}\ n_i > 0) \tag{3.12}$$

isthe continued fraction expansion of α, then the ith iterate α_i is given by $\alpha_i = (-1)^i(2n_i, 2n_{i+1}, 2n_{i+2}, \ldots)$. Our renormalization argument requires all α_i mod 2 to be sufficiently small. It follows that all numbers α whose continued fractions are composed of sufficiently large even integers will lead to infinitely renormalizable patterns $\{S_N\}$ in $\mathbb{C}$. If α is also a quadratic irrational, namely, if its continued fraction becomes periodic with period k eventually, then the pattern generated by $\{S_N\}$ in $\mathbb{C}$ is eventually selfsimilar of order k. The special case $n = n_0 = -n_1 = n_2 = -n_3 = \ldots$, which implies $\alpha = \pm n-(n^2-1)^{1/2}$, leads to a pattern that is selfsimilar; see Fig. 2(a). The case $n = n_0 = n_1 = n_2 = \ldots$, which implies $\alpha = (n^2 + 1)^{1/2} \pm n$, leads to a pattern that is anti-selfsimilar, i.e., a pattern whose spirals are inverted by a reflection at each iteration; see Fig 2(b).

The similarity observed is precisely this: Let $v_{\ell+1}$ be the vector from the origin to the center of the first spiral of order $\ell + 1$ formed by the spirals of order ℓ, and let the length of v_ℓ be $\|v_\ell\|$. Then $v_{\ell+1}$ makes an angle of $\pi/4$ to v_ℓ (clockwise) and $\alpha^{1/2}\|v_{\ell+1}\| = \|v_\ell\|$. If we rotate the plane by $-\pi/4$ and contract by $\alpha^{1/2}$ with respect to the origin then the images of the points of inflection of the spirals of order ℓ forming the first spiral of order $\ell + 1$ almost coincide with the points of inflection of of the spirals of order $\ell - 1$ that form the first spiral of order ℓ. In the case of anti-selfsimilarity $v_{2\ell}$ makes an angle of $-\pi/4$ with $v_{2\ell-1}$; $v_{2\ell+1}$ makes an angle of $\pi/4$ with $v_{2\ell}$. The selfsimilarity is never exact (α is irrational), but it improves as $\alpha \to 0$.

To include other irrational numbers, we allow negative integers in their continued fraction expansions. Since it can be shown that every irrational number has a unique expansion of the form (3.15) if we allow negative integers n_i, the statements made above include any irrational in $(-1, 1)$. Thus, the numbers $\alpha = (2n, -2n, 2n, -2n, \ldots) = 2n-1/\alpha$, correspond to patterns generated by $\{S_N\}$ in $\mathbb{C}$ that are selfsimilar.

We close this section with a generalization of selfsimilarity to multi-indexed sums. Jacobi's θ_3 has been generalized and used by number theorists in a number of ways [2, Chapt.XI]. This enables us to generalize (1.1) and, heuristically, its renormalization for $p = 2$. Let $Q(x) = \sum_{k,l=1}^{m} \alpha_{kl}\, x_k x_l$ be a positive-definite quadratic form associated with the symmetric, positive-definite, real matrix $(\alpha_{k,l})$ with determinant D. Let $(\alpha'_{k,l})$ be the inverse of $(\alpha_{k,l})$, and let $Q'(x)$ be the quadratic form associated with $(\alpha'_{k,l})$. Define

$$\theta(\tau, Q) = \sum_{n_1,\ldots n_m=1}^{\infty} \exp(i\pi\tau Q(n_1, \ldots n_m) \qquad (\mathrm{Im}(\tau) > 0).$$

Then $\theta(\tau,Q) = [\theta_3(0,\tau)]^m$, and $\theta(-1/\tau,Q) = (\tau/i)^{m/2} D^{-1/2} \theta(\tau,Q')$. For $\tau = \alpha = m - (m^2-1)^{1/2}$ and $(\alpha_{k,l})$ the identity matrix we obtain the formal identity $\theta(-1/\alpha,Q) = (\alpha/i)^{m/2}\theta(\alpha,Q)$. It should correspond to an approximate functional equation, analogous to (3.10), for multidimensional partial sums for real τ. Iterations of the map $z_{n+1} = z_n + \exp(i\pi\alpha n^2)$, where now n is a multi-index with components $n_1, \ldots, n_m$ and $n + 1$ means add 1 to one component of n, yield graphs exhibiting selfsimilarity or anti-selfsimilarity for α a quadratic irrationality; see Fig. 5.

4. The Case $p > 1$, $p \neq 2$.

In close analogy to that given above for $p = 2$ one may derive a functional equation for sums of the form

$$S_{a,b} = \Sigma_{a \le n \le b} \exp(i\pi\alpha n^p) \qquad (p \neq 2,\ p > 1). \qquad (4.1)$$

Our result (4.6-7) below in which $p \mapsto p' = p/(p-1)$ agrees with the appoximate functional equation given by van der Corput [11] for general sums of the form $\Sigma_{a \le n \le b}\, g(n)e^{2\pi i f(n)}$ (when it is specialized to the sums (4.1) we consider). We also apply the method of stationary phase to reduce our functional equation, if the range of n in (4.1) is further limited, to the case $p \mapsto 2$ so that we obtain C-spirals of C-spirals of points for these sums. The self-similarity obtained is only approximate, and we were able to find it only up through level 2: the C-spirals of C-spirals of points do not form C-spirals.

Our result is the approximate functional equation

$$\sum_{1+n_{2\ell_0+1}}^{n_{2\ell_1+1}} \exp[i\pi\alpha n^p] \approx \sum_{\ell=\ell_0}^{\ell_1} \mathcal{B}_\ell, \tag{4.2}$$

where

$$\mathcal{B}_\ell \approx \{[\gamma/(p-1)]\ell^{(2-p)/(p-1)}\}^{1/2} e^{i\pi/4} \exp\{-i\pi\alpha'\ell^{p'}\}. \tag{4.3}$$

$$\alpha' = -2[(p-1)/p]\gamma, \quad p' = p/(p-1), \text{ and } \quad r_\ell = \{[\gamma/(p-1)]\ell^{(2-p)/(p-1)}\}^{1/2} \tag{4.4}$$

Note that one must understand that (4.2) holds only for large $\ell_0 < \ell_1$. We also observe that our block summation process has led to a new sum, the right-hand side of (4.2), in which α and p have been replaced by α' and p', respectively, while a change of phase by $\pi/4$ and a scale factor r_ℓ have been introduced. The search for nearly selfsimilar patterns is thus complicated. Since α' is not necessarily small, the renor- malized sum will exhibit regular behavior if p' is an integer and α' is rational. In this case the graph of the right-hand side of (4.2) is a portion of a pseudo-periodic graph of C-spirals. The graphs produced by (4.1) will have spirals growing in size, according to (4.4), if $1 < p < 2$ and decreasing in size for $p > 2$. A series of patterns produced for rational p's and special α's yielding $p' = 3, 4, \ldots, 8$, $\gamma = 1$, and α' is rational are shown in Fig. 4.

To find selfsimilar behavior in sums of the form (4.1) it turns out to make sense to look for selfsimilar patterns locally in n, that is, for n close to a fixed n_c. Let n_c be a large integer and consider a sum

$$S = \sum_{n_c-a}^{n_c+b} \exp(i\pi\alpha n^p)$$

$$\approx \exp(i\pi\alpha n_c^p) \sum_{n_c-a}^{n_c+b} \exp\{i\pi\alpha[p n_c^{p-1} k + \tfrac{1}{2} p(p-1) n_c^{p-2} k^2]\} \tag{4.5}$$

In (4.5) the first neglected term in the exponent is of order $k^3 n_c^{p-3}$. Now let $\alpha p n_c^{p-1} = 2\|\frac{1}{2}\alpha p n_c^{p-1}\| + r$, where $\|\cdot\|$ denotes the nearest integer function. Then, after we complete the square in the exponent of the terms in the sum (4.5) and neglect even multiples of π, it becomes

$$S \approx \exp(i\pi[\alpha n_c^{p}-\delta']) \sum_{n_c-a}^{n_c+b} \exp\{i\pi[(p-1)/n_c][\tfrac{1}{2}\alpha p n_c^{p-1}](k + \delta)^2]\} \qquad (4.6)$$

where

$$\delta' = r^2/(2\alpha p(p-1)n_c^{p-2}) \quad \text{and} \quad \delta = r/(\alpha p(p-1)n_c^{p-2}) \qquad (4.7)$$

The sum in (4.6) will behave like a quadratic sum

$$\sum_{n_c+a}^{n_c+b} \exp\{i\pi\alpha' k^2\} \qquad (4.8)$$

rovided δ can be made sufficiently small. This will be so provided n_c^{p-1} is sufficiently close to the denominator of a convergent to $\frac{1}{2}\alpha p$ in its continued fraction expansion. For, then by the best approximation property, one can always find a fixed constant c such that $|\frac{1}{2}\alpha p - m/n_c^{p-1}| < c/n_c^{p-1}$. This means $r \approx c/n_c^{p-1}$, and this in turn implies

$$\delta \approx c/(\tfrac{1}{2}\alpha p(p-1)n_c^{2p-3}) \qquad (4.9)$$

Since in (4.9), $\frac{1}{2}\alpha p$, although it might be small, is fixed, and c is fixed (like 1/5, if we use Hurwitz's theorem [7]), δ can be made arbitrarily small by choosing n_c large enough provided $p > 3/2$. However, in terms of observing patterns arising from (4.2) this formula is not useful for $p < 2$ except for special values of α. If $\alpha = m - (m^2 - 1)^{1/2}$ ($m \in \mathcal{J}^+$) is small, then $-1/\alpha = \alpha \bmod 2$ and δ is small. For, if $\alpha = K[m - (m^2 - 1)^{1/2}]$ and we choose K so that the coefficient of $(k+\delta)^2$ is exactly $i\pi[m-(m^2-1)^{1/2})]$, then (4.5) should produce selfsimilarity for a and b small, provided the hypotheses for (4.2) to hold are fulfilled. Three 2nd order spirals for p's $\neq$ 2, found by using the α's determined from (4.4) are shown in Fig. 3.

The estimate (4.9) and (4.6) also have an interesting implication if $p = 2$: (4.6) with δ as in (4.9) implies that the pattern centered at n_c (i.e., n is close to n_c) will be arbitrarily close to the pattern centered at $n = 0$ if n_c is a convergent denominator for α.

For some $p \neq 2$ we can also use the result (4.8) as follows. Let $p > 2$ be an integer, and let

$$\tfrac{1}{2}\alpha p = [\,0; n_0^{p-1}; n_1^{p-1}; \ldots] \qquad (n_i \in Z^+) \qquad (4.10)$$

Choose n_0 not too large and choose n_1 very large. Let $n_c = n_0 n_1$. Then the second convergent's denominator $q = n_c^{p-1} + 1 \approx n_c^{p-1}$, so that by

(4.7), we get (4.8) with $\alpha' = [(p-1)/n_c][n_1^{p-1} + O(n_0^{1-p})]$. In this case α' is nearly equal to $(p-1)n_1^{p-1}/n_c$, and we expect ordered behavior in a range of n for $\frac{1}{2}\alpha p$'s whose continued fractions contain integers that are in some sense close to numbers of the form n^{p-1}. This leads to questions of Diophantine approximation which we do not pursue.

We also conjecture that if $\alpha = -1/\alpha \bmod 2$ and one approximates α by a finite portion of its continued fraction, then there exists similarity up to a corresponding level in the graph generated by (1.1). If such a magic α is replaced by $2^m\alpha$ and $\beta = 0$, then (1.1) yields a lattice of points building spirals building spirals of spirals; see Fig. 2. Further, there is a variety of cases for which (1.1) is periodic. For example, if $\alpha = s/q$, $\beta = r/q$, and $s \geq 2$, q and r are positive integers, with $(s, q) = s$, $(r, q) = 1$, then the graph of (1.1) is periodic with period at least as small as $2q$ is a periodic lattice; see Fig. 1. The case $s = 1$ and $r = q = 2m$ is associated with generalized Gauss sums [2, p. 144].

Finally, we observe that the S_n of (1.1) can also be interpreted to be special solutions of Schrödinger's equation with $x \in [0, 1]$, $t > 0$ and periodic boundary conditions. If $u_{xx} = iu_t$, then $\exp i(n\pi x + n^2\pi^2 t)$ is a solution; and setting $\alpha = \pi^2 t$ and $\beta = \pi x$, we see that S_n is a sum of such solutions. Indeed, S_∞ is the (formal) solution of this problem satisfying the initial condition $u(x,0) = \delta(x)$; that is, S_∞ is formally the causal Green's function for Schrödinger's equation. For a map like (1.1), but involving quartic, quadratic, and first powers of n, S_∞ is formally the Green's function for the linear part of the Kuramoto-Velarde operator $u_t + 4u_{xxxx} + \alpha(u_{xx} + \frac{1}{2}(u_x)^2 + (uu_x)_x)$.

Acknowledgements. The authors gratefully acknowledge the help of Dr. Andrew Mulhaupt, who introduced them to [5] and the connection between the map (2.1) and Beatty sequences and $\theta_3(v,\tau)$. We thank Dr. E. Bombieri for introducing us to the map (1.1).

REFERENCES

1. de Bruijn, N. G. (1981) Sequences of zeros and ones generated by special production rules, Mathematics, Proc. A. 84, 27-37.

2. Chandrasekharan, K. (1985) *Elliptic Functions*, vol. 281 in *Grundleheren*

der mathematischen Wissenschaften, Springer-Verlag, Berlin,.

3. Coutsias E. E. and Kazarinoff, N.D. (1987) Disorder, renormalization, theta functions and Cornu spirals, Physica D, to appear.

4.Feynman, R.C., Leighton, R. B. and Sands, M. (1963) *The Feynman Lectures on Physics*, Addison Wesley, Reading, Mass., vol. 1,Section 30-6.

5. Hardy, G.H. and Littlewood, J. E. (1914) Some problems of Diophantine approximation, Acta Math. 37, 193-238.

6. Hardy, G. H. and Wright, E. M. (1979) *An Introduction to the Theory of Numbers,* Fifth Ed., Oxford Univ. Press, Oxford, see Thms. 193-195, p. 164.

7. Mendès-France,M. (1984) Entropy of curves and uniform distributions, *Topics in Classical Number Theory,* G. Halasz, ed., vol II, (Budapest, 1981), 1051-1068, Colloq. Math. Soc. Janos Bolyai, 34, , North Holland Pub. Co., Amsterdam-New York; (1983) Entropie, dimension et thermodynamique des courbes planes, *Seminar on Number Theory,* Paris 1981-82, 153-177, Prog. Math. 38 Birkhaüser, Boston, Mass..

8. Mordell, L. J. (1927) The approximate functional formula for the theta function. J. London Math. Soc. 2, 68-72.

9. Mullhaupt, A. P. (1987) Discrete self-similarity, Preprint.

10.Stolarsky, K. B. (1976) Beatty sequences, continued fractions, and certain shift operators, Canad. Math. Bull. 17, 473-482.

11. van der Corput, J. D. (1928) Beweis einer approximativen Functional gleichung, Math. Z. 28, 238-300.

Dr. Nicholas D. Kazarinoff, Department of Mathematics, State University of New York at Buffalo, 106 Diefendorf Hall, Main St. Buffalo, New York, 14214-3093, U.S.A.

International Series of
Numerical Mathematics, Vol. 79

OPTIMIZATION IN BIFURCATION PROBLEMS USING A CONTINUATION METHOD

J. P. Kernévez

Génie Informatique, UTC, Compiègne, France

E. J. Doedel

Applied Mathematics, Caltech, Pasadena CA, USA

1. Statement of the Problem.

This paper deals with equations

$$(1.1) \qquad f(\lambda, u) = 0,$$

where u denotes a *state variable*, λ a *control variable*, and to which we want to find solution pairs (λ_0, u_0) that maximize an objective functional $g(\lambda, u)$. Our specific objectives are (i) optimization in systems that have bifurcations, and (ii) control of the bifurcation phenomena themselves. Above $g : \Lambda \times U \to R$ and $f : \Lambda \times U \to Y$ are continuously Fréchet differentiable mappings with Λ, U, and Y real Banach spaces. In particular Λ may be R^m. Let $X \equiv \Lambda \times U$, with element $x = (\lambda, u)$. We state a general result on the validity of an optimality system and we give a continuation algorithm for finding optimal solutions.

Besides the equality constraint (1.1) we may have inequality constraints

$$(1.2) \qquad h_i(\lambda, u) \leq 0, \quad i = 1, \cdots, n_h,$$

where $h_i : X \to R$ is C^1. The constraints (1.2) are often needed for existence of an optimum solution $x_0 = (\lambda_0, u_0)$, since they introduce boundedness.

Only active constraints, i.e., those that satisfy $h_i(\lambda_0, u_0) = 0$, $i \in a \equiv \{\text{indices of active constraints}\}$, have to be taken into account in the optimality system and they can be considered as included in $f(\lambda, u) = 0$. Define $X_{ad} \equiv \{x \in X \ : \ h_i(x) \leq 0, \quad 1 \leq i \leq n_h\}$, and suppose that the h_i are such that X_{ad} is a closed convex subset of X. This is the case, for example, if the constraints h_i are of the form $m_j - \lambda_j \leq 0$ or $\lambda_j - M_j \leq 0$. We define $A \equiv \{x \in X_{ad} : f(x) = 0\}$, and we assume that: (i) A is a nonempty, bounded and weakly closed subset of X, (ii) X is a reflexive Banach space, (iii) $g : A \to R$ is weakly lower semi-continuous. Then the following is well known [1]:

THEOREM 1.1. *With the above assumptions g attains its minimum on A, i.e., there exists $x_0 \in A$ such that $g(x_0) = min_A \ g$.* □

REMARK. The main work in verifying that the theorem applies is to show that A is bounded in X and that $f^{-1}(0)$ is weakly closed.

2. Validity of the Optimality System.

Here we give necessary conditions for an optimal solution $x_0 \in A$, i.e., $f(x_0) = 0$, $g(x_0) = min_A \ g$. We suppose that equalities $h_i(x) = 0$ are included in $f(x) = 0$. Then we have

THEOREM 2.1. *Let x_0 be an optimal solution. Let $X = X_1 \oplus X_2$, where X_1 and X_2 are two closed complementary spaces in X, with the restriction of $Df(x_0)$ to X_2 a linear homeomorphism on Y. Then there exists $(p, q) \in Y^* \times R$ such that*

$$(Df(x_0))^* p + (Dg(x_0))^* q = 0,$$

$$| \, p \, |_{Y^*}^2 + q^2 = 1. \quad \square$$

REMARKS.

(i) Theorem 2.1 follows from the Implicit Function Theorem. We omit a proof here. Often one uses a penalty method to obtain an optimality system [7]. However, in our general setting this is less convenient and not necessary.

(ii) Let T denote $Df(x_0)$. Theorem 2.1 applies in particular when

$$Range(T) = Y \quad and \quad Null(T) \equiv T^{-1}(0) \ is\ complemented\ in\ X,$$

which happens in particular if T admits a right inverse $S \ : \ T \circ S = Id_Y$. It is well known [1] that in this situation $X = X_1 \oplus X_2$ with $X_1 = T^{-1}(0)$ and $X_2 = Range(S) = S(Y)$. $Null(T)$ is also complemented when it is finite dimensional. The main work when applying Theorem 2.1 is then to show that T is surjective, and either finding S or proving that $dim\ Null(T) < \infty$.

(iii) In the classical case, when $D_2 f(x_0) = f_u^0 \ : \ U \to Y$ is a linear homeomorphism, i.e., when the state u_0 is *regular*, then we can take $X_1 = \{(l,0) \ : \ l \in \Lambda\}$ and $X_2 = \{(0,v) \ : \ v \in U\}$.

(iv) If $m \equiv dim(\Lambda) < \infty$ then the optimality equations split into

$$(2.1) \qquad \begin{aligned} &(D_u f)^* p + (D_u g)^* q = 0, \\ &(D_{\lambda_i} f)^* p + (D_{\lambda_i} g) q = 0, \quad i = 1, \cdots, m. \end{aligned}$$

(v) If we distinguish the active constraints $h_i(\lambda, u) = 0, \quad i \in a$, from the state equation $f(\lambda, u) = 0$, then we have the optimality system

$$\begin{aligned} &f(\lambda, u) = 0, \\ &h_i(\lambda, u) = 0 \quad i \in a, \\ &(D_u f)^* p + (D_u g)^* q + (D_u h_a)^* r = 0, \\ &(D_\lambda f)^* p + (D_\lambda g)^* q + (D_\lambda h_a)^* r = 0, \\ &| p |_{Y^*}^2 + q^2 + | r |^2 = 1. \end{aligned}$$

3. Numerical Method: Successive Continuation.

We do not assume that the state equation (1.1) uniquely defines the state u as a function of the control parameters λ. The state of the system may depend also on the path followed to the final values of the control variables. In fact, our main interest is precisely in applications where (1.1) has bifurcation phenomena as the control parameters are varied.

To deal with this type of optimization problem we have experimented with descent methods, such as a projected gradient technique, and penalty function methods. We couple these to numerical continuation techniques (See [9] for an analysis of the penalty function continuation method) and numerical bifurcation techniques [2,5]. Below we describe another method that we have successfully used. It is basically a systematic continuation procedure for the optimality equations (See [8] for a classification of singularities that may arise). Our *successive continuation* method is most useful when the manifolds defined by the state equation (1.1) have a complicated geometry.

The optimality system (2.1) gives rise to the system

$$
\begin{aligned}
&f(\lambda,u)=0,\\
&g(\lambda,u)-\omega=0,\\
&(D_u f)^* p+(D_u g)^* q=0,\\
&(D_{\lambda_i} f)^* p+(D_{\lambda_i} g)^* q=\alpha_i, \quad i=1,\cdots,m,\\
&\mid p\mid^2+q^2=1,
\end{aligned}
\tag{3.1}
$$

where we have introduced new parameters $\alpha_1,\cdots,\alpha_m$. The successive continuation algorithm is then as follows:

(1) INITIALIZATION: Find a solution (λ,u) to $f(\lambda,u)=0$. Calculate $\omega=g(\lambda,u)$.

(2) STARTING PROCEDURE: Continue a branch of solutions to $f(\lambda,u)=0$, $g(\lambda,u)-\omega=0$, with all control parameters λ_i fixed, except one, say λ_k. For notational

simplicity we may take $k = 1$. Each point on the branch then consists of (λ_1, u, ω). Try to locate a quadratic extremum of ω. Such a point can also be thought of as a limit point (fold) with respect to ω on the branch.

(3) GENERATION OF STARTING POINT: At each extremum from Step (2) we can find p and q, not both zero, and $\{\alpha_i,\ i = 1, \cdots, m\}$ with $\alpha_1 = 0$, satisfying the last three equations in (3.1).

(4) MAIN ALGORITHM: Free another λ_i, say λ_2, and compute an entire branch of solutions to (3.1). Each point consists of $(\lambda_1, \lambda_2, u, p, q, \alpha_2, \cdots, \alpha_m, \omega)$. Locate zero intercepts of the remaining α_i. Return to each of these, freezing the proper α_i at zero and free one of the remaining λ_i. Continue while intercepts are still found and while there are still frozen λ_i left.

REMARKS.

(i) Computation of solution branches is done using pseudo arc-length continuation and branch switching [2,5].

(ii) Along solution branches in the continuation procedure we monitor the quantities $\beta_i \equiv h_i(\lambda, u)$. Zero intercepts of the β_i are located accurately. Starting from these points the algorithm can be repeated with the corresponding active inequality constraint included in the state equation $f(\lambda, u) = 0$.

(iii) We do not address the question of implementation here, except to note that much efficiency can be gained by making use of ideas in [3] or [4].

4. Application.

The successive continuation algorithm has been applied to the optimization of boundary value problems and Hopf bifurcation problems [2] and to the control of folds, period doubling bifurcations and traveling waves [10]. Most applications concern enzyme models from [6].

As an example of the method we consider here the following system of differential equations

$$
\begin{aligned}
u_1'(t) &= u_2(t), \\
u_2'(t) &= -\lambda_1 e^{p(u_1,\lambda_2,\lambda_3)},
\end{aligned} \tag{4.1}
$$

where $p(u_1, \lambda_2, \lambda_3) \equiv u_1 + \lambda_2 u_1^2 + \lambda_3 u_1^4$. The boundary conditions are

$$
\begin{aligned}
u_1(0) &= 0, \\
u_1(1) &= 0,
\end{aligned} \tag{4.2}
$$

and as objective functional we take

$$
\omega = \int_0^1 \left[u_1(t) - 1\right]^2 dt + \frac{1}{10} \sum_{k=1}^{3} \lambda_k^2. \tag{4.3}
$$

For the problem (4.1)-(4.3) the optimality equations (3.1) can be written as

$$
\begin{aligned}
u_1'(t) &= u_2(t), \\
u_2'(t) &= -\lambda_1 e^{p(u_1,\lambda_2,\lambda_3)}, \\
w_1'(t) &= \lambda_1 e^{p(u_1,\lambda_2,\lambda_3)} p_{u_1} w_2(t) + 2\gamma(u_1(t) - 1), \\
w_2'(t) &= -w_1(t),
\end{aligned}
$$

where

$$
p_{u_1} \equiv \frac{\partial p}{\partial u_1} = 1 + 2\lambda_2 u_1 + 4\lambda_3 u_1^3,
$$

$$
\begin{aligned}
&u_1(0) = 0, \qquad w_1(0) - \beta_1 = 0, \qquad w_2(0) = 0, \\
&u_1(1) = 0, \qquad w_1(1) + \beta_2 = 0, \qquad w_2(1) = 0,
\end{aligned}
$$

$$
\int_0^1 \left[\omega - (u_1(t) - 1)^2 - \frac{1}{10} \sum_{k=1}^{3} \lambda_k^2\right] dt = 0,
$$

$$
\int_0^1 \left[w_1^2(t) - 1\right] dt = 0,
$$

$$\int_0^1 \left[-e^{p(u_1,\lambda_2,\lambda_3)} w_2(t) - \frac{1}{5}\gamma\lambda_1 - \alpha_1\right] dt = 0,$$
$$\int_0^1 \left[-\lambda_1 e^{p(u_1,\lambda_2,\lambda_3)} u_1(t)^2 w_2(t) - \frac{1}{5}\gamma\lambda_2 - \alpha_2\right] dt = 0,$$
$$\int_0^1 \left[-\lambda_1 e^{p(u_1,\lambda_2,\lambda_3)} u_1(t)^4 w_2(t) - \frac{1}{5}\gamma\lambda_3 - \alpha_3\right] dt = 0.$$

The coefficients α_i above correspond to those in (3.1). Numerical results are summarized in Figure 1. For a more detailed description see [2].

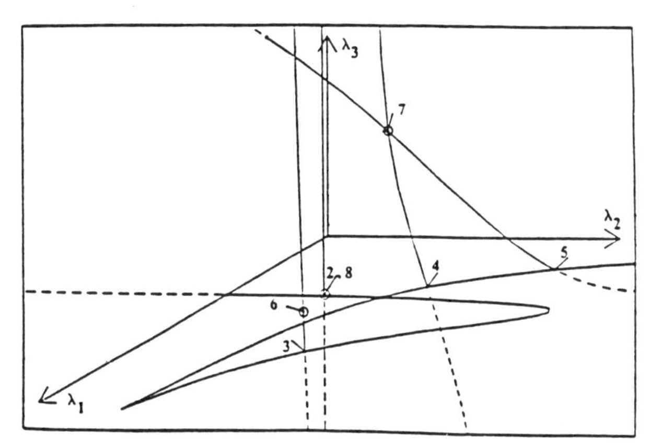

Figure 1 *The final 3-parameter curves of 2-parameter extrema of (4.3) subject to (4.1),(4.2). The circled points denote 3-parameter extrema. Also shown is a curve of 1-parameter extrema in the* $\lambda_3 = 0$ *plane. Note that the 3-parameter extremum with label* 7 *has been reached starting from both of the 2-parameter extrema* 4 *and* 5. *The value of the objective functional at points* 6, 7 *and* 8 *equals* 10.61, 0.2845 *and* 0.9234 *respectively. All are local minima. Thus the global minimum is attained at point* 7. *At this point* $\lambda_1 = 0.3772$, $\lambda_2 = 0.2378$, *and* $\lambda_3 = 0.4676$.

References

[1] H. Brezis, Analyse Fonctionelle, Théorie et Applications, Masson, Paris, 1983.

[2] E. J. Doedel, J. P. Kernévez, AUTO: Software for continuation and bifurcation problems in ordinary differential equations, Applied Mathematics Technical Report, California Institute of Technology, 1986, 226 pages, (Includes the AUTO 86 User Manual).

[3] J. M. Fier, Fold Continuation and the flow between rotating, coaxial disks, Thesis, Part I, California Institute of Technology, Pasadena, Ca., 1985.

[4] A. Griewank, G. W. Reddien, Characterization and computation of generalized turning points, SIAM J. Numer. Anal. 21, No. 1, 1984, 176-185.

[5] H. B. Keller, Numerical solution of bifurcation and nonlinear eigenvalue problems, in: Applications of Bifurcation Theory, P. H. Rabinowitz, ed., Academic Press, 1977, 359-384.

[6] J. P. Kernévez, Enzyme Mathematics, North-Holland Press, 1980.

[7] J. L. Lions, Contrôle optimale de systèmes distribués singuliers, Dunod, Paris, 1983.

[8] A. B. Poore, C. A. Tiahrt, Bifurcation problems in nonlinear parametric programming, Preprint, Dept. of Math., CSU, Fort Collins CO, 1986.

[9] A. B. Poore, The expanded Lagrangian system for constrained optimization problems, Preprint, Dept. of Math., CSU, Fort Collins CO, 1986.

[10] A. Trubuil, Thèse, Université de Technologie de Compiègne, 1986, (to appear).

J. P. Kernévez
Génie Informatique
Université de Technologie de Compiègne
BP 233
60206 Compiègne, France

E. J. Doedel
Applied Math 217-50
California Institute of Technology
Pasadena CA, 91125 USA
(On leave from Concordia University,
Montréal, Canada)

International Series of
Numerical Mathematics, Vol. 79

ON THE NUMERICAL STUDY OF BIFURCATION PROBLEMS

Edwin J. Kreuzer
Institute B of Mechanics, University of Stuttgart, Stuttgart, FRG

Abstract: Lyapunov exponents are normally used to characterize the behavior of dynamic systems, either if the system is continuous or discrete. It is shown that Lyapunov exponents are equally applicable for the study of bifurcation problems to obtain both bifurcation diagrams and stability charts.

1. Introduction

Nonlinear dynamic problems in engineering are usually described by a set of coupled nonlinear differential equations and contain several physical parameters which vary continuously. Variation of parameters may often change behavior so essential - multiple steady states, oscillatory, or even chaotic solutions may appear - that trial and error methods are not sufficient to study all kinds of phenomena.

Most technical dynamical systems are dissipative and there long-term behavior is described by limit sets, so-called attractors. The qualitative change of behavior and related attractors occuring due to parameter changes is a result of a sequence of bifurcations, thereby, the motion may change from regular to chaotic and vice versa.

Studying bifurcations in arbitary systems by means of analytical methods is often difficult. In engineering practice numerical techniques to determine both the bifurcation points as well as the dependence of solutions on parameters are often sufficient. With efficient numerical algorithms modern computer technology allows us to simulate systems and to study bifurcation problems with relative ease.

The principle objective of this paper is to show that the dynamic change from one behavior to another as a parameter is varied can be examined by means of Lyapunov exponents. Bifurcation diagrams and stability charts of systems described by ordinary differential equations or difference equations may be produced. To obtain a global picture of the domains of attraction of different types of behavior observed the cell mapping approach may be used to supplement the studies of dynamic bifurcations. Applying the methods described to an example will demonstrate what result may be obtained.

2. Lyapunov Exponents

Lyapunov exponents are of interest in the study of dynamical systems in order to characterize quantitatively the average exponential divergence or convergence of nearby trajectories. Since they can be computed either for a model or from experimental data, they are widely used for the classification of attractors. Negative exponents, besides the one along the flow being zero, signal periodic orbits, whilst at least one positive exponent indicates a chaotic orbit and the divergence of initially close trajectories. In the periodic region bifurcations in periodic solutions coincide with one exponent being just zero, as one orbit loses stability and another gains it.

A continuous dynamical system may be governed by the set of differential equations

$$\dot{\mathbf{x}} = \mathbf{f}(\mathbf{x}) \quad , \tag{1}$$

where $\mathbf{x} = \mathbf{x}(t) \varepsilon \mathbf{R}^n$ is a vector valued function of time t and and $\mathbf{f}: U \rightarrow \mathbf{R}^n$ is a smooth function defined on $U \subseteq \mathbf{R}^n$. The vector field $\mathbf{f}$ generates a flow $\varphi_t(\mathbf{x})$, which is a smooth function defined for all $\mathbf{x}$ in U and $t \varepsilon I \leq \mathbf{R}$ satisfying (1). In general $\varphi_t(\mathbf{x})$ generates a flow on a manifold M.

2.1 One-dimensional Lyapunov Exponents

To give a precise quantitative definition of exponential divergence, we consider on an n-dimensional manifold M a trajectory and a nearby trajectory with initial conditions $\mathbf{x}_o$ and $\mathbf{x}_o + \Delta\mathbf{x}_o$, respectively (Fig. 1). If $\Delta\mathbf{x}$ is small, one is naturally led to introduce the tangent vector $\mathbf{w} \varepsilon T_x M$, where $T_x M$ is the tangent space to M at $\mathbf{x}$. Under the flow φ_t the system evolves and thereby yielding the tangent vector $\mathbf{w}(t)$, (Fig. 1).

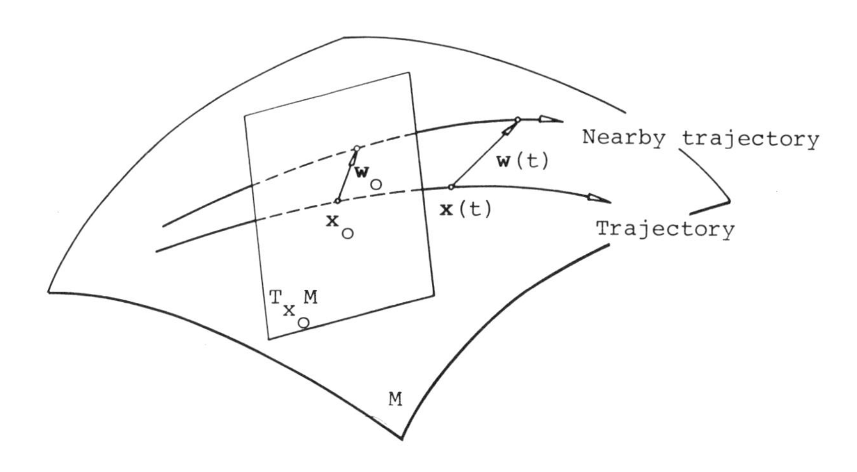

Fig. 1. Trajectory and nearby trajectory on a manifold

The time evolution for $\mathbf{w}$ is described by linearizing (1) to obtain

$$\dot{\mathbf{w}} = Df(\mathbf{x}(t))\mathbf{w} \tag{2}$$

where

$$Df(\mathbf{x}(t)) = \left.\frac{\partial \mathbf{f}(\mathbf{x})}{\partial \mathbf{x}}\right|_{\mathbf{x} = \mathbf{x}(t)}$$

is the Jacobian matrix for $\mathbf{f}$ evaluated at $\mathbf{x}$. The one-dimensional Lyapunov exponent is then defined as

$$\sigma(\mathbf{x}_o,\mathbf{w}_o) = \lim_{t\to\infty} \left\{ \frac{1}{t} \ln \frac{\|\mathbf{w}(t)\|}{\|\mathbf{w}\|} \right\} , \tag{3}$$

with $\|\bullet\|$ being the euclidean norm. OSELEDEC and BENETTIN et al. (1976) have shown that the limit (3) exists and is finite under very general conditions. The numbers σ defined in (3) depend on $\mathbf{x}_o \varepsilon M$ and $\mathbf{w}_o \varepsilon T_x M$. Equation (2) represents a homogeneous time-invariant system of equations the behavior of which is determined by a state transition matrix. Hence there exists an n-dimensional orthonormal basis $\{\mathbf{e}_i(\mathbf{x})\}$, $i=1,\ldots,n$, of $\dot{\mathbf{w}}$ such that for any $\mathbf{w}$ the number σ takes on one of the n, not necessarily distinct, values

$$\sigma_i(\mathbf{x}_o) \equiv \sigma(\mathbf{x}_o,\mathbf{e}_i) . \tag{4}$$

These so-called Lyapunov exponents can be ordered as $\sigma_1 \geq \sigma_2 \geq \ldots \geq \sigma_n$, with $\sigma_1 = \sigma_{max}$ being the largest exponent, and one of the exponents, representing the direction along the flow, being zero. This follows because in the direction along the flow, $\mathbf{w}$ grows only linearly with time. If one takes at random a vector $\mathbf{w}_o$ in $T_x M$ then $\sigma(\mathbf{x}_o,\mathbf{w}_o)$ tends to σ_1 with probability one.

The Lyapunov exponents apply to maps as well as flows, including the Poincaré map for a given flow. The n-1 exponents σ_i of the Poincaré map are proportional to the n exponents of the flow with the zero exponent deleted.

2.2 Multi-dimensional Lyapunov Exponents

We have considered the Lyapunov exponents of the vector $\mathbf{w}$, which are also called exponents of order 1. OSELEDEC has generalized the concept to describe the mean rate of exponential growth of an l-dimensional volume in the tangent space.

Let $\mathbf{w}^1, \ldots, \mathbf{w}^l$ be a system of l linearly independent vectors of $T_{\mathbf{x}}M$ $(1 \leq l \leq n)$, and $\mathbf{w}^{(l)}$ the l-dimensional parallelepiped whose edges are the vectors $\mathbf{w}^i$. Then

$$\sigma(\mathbf{x}_o, \mathbf{w}_o^{(l)}) = \lim_{t \to \infty} \{ \frac{1}{t} \ln \frac{\|\mathbf{w}^{(l)}(t)\|}{\|\mathbf{w}_o^{(l)}\|} \} \tag{5}$$

defines a Lyapunov exponent of order l. It can be shown that $\sigma(\mathbf{x}_o, \mathbf{w}_o^{(l)})$ equalls the sum of l Lyapunov exponents of order 1. For an arbitrary choice of $\mathbf{w}^{(l)}$ we find, just as in the one-dimensional case, that $\sigma(\mathbf{x}_o, \mathbf{w}_o^{(l)})$ is the sum of the l largest Lyapunov exponents

$$\sigma(\mathbf{x}_o, \mathbf{w}_o^{(l)}) \equiv \sigma^{(l)}(\mathbf{x}) = \sum_{i=1}^{l} \sigma_i(\mathbf{x}) \quad . \tag{6}$$

The motion for dissipative systems contracts, on the average, volume in phase space. For l=n we obtain the mean exponential rate of growth of the phase space volume as

$$\sigma^{(n)}(\mathbf{x}_o) = \sum_{i=1}^{n} \sigma_i(\mathbf{x}_o) \quad . \tag{7}$$

We can generally assume that all trajectories that eventually move on the same attractor have the same spectrum of exponents, though distinct attractors generally have different exponents.

3. Numerical Techniques to Compute Lyapunov Exponents

Simple application and integration of eqs. (1) to (3) leads after a sufficiently large time to computer overflow as the norm of $\mathbf{w}$ increases exponentially with t. The difficulty is overcome making use of the linearity of (2). We choose an initial vector $\mathbf{w}$ of norm 1 and renormalize the evolved vectors to a norm of unity at arbitrary but fixed time intervals τ. Thus we iteratively compute

$$\sigma(\mathbf{x}_o,k) = \frac{1}{k\tau} \sum_{j=1}^{k} \ln \alpha_j \tag{8}$$

where α_j is the renormalizing factor at time $j\tau$. For τ not too large, it can be shown that

$$\sigma_\infty = \lim_{k\to\infty} \sigma(\mathbf{x}_o,k) = \sigma_1 \tag{9}$$

exists and is independent of τ. Since τ is arbitrary, the same trick can be applied to calculate σ for maps as well as flows, FROESCHLE.

Using the definition of Lyapunov exponents of order l and relation (6), BENETTIN et al. (1980) have shown how to compute the entire set of Lyapunov exponents for a n-dimensional flow. In this case a further difficulty occurs, due to numerical errors, the angles between any two tangent vectors generally becomes too small for numerical computations. To overcome this difficulty the above renormalization procedure is extended. Thus given $\mathbf{w}^1,\ldots,\mathbf{w}^l$ orthonormal, we must replace at each time interval τ the l evolved vectors by a new set of orthonormal vectors, using the Gram-Schmidt procedure, spanning the same l-dimensional subspace.

A difficulty still remains and has to be mentioned at the end of this section. In numerically evaluating the Lyapunov exponents, there is no a priori condition for determining the number of iterations that must be used. Thus the results have to be supplemented by other techniques like Poincaré maps to clarify the bifurcation picture obtained by Lyapunov exponents.

4. Application

We will now study the dynamic bifurcation of an equation of Duffing type

$$\ddot{x} + d\dot{x} - x + x^3 = a \cos \omega t, \tag{10}$$

which describes a mechanical model of a sinusoidally forced structure with a linear and cubic stiffness. For such a driven nonlinear oscillator, closed-form analytical solutions are not available and recourse must be made to numerical methods.

Equations (10) rewritten as an autonomous system

$$\left.\begin{aligned} \dot{x}_1 &= x_2 \,, \\ \dot{x}_2 &= x_1 - x_1^3 - dx_2 + a \cos \omega\theta, \\ \dot{\theta} &= 1 \,, \end{aligned}\right\} (x_1, x_2, \theta) \varepsilon R^2 x S^1, \tag{11}$$

where $S^1 = \mathbf{R}/T$ is a circle of length $T = 2\pi/\omega$. In (Fig. 2) we show for fixed values of $\omega = 1.0$, $a = 0.3$ the spectrum of Lyapunov exponents depending on d. Different types of qualitative behavior occur and we note that chaotic motions exist for relatively wide sets of parameter values. The enlargement shows a sequence of bifurcations. Starting with a subharmonic of period 3 reducing a results in a bifurcation with symmetry breaking, that is, another period 3 motion coexists. After a further bifurcation, period doubling occured and we observe period 6 solutions. If we continue to decrease a , further period doubling bifurcations occur. These accumulate at a point at which transition from periodic to chaotic motion occurs.

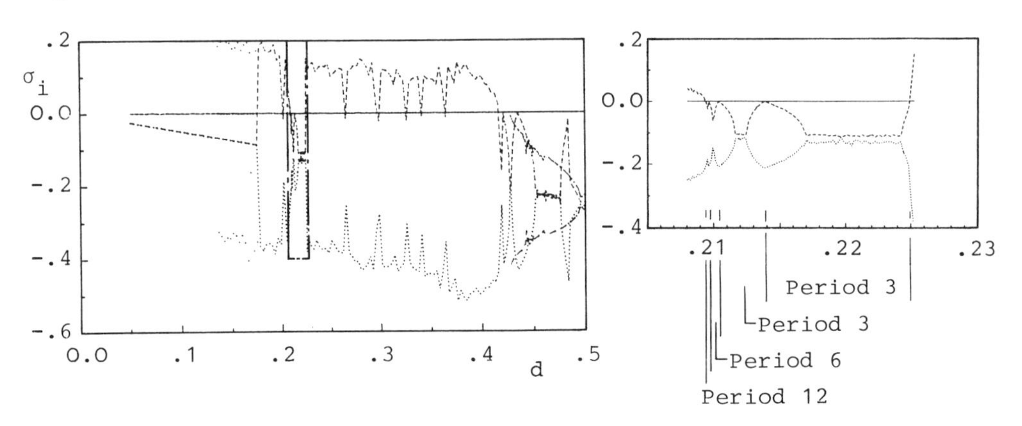

Fig. 2. Spectrum of Lyapunov exponents, ω=1.0, a=0.30

Summarizing many bifurcation diagrams like (Fig. 2) we end up with the comprehensive diagram shown in (Fig. 3).

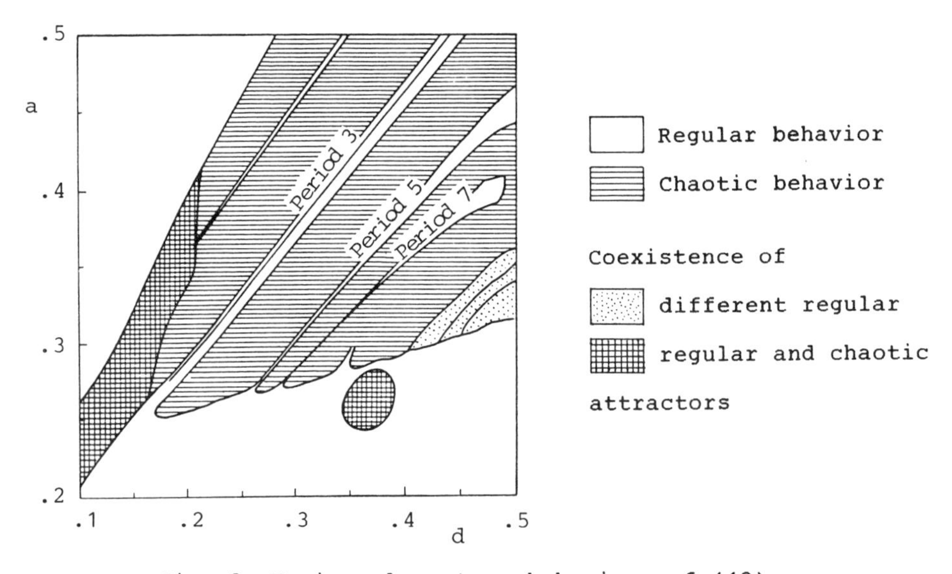

Fig. 3. Various long-term behaviors of (10)

In (Fig. 3) the damping d and the forcing magnitude a are treated as parameters and the d-a plane is divided into regions where various attractors exist. We note that the regions of multiple attractors and of chaotic attractors are quite extensive and by no means untypical or pathological. The results were obtained by following many trajectories simultaneously using a vector computer, KLECZKA. This is a very efficient way because the numerical scheme outlined above lends itself to vectorization. A factor of 10 improvement and more was observed.

5. Global Dynamics in Phase Space

Having obtained a bifurcation diagram of the system under consideration we turn to the qualitative description of dynamic behavior which begins with the identification of all possible attractors in phase space. In nonlinear systems, multiple attractors are common, and indeed there may be periodic and chaotic attractors coexisting for the same values of parameters. Having mapped out the locations of attractors, the type of each is known from their Lyapunov exponents, one would like to know the ensemble of initial conditions that settle to it. This set forms the basin of attraction. The basins can be identified by evaluating many starting conditions, but a more application oriented method is the cell mapping approach.

5.1 Cell Mapping Approach

This method describes the evolution of a system based upon a large collection of very small cells, HSU (1980, 1981). Cell mappings have been applied successfully to find the locations in phase space of attractors and basins. Especially the generalized cell mapping is tailormade for the analysis of dissipative dynamical systems. It describes the behavior of a system in a probabilistic sense and is essentially a Markov chain analysis of dynamical systems. The attractors are represented as persistent groups of the Markov chain.

To study periodic solutions and in particular chaotic solutions Poincaré maps are often used. From such a mapping we can also construct a cell mapping and analyze the dynamic behavior in a very efficient manner. Periodic solutions of the Poincaré map are represented by periodic cells or groups of periodic cells of the same period. A chaotic or strange attractor is replaced by a cluster of cells covering the same domain of the surface of section as the points of the Poincaré map. For more details see also KREUZER.

5.2 Applications

We now present the results obtained from the cell mapping approach for the model system (10). We pick a cross section $\Sigma = \{(x_1,x_2,\theta)|\theta=0\}$ and consider the Poincaré map $P: \Sigma\to\Sigma$. The generalized cell mapping was constructed from Σ for $-2.0\leq x_i\leq 2.0$, $i=1,2$, using 100x100 cells. In (Fig. 4) the results for $\omega=1.0$, $a=0.3$, $d=0,15$ are shown. There is a period 1 motion and a strange attractor supplemented by their basins of attraction. Similar portraits for other values of parameters can be obtained easily by using the cell mapping approach.

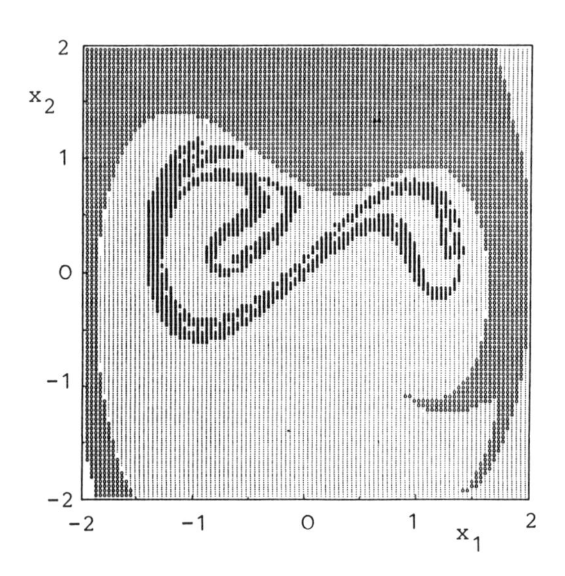

Fig. 4. Attractors with their basins of attraction

References

Benettin, G.; Galgani, L.; Strelcyn, J.-M. (1976) Kolmogorov entropy and numerical experiments. Phys. Rev. A 14, 2338-2345.

Benettin, G.; Galgani, L.; Giorgilli, A.; Strelcyn, J.-M. (1980) Lyapunov characteristic exponents for smooth dynamical systems and for Hamilton systems; a method for computing all of them. Mechanica 15, 9-30.

Froeschle, C. (1984) The Lyapunov characteristic exponents and applications. J. Méc. thér. appl., 101-132.

Hsu, C.S. (1980) A theory of cell-to-cell mapping dynamical systems. J. Appl. Mech. 47, 931-939.

Hsu, C.S. (1981) A generalized theory of cell-to-cell mapping for nonlinear dynamical systems. J. Appl. Mech. 48, 634-642.

Kleczka, M. (1986) Numerische Untersuchung nichtlinearer dynamischer Systeme mit dem Vektorrechner, DIPL-14, Inst. B für Mech., Univ. Stuttgart, Stuttgart.

Kreuzer, E. (1986) Numerische Untersuchung nichtlinearer dynamischer Systeme, (Springer, Berlin).

Oseledec, V.I. (1968) A multiplicative ergodic theorem: Ljapunov characteristic numbers for dynamical systems. Trans. Moscow math. Soc. 19, 197-231.

Prof. Dr.-Ing. Edwin J. Kreuzer, Institute B of Mechanics, University of Stuttgart, Pfaffenwaldring 9, D-7000 Stuttgart, FRG.

International Series of
Numerical Mathematics, Vol. 79

EXTERNAL FORCING IN A GLYCOLYTIC MODEL

Magnus Küper , Universität Dortmund

Introduction

In this paper we consider an external forced Sel'kov model. Many of the described phenomea can also be found in higher complicated models, which are in preparation. Consider the simple Sel'kov model [1]

$$\dot{x} = v - xy^2$$
$$\dot{y} = xy^2 - y$$

where x, y denote concentration and v the input velocity.
This system has a *Hopf-bifurcation* point at $v=1$, i.e. for $v<1$ it has a stable periodic solution $\varphi(t)$ and for $v>1$ a stable steady-state solution. Furthermore Sel'kov has shown that there exists another solution depending on the initial value with $x(t) \to \infty$, $y(t) \to 0$. Further particulars can be found in [1,2,3].

A short summary of the theory

Consider the perturbed system

$$\begin{aligned} \dot{x} &= v + v\cdot\varepsilon \sin \omega t - xy^2 \\ \dot{y} &= xy^2 - y \end{aligned} \tag{1}$$

i.e. the input velocity is periodicly varied ($0 \le \varepsilon \le 1$). Using vector notation one can rewrite this system as

$$\dot{x} = F(x) + \varepsilon f(t, \varepsilon) \tag{2}$$

where $x \in \mathbb{R}^n$, $F: \mathbb{R}^n \to \mathbb{R}^n$, ε a real parameter and $f(t,\varepsilon)$ is periodic of period T, and for $\varepsilon = 0$ the system (2) has a periodic solution $\varphi(t)$ of period T_0.

Question :
Does the equation (2) have any ***almost periodic*** solution "near" $\varphi(t)$ if $f(t,\varepsilon)$ is of period T in t ?

Under some restrictions one can expect that (see [4])

(i) if $T = \frac{n}{m} T_0 \quad n, m \in \mathbb{N}$, the solution of system (2) is periodic of period $m\, T_0$ for small ε.

(ii) otherwise (T and T_0 rational independent), the solution of system (2) is quasi-periodic for small ε .

It is also possible that on the branches the solutions bifurcate into ***subharmonic*** or into ***quasiperiodic*** solutions (see [5]).

Numerical investigation of the external forced Sel'kov model

For this problem I developed a program *NONAUTO* based on the program *DERPER* [6] . This program can continue the solution branches in one or two parameter continuation. I consider the system

$$\dot{x} = 0.99 + 0.99 \cdot \varepsilon \sin \omega t - xy^2$$

$$\dot{y} = xy^2 - y$$

which has a periodic solution of period $T_0 = 6.4585077$ or of frequency $\omega_0 = 2\pi / T_0$.

Figure 1 :
x-axis denotes ω / ω_0 and y-axis the amplitude of the sinus term.

The curves marked by a number mean :

1,2	Torus bifurcation	$\|\lambda_i\|$	=	1
3,4,5,6	period-doubling	λ_i	=	-1
7,8	"limit points"	λ_i	=	1

between			existence of
1	and	3	T-periodic solution
2,4	and	7	2T-periodic solution
6	and	8	3T-periodic solution
4	and	5	4T-periodic solution
	above	5	there is a cascade of period doubling points converging into **chaos** (see figure 3)

Note that there are tongues of 2T (3T) periodic solutions in the region of the T (2T) solutions, i.e. there is co-existence of stable periodic solutions.

Figure 2 :
These are stroboscopic plots ($\omega / \omega_0 = 1.5$ and $\varepsilon = 0.28$) of the T-periodic solution (a) , chaos (b) and the "trivial" solution (c) depending on the initial value. But for $\varepsilon = 0.3$

we find only T-periodic and the "trivial" solutions.

Figure 3:

A cascade of period doubling points (ω/ω_0=2.0) . Note that the sequence of the bifurcation points satisfies the Feigenbaum number δ [7].

$$\delta = \lim_{n \to \infty} \frac{b_{n+1} - b_n}{b_{n+2} - b_{n+1}} = 4.669201 \cdots$$

	T	b	δ
1	2	.56875880	
2	4	.62840943	
3	8	.64106133	4.71477360
4	16	.64377590	4.66071600
5	32	.64378145	4.66385258

Figure 4:

An interesting fact is that both points #1 and #2 of the branch "---" (period doubling continuation of T periodic solution) have double multiplier -**1**. On the other hand #2 results in having double multiplier **+1** when reaching the same point #2 on the Torus bifurcation curve 3.

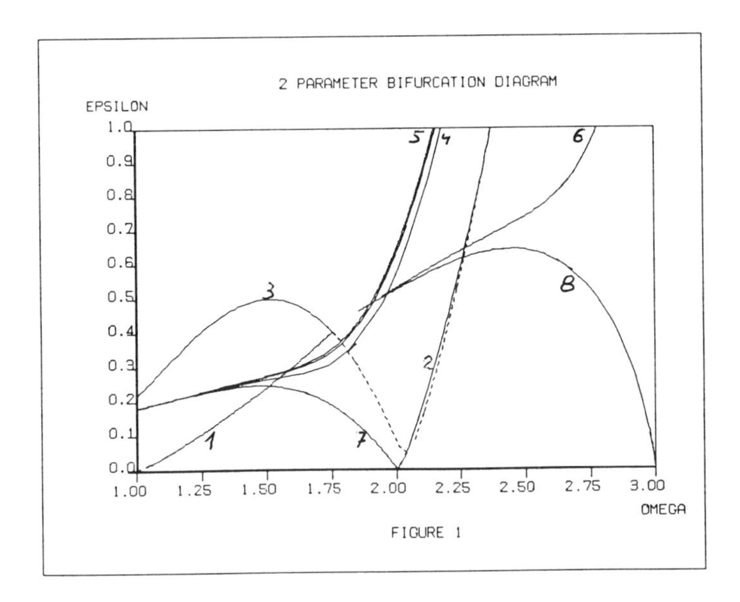

FIGURE 1

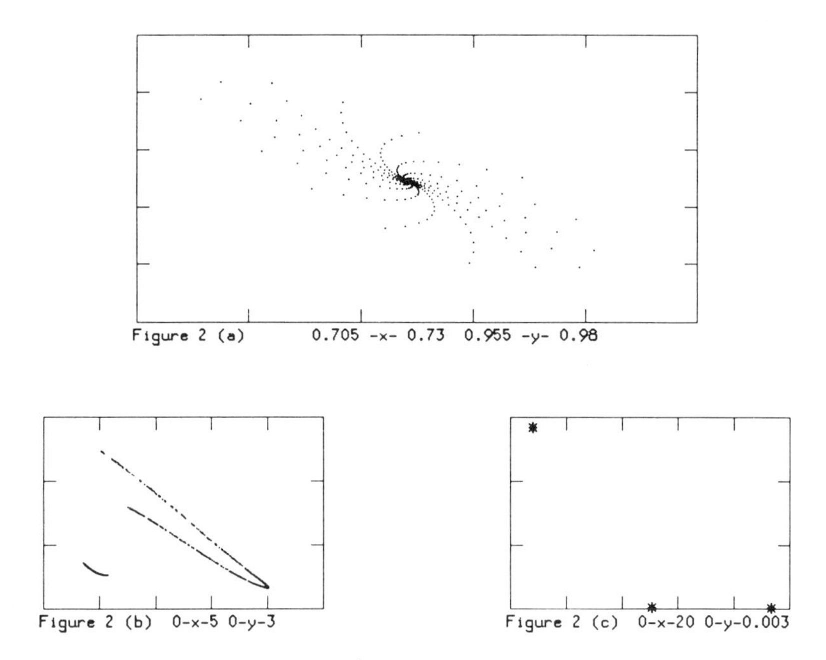

Figure 2 (a) 0.705 -x- 0.73 0.955 -y- 0.98

Figure 2 (b) 0-x-5 0-y-3

Figure 2 (c) 0-x-20 0-y-0.003

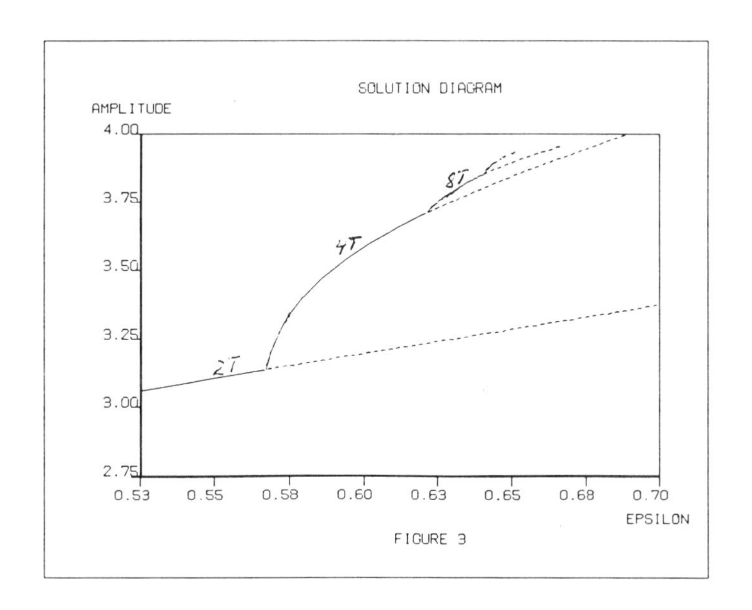

FIGURE 3

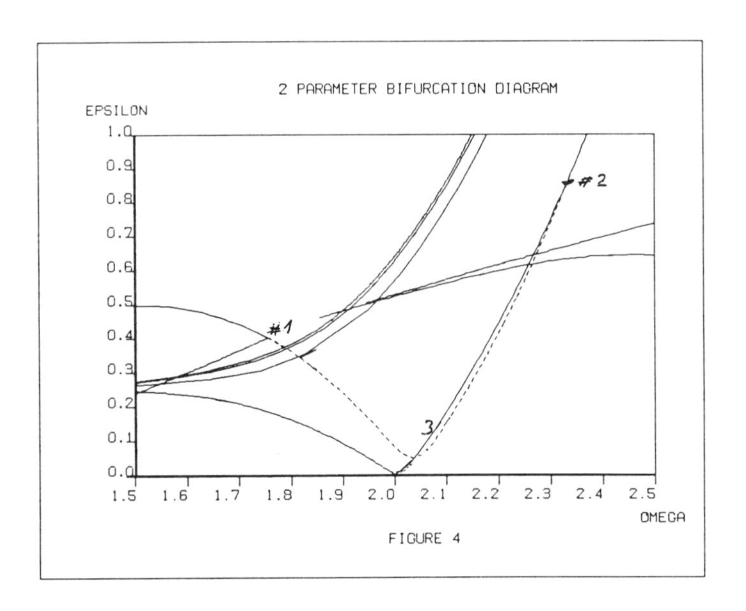

FIGURE 4

References :

[1] Sel'kov E.E. , Model of Glycolytic Oscillations, European J. Biochem. , Vol. 4, 79-86 (1968)

[2] Tyson, J. and Kauffman, S. , Control of mitosis by a continuous biochemical oscillation, J. Math. Biol., Vol.1, 289-310 (1975)

[3] Ashkenazi, M. and Othmer, H., Spatial patterns in coupled oscillators, J. Math. Biol., Vol.5, 305-350 (1978)

[4] Diliberto, S.P. and Hufford, G. , Perturbation theorems for nonlinear ordinary differential equations, Contributions to the theory of nonlinear oscillations, Vol III , 207, Annals of Math. Studies 36 (1956)

[5] Iooss, G. and Joseph, D.D., Elementary Stability and Bifurcation Theory, Springer (1980)

[6] Holodniok, M. and Kubicek, M. , An algorithm for continuation of periodic solution in ordinary differential equations, J. Comp. Phys. Vol.55, 254-267 (1984)

[7] Feigenbaum, M.J., Quantitative universality for a class of nonlinear transformations, J. Statistical Physics Vol.19, 25-52 (1978)

Magnus Küper

Fachbereich Mathematik, Universität Dortmund

D-4600 Dortmund 50, West-Germany

International Series of
Numerical Mathematics, Vol. 79

SOME REMARKS ON THE MORPHOLOGY OF NON-UNIQUE SOLUTIONS IN NONLINEAR ELASTOSTATICS

Franz Karl Labisch
Lehrstuhl für Mechanik II, Ruhr-Universität Bochum, FRG

1. Introduction

Elastostatics is concerned with the study of equilibrium states of elastic structures and may be regarded as a subtheory of elastodynamics, when velocities and accelerations are zero. A reliable prediction of the non-unique response of a suitably supported structure subject to outer loads, tractions and other boundary conditions relies decisively on the consideration of all possible solution points corresponding to a whole set (domain) of outer control quantities. The required prediction of the ambiguous behaviour of the structure and the nonlocal character of some singular phenomena, such as e. g. snap-through from one to another stable equilibrium state illustrate the urgent necessity to develop mathematical methods which are suitable for a rigorous nonlocal qualitative and quantitative analysis of multivalued solutions. In this paper such a method called morphology analysis is briefly outlined and applied to a shallow arch and to a shallow shell problem.

1.1 Abstract Formulation

In the nonlocal morphology analysis a b. v. p. (boundary value problem) stands for a triple (U_F, P, $\mathbf{F}$) where

U_F is the set of admissible state quantities $\mathbf{u}$,
P is the set of outer control quantities $\mathbf{p}$ and
$\mathbf{F} \in C^r$; $r \geq 1$ is a sufficiently smooth operator mapping

$$F: \begin{cases} U_F \times P \to Z \\ \mathbf{F}(\mathbf{u},\mathbf{p}) = \mathbf{0}. \end{cases} \tag{1.1}$$

Here Z is a suitable topological vector space usually definded by the operator **F** and the set P. "Admissible" should be interpreted as restricted by boundary conditions and smoothness demands. In elastostatics the operator **F** stands frequently for the equilibrium equations. Indispensable in a nonlocal morphology analysis are the following sets:

the complete solution set

$$S_F := \{(\mathbf{u},\mathbf{p}) \in U_F \times P,\ \mathbf{F}(\mathbf{u},\mathbf{p}) = \mathbf{0}\}, \tag{1.2}$$

the complete degeneracy set

$$D_F := \{(\mathbf{u},\mathbf{p}) \in S_F,\ \frac{\partial \mathbf{F}(\mathbf{u},\mathbf{p})}{\partial \mathbf{u}} \text{ is singular}\}, \tag{1.3}$$

the complete degenerate state set

$$DS_F := \{\mathbf{u} \in U_F,\ (\mathbf{u},\mathbf{p}) \in D_F\} \tag{1.4}$$

and the complete degenerate control set

$$DC_F := \{\mathbf{p} \in P,\ (\mathbf{u},\mathbf{p}) \in D_F\}. \tag{1.5}$$

The partial derivative in the definition of D_F is the Frechet derivative and "complete" means that all elements of the defined set corresponding to any $\mathbf{p} \in P$ are taken into account.

The nonlocal morphology of a solution is said to be known if

the complete solution set S_F,

the complete degeneracy set D_F together with

the complete degenerate state set DS_F

the complete degenerate control set DC_F

and all physically relevant properties and peculiarities of each solution point, such as e. g. stability are known.

2. The Shallow Arch Problem

Well suited for a demonstration of some of the various abilities of a nonlocal morphology analysis is a shallow arch problem. In a dimensionless Lagrangean description the b. v. p. of the considered simply supported arch is given in $\Omega := \{\xi;\ 0 < \xi < 1\}$ by the equilibrium equations

$$-(n_1' + q_1) = 0\ ,\ -(m_1'' + n_{13}' + q_3) = 0 \tag{2.1}$$

and by the geometric and static boundary conditions

$$u_1(0)=u_1(1)=u_3(0)=u_3(1)=0\ ,\ m_1(0)=m_1(1)=0. \tag{2.2}$$

The moment, stress and vertical stress resultants m_1, n_1, n_{13} appearing in the equilibrium equations are replaced by means of the constitutive relations

$$m_1=-Ku_3'',\ n_1=u_1' + z'u_3'+ \frac{1}{2}(u_3')^2,\ n_{13}= n_1(u_3'+z') \tag{2.3}$$

by known functions in terms of derivatives of the unknown displacements u_1 and u_3. Here K is a physical constant, $z = z(\xi)$ describes the middle line of the undeformed arch and $' = \frac{d}{d\xi}$, $'' = \frac{d^2}{d\xi^2}$.

Growth condition applied to the constitutive relations lead for smooth enough $z(\xi)$ and for $q_1(\xi)\in L_2\ni q_3(\xi)$ together with the boundary conditions (2.2) to

$$\begin{aligned}&\mathbf{u} =(u_1,\ u_3)\in U_F= U_{F1} \times U_{F3}\\ &U_{F1}=\{u_1\in W_{2,2}\cap \overset{\circ}{W}_{1,2}\}\\ &U_{F3}=\{u_3\in W_{4,2}\cap \overset{\circ}{W}_{1,2},\ u_3''(0)=u_3''(1)=0\}.\end{aligned} \tag{2.4}$$

Thus U_F is a separable normed space and the set

$$B:=\{\sin i\pi\xi,\ 0 < \xi < 1,\ i=1,\ 2,\ \ldots\} \tag{2.5}$$

is a basis in both U_{F1} and U_{F3}.

The simplicity of the particular problem considered below provides a deep insight into the complicated morphology of multi-valued solutions.

<u>Theorem:</u> For the arch problem arising for $z = \sum_{i=1}^{k} \varepsilon_i \sin i\pi\xi$, $q_1=0$, $q_3= \sum_{i=1}^{k} p_i \sin i\pi\xi$, $\sum_{i=1}^{k} \theta_i^2< 4(k+1)^2K$, $\theta_i=i\varepsilon_i$ the complete, strong and exact solution is given by

$$u_1= \sum_{i=1}^{2k} \alpha_i \sin i\pi\xi,\ u_3= \sum_{i=1}^{k}\frac{\phi_i-\theta_i}{i} \sin i\pi\xi\ , \tag{2.6}$$

where the active coordinates ϕ_i, $i=1,\ 2,\ \ldots,\ k$ are given by

$$\Big(\sum_{j=1}^{k} (\phi_j^2-\theta_j^2) + 4i^2K\Big)\phi_i- C_i=0, \tag{2.7}$$

α_i, $i=1,\ 2,\ \ldots,2k$ are passive coordinates and can be eliminated by means of the equations

$$i\alpha_i+ \frac{\pi}{4}\sum_{l=1}^{k} \delta_i^{2l}(\phi_{2l}^2-\theta_{2l}^2)+ \frac{\pi}{2}\sum_{l=2}^{k}\ \sum_{m=1}^{l} (\phi_l\phi_m-\theta_l\theta_m)(\delta_i^{l-m}+\delta_i^{l+m})= 0 \tag{2.8}$$

and $\phi_i=0$, $i > k$; $\alpha_i=0$, $i > 2k$.

Here $C_i= \frac{4}{i\pi^4} p_i+ 4i^2K\theta_i$ and δ_i^{2l} denotes the Kronecker delta function.

The proof results from a coefficient comparison and is a simple consequence of some trivial trigonometric identities. The nonlocal morphology analysis is therefore in a rigorous way reduced to an analysis of the nonlocal solution of the algebraic system

$$F_i=(N+4i^2K)\phi_i - C_i=0,\ i=1,\ 2,\dots,\ k,\ N=\sum_{j=1}^{k}(\phi_j^2-\theta_j^2). \tag{2.9}$$

The degeneracy set D_F is given by

$$\det \frac{\partial F_i}{\partial \phi_j}=0,\ i,\ j=1,\ 2,\dots,\ k \text{ where } \frac{\partial F_i}{\partial \phi_j}=2\phi_i\phi_j+\delta_i^j(N+4i^2K). \tag{2.10}$$

For the loads considered here this is the expression for the degenerate state set DS_F, too. The degenerate control set DC_F is obtained by an elimination of ϕ_i from the equations

$$\det \frac{\partial F_i}{\partial \phi_j}=0,\ (N+4i^2K)\phi_i - C_i=0,\ i,\ j=1,\ 2,\dots,\ k. \tag{2.11}$$

In a linearized problem $\sum_{j=1}^{k}\phi_j^2$ is omitted. Then the solution satisfies

$$F_i=[4K(i^2-\chi)]\phi_i - C_i=0,\ i=1,\ 2,\dots,\ k \tag{2.12}$$

and the degeneracy set D_F is given by

$$\det \frac{\partial F_i}{\partial \phi_i}=\prod_{i=1}^{k}[4k(i^2-\chi)]=0. \tag{2.13}$$

The set DS_F is reduced to the isolated points

$$\sum_{j=1}^{k}\phi_j^2=0,\ \chi=i^2,\ i<k+1 \tag{2.14}$$

and the set DC_F to the isolated points

$$\sum_{j=1}^{k}C_j^2=0,\ \chi=i^2,\ i<k+1. \tag{2.15}$$

The linearized set DS_F describes through the ratio

$$\chi=\frac{\sum_{j=1}^{k}\theta_j^2}{4K} \tag{2.16}$$

those discrete lengths of the undeformed arch for which a snap-through with respect to the involved modes is predicted by the linear theory.

The morphology of the nonlinear problem is far more complicated. Any element $\frac{\partial F_i}{\partial \phi_j}$ is a quadratic function in terms of the ϕ_i, i=1, 2, ..., k . For each active mode sin $i\pi\xi$ linked to ϕ_i, $i^2<\chi$ there exists a (k-1) dimensional hypersurface in the $\{\phi_i\}$ space as elementary degenerate state set and a (k-1) dimensional hypersurface in the $\{C_i\}$ space as elementary degenerate control set. Elementary degenerate state and control sets corresponding to passive modes are empty. The complete degenerate state set and the complete degenerate control set are given by the set of all nonempty elementary degenerate state and elementary degenerate control sets, respectively. All phenomena which are singular with respect to a given active mode, such as e. g. appearance of limit, bifurcation or snap-through points, gain or loss of stability (with respect to the involved mode) have to occur at state quantities prescribed by the elementary degenerate state set for control quantities prescribed by the associated elementary degenerate control set. A solution point is stable if the matrix with the elements $\frac{\partial F_i}{\partial \phi_j}$ is positive definite. For a thorough mathematical morphology analysis of the considered shallow arch problem, including essentially new results concerning a dual variational approach, which remains equivalent to the primary b. v. p. on the complete degeneracy set, too, the reader is referred to LABISCH [2].

3. The Shallow Shell Problem

The morphology of a shallow shell problem is far more complicated. The b.v.p. considered here is in a dimensionless Lagrangean formulation given in $\Omega:=\{(\xi,\eta),0 < \xi < 1,0 < \eta < 1\}$ by the equilibrium equations

$$\begin{aligned}&-(n_{11,1}+n_{21,2})=0,\ -(n_{12,1}+n_{22,2})=0\\&-(m_{11,11}+m_{22,22}+2m_{12,21}+n_{13,1}+n_{23,2}+q_3)=0\ ,\end{aligned}\tag{3.1}$$

by the geometric boundary conditions

$$\begin{aligned}&u_1(0,\eta)=u_1(1,\eta)=0,\ u_2(\xi,0)=u_2(\xi,1)=0\\&u_3(0,\eta)=u_3(1,\eta)=u_3(\xi,0)=u_3(\xi,1)=0\end{aligned}\tag{3.2}$$

and by the static boundary conditions

$$\begin{aligned}&m_{11}(0,\eta)=m_{11}(1,\eta)=0,\ m_{22}(\xi,0)=m_{22}(\xi,1)=0\\&n_{12}(0,\eta)=n_{12}(1,\eta)=0,\ n_{21}(\xi,0)=n_{21}(\xi,1)=0\ .\end{aligned}\tag{3.3}$$

By means of the constitutive relations are moment resultants

$$\begin{aligned}&m_{11}=-D(u_{3,11}+\nu u_{3,22}),\ m_{22}=-D(u_{3,22}+\nu u_{3,11})\\&m_{12}=-D(1-\nu)\ u_{3,12}\ ,\end{aligned}\tag{3.4}$$

stress resultants

$$n_{11}= D_1(e_{11}+\nu e_{22}),\ n_{22}= D_1(e_{22}+\nu e_{11})$$
$$n_{12}= n_{21}= D_1(1-\nu)e_{12} \tag{3.5}$$

and vertical stress resultants

$$n_{13}= n_{11}(u_{3,1}+z_{,1})+n_{12}(u_{3,2}+z_{,2})$$
$$n_{23}= n_{12}(u_{3,1}+z_{,1})+n_{22}(u_{3,2}+z_{,2}) \tag{3.6}$$

appearing in the equilibrium equations replaced by known functions in terms of derivatives of the unknown displacements u_1, u_2 and u_3.

Here ν denotes Poisson's ratio, $D = \frac{1}{12} D_1$, $D_1= \frac{1}{1-\nu^2}$ and the strains are given by

$$e_{11}= u_{1,1}+ \frac{1}{2} (u_{3,1}+z_{,1})^2- \frac{1}{2} z_{,1}^2$$

$$e_{22}= u_{2,2}+ \frac{1}{2} (u_{3,2}+z_{,2})^2- \frac{1}{2} z_{,2}^2 \tag{3.7}$$

$$2e_{12}= u_{1,2}+ u_{2,1}+ (u_{3,1}+z_{,1})(u_{3,2}+z_{,2})-z_{,1}z_{,2},$$

where $u_{1,1}= \frac{\partial u_1}{\partial \xi}$, $u_{2,2}= \frac{\partial u_2}{\partial \eta}$, $m_{12,21}= \frac{\partial^2 m_{12}}{\partial \xi \partial \eta}$ et cetera and $z=z(\xi,\eta)$ describes the middle surface of the undeformed shell. For a derivation of this dimensionless formulation of the b. v. p. see LABISCH [1].

Consider now the particular problem arising for the undeformed middle surface

$$z= \sum_{i=1}^{k} \sum_{j=1}^{k} \theta_{ij}\sin i\pi\xi \sin j\pi\eta \tag{3.8}$$

and the load

$$q_3= \sum_{i=1}^{k} \sum_{j=1}^{k} p_{ij}\sin i\pi\xi \sin j\pi\eta \tag{3.9}$$

where k is an integer. For this case an exact solution cannot be calculated. However, it is possible to construct the intersection of the exact solution with a space spanned by any finite number of active modes.

With $\varphi_{ij}= \gamma_{ij}+ \theta_{ij}$, $C_{ij}= p_{ij}+ \frac{(i^2+j^2)^2}{12} \theta_{ij}$

for the first k^2 active modes all appearing in

$$u_3= \sum_{i=1}^{k} \sum_{j=1}^{k} \gamma_{ij} \sin i\pi\xi \sin j\pi\eta, \tag{3.10}$$

a coefficient comparison and trivial trigonometric identities lead to the algebraic system

$$F_{ij}= \sum_{l=1}^{k} \sum_{m=1}^{k} [\delta_l^m \frac{(l^2+m^2)^2}{12} + R_{lm}-\overset{o}{R}_{lm}] \varphi_{lm}-C_{ij}=0,\ i,\ j=1,\ 2,\ \ldots,\ k \tag{3.11}$$

where $R_{lm}=R_{lm}(\varphi_{ij})$ are quadratic functions of the unknown φ_{ij} and $\overset{o}{R}_{lm}=R_{lm}(\theta_{ij})$ are these functions with φ_{ij} replaced by θ_{ij}. If k=2 then the exact intersection with the space spanned by the active modes sin iπξ sin jπη, i,j=1, 2 is obtained for

$$u_1=\sum_{i=1}^{4}\sum_{j=0}^{4}\alpha_{i\gamma}\sin i\pi\xi\cos j\pi\eta\ ,\ u_2=\sum_{i=0}^{4}\sum_{j=1}^{4}\beta_{ij}\cos i\pi\xi\sin j\pi\eta$$
$$u_3=\sum_{i=1}^{2}\sum_{j=1}^{2}\gamma_{ij}\sin i\pi\xi\sin j\pi\eta. \tag{3.12}$$

All α_{ij} and β_{ij} are passive and can be eliminated. The formalism of the morphology analysis is exactly the same as for the arch problem.

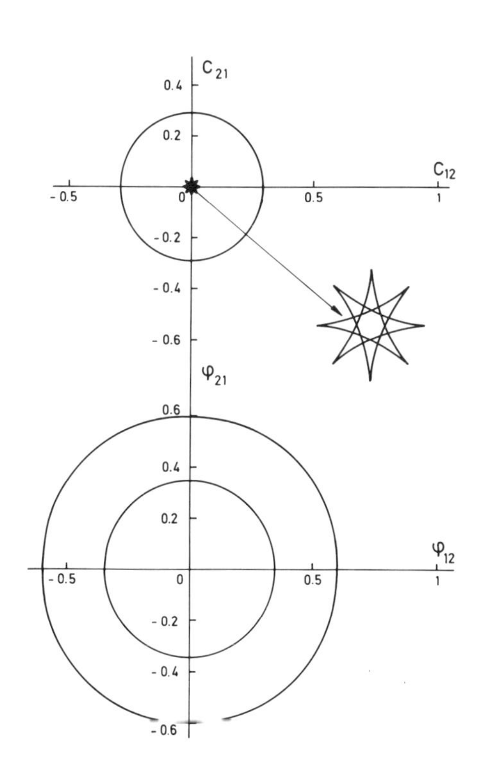

Fig. 1 Essential intersection of DC_F with the (C_{12}, C_{21}) plane and of DS_F with the $(\varphi_{12}, \varphi_{21})$ plane. $\theta_{11}^2= 4$.

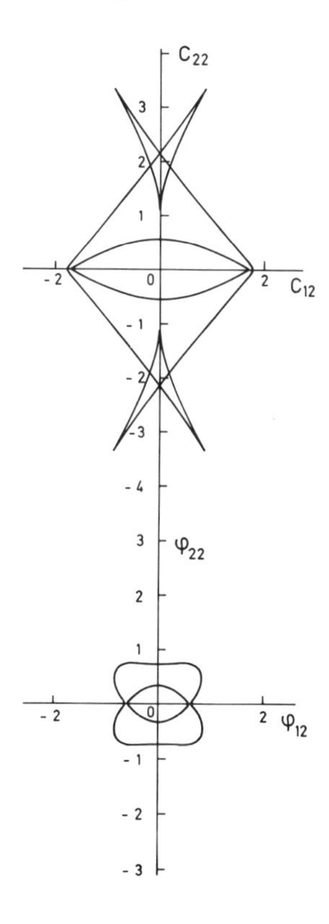

Fig. 2 Essential intersection of DC_F with the (C_{12}, C_{22}) plane and of DS_F with the $(\varphi_{12}, \varphi_{22})$ plane. $\theta_{11}^2= 6$.

For each active mode there exist an elementary degenerate state set and an elementary degenerate control set. In contrast to the arch problem where all eigenvalues of the linearized problem are simple, the linearized shell problem considered here has double eigenvalues, too. If e. g. $\theta_{12}=\theta_{21}=\theta_{22}=0$, then $\theta_{11}^2=0.7597340931$ corresponds to a simple eigenvalue with respect to φ_{11}, $\theta_{11}^2=2.003205128$ to a double eigenvalue with respect to φ_{12} and φ_{21} and $\theta_{11}^2=4.102564103$ to a simple eigenvalue with respect to φ_{22}. The essential intersection of the elementary degenerate sets of the shell problem with the $(\varphi_{11}, \varphi_{22})$ plane and with the (C_{11}, C_{22}) plane are topologically equivalent to the corresponding intersections of the arch problem. The intersections of the elementary degenerate sets with the $(\varphi_{12}, \varphi_{21})$ plane and the (C_{12}, C_{21}) plane or with the $(\varphi_{12}, \varphi_{22})$ plane and the (C_{12}, C_{22}) plane, shown in Fig. 1 and Fig. 2 for $\theta_{12}=\theta_{21}=\theta_{22}=0$, are more complicated.

4. Conclusion

In a nonlocal qualitative and quantitative analysis of nonlinear problems is the complete degeneracy set a natural and indispensable generalization of the spectrum of the corresponding linearized problem. Singular state points and singular control points (eigenvalues) of the linear problem are the nuclei of elementary degenerate state sets and the nuclei of elementary degenerate control sets of the corresponding nonlinear problem. The considered two simple examples lead to the conclusion, that any nonlocal qualitative and quantitative analysis of a concrete nonlinear problem would be incomplete if the complete degenerate control set and the complete degenerate state set are not taken into account.

5. References

1. Labisch, F.K. (1985) Grundlagen einer Analyse mehrdeutiger Lösungen nichtlinearer Randwertprobleme der Elastostatik mit Hilfe von Variationsverfahren, Mitt. Inst. für Mech. **47** Ruhr-Universität Bochum.

2. Labisch, F.K. (1986) On the morphology of multi-valued solutions of a simple shallow arch problem, ZAMM (in press).

Dr. Franz Karl Labisch, Lehrstuhl für Mechanik II, Ruhr-Universität Bochum, Universitätsstraße 150, Postfach 10 21 48, 4630 Bochum 1, FRG.

International Series of
Numerical Mathematics, Vol. 79

LUSTERNIK–SCHNIRELMANN CRITICAL VALUES AND BIFURCATION PROBLEMS

Ari Lehtonen

University of Jyväskylä, Department of Mathematics, Finland

1. Introduction

We present a method to calculate bifurcation branches for nonlinear two point boundary value problems of the following type

$$\text{(1.1)} \qquad \begin{cases} -u'' = \lambda G'(u) \\ u(a) = u(b) = 0, \end{cases}$$

where $G : \mathbf{R} \to \mathbf{R}$ is a smooth mapping. This problem can be formulated equivalently as

$$\text{(1.2)} \qquad g'(u) = \mu\, u,$$

where

$$\text{(1.3)} \qquad g(u) = \int_a^b G(u(t))\, dt$$

and $\mu = 1/\lambda$. Solutions of this problem can be found by locating the critical points of the functional $g : H \to \mathbf{R}$ on the spheres $S_r = \{x \in H \mid \|x\| = r\}$, $r > 0$. (The Lagrange multiplier theorem.)

Calculating critical points x_r for $r > 0$ can be used to obtain bifurcation branches of (1.2). Using the Lusternik–Schnirelmann critical point theory we can find saddle points of (1.1) and also higher order bifurcation branches.

Very general results on nonlinear eigenvalue problems in Banach spaces can be found in H. Amann [1] and in E. Zeidler [11], which also contains an extensive bibliography on critical point theories. Iterative methods for the construction of all Lusternik–Schnirelmann critical values and critical vectors has been presented by J. Nečas [7] and by A. Kratochvíl and J. Nečas [4], [5]. In this paper we shall give an extension of the method used in [5] to study the eigenvalue problem (1.2). Also, we give examples of problems that have been tested numerically.

The results to be presented are joint work with Jindrich Nečas (Prague) and Pekka Neittaanmäki (Jyväskylä, Finland). The author is indebted to Mr. Timo Männikkö for his assistance in numerical tests. Proofs of the following results can be found in [8]; in [9] we present a generalization of the method which works in Banach spaces.

2. Algorithm for one Lusternik–Schnirelmann critical value

Let H be real Hilbert space with inner product $\langle\cdot,\cdot\rangle$ and norm $\|\cdot\|$. Furthermore, we set $S = \{x \in H \mid \|x\| = 1\}$. Let g be a continuously differentiable functional on H such that the derivative g' is strongly continuous, i.e. for each sequence $(x_n)_{n=1}^{\infty} \subset H$ converging weakly to $x_0 \in H$, the sequence $(g'(x_n))_{n=1}^{\infty}$ converges to $g'(x_0)$.

Assume that the following conditions are fulfilled for some constant $c > 0$:

(2.1) $$g(0) = 0, \quad g'(0) = 0;$$

(2.2) if $g(x) \neq 0$ and $\|x\| \leq 1$ then $g'(x) \neq 0$;

(2.3) $$\langle g'(x) - g'(y), x - y\rangle \geq -c\|x-y\|^2 \quad \text{for all } x, y \in H,\ \|x\| \leq 1,\ \|y\| \leq 1.$$

Theorem 2.1. *Let the above hypothesis be satisfied. Assume that $x_1 \in S$ and $g(x_1) > 0$. Let $0 < \theta < 1/c$. Define a sequence $(x_n)_{n=1}^{\infty}$ by*

(2.4) $$x_{n+1} = \frac{x_n + \theta g'(x_n)}{\|x_n + \theta g'(x_n)\|}.$$

Then there exists a subsequence of $(x_n)_{n=1}^{\infty}$ converging to a solution of (1.2). Hence, there exists a point $x_0 \in S$ and a number $\mu \in \mathbf{R}$ such that

(2.5) $$g'(x_0) = \mu x_0.$$

3. Higher order critical points

To find higher order critical poins we use the notion of the order of a set, cf. [2]. Let K be a symmetric closed set in H. We say that ord $K = 0$ if K is empty; that ord $K = 1$ if $K = K_1 \cup K_2$, where the K_i are closed subsets of K and neither K_1 nor K_2 contains antipodal poins. In general, ord $K = n$ if $K = \cup_{i=1}^{n+1} K_i$, where the K_i are closed subsets of K not containing antipodal points and n is the least possible number. Finally, ord $K = \infty$ if no such n exists.

Let $V_k^{\pm} = \{K \subset S \mid K$ is symmetric, compact, $\operatorname{ord} K \geq k$ and $\pm g(x) > 0$ on $K\}$. Denote

$$\pm\gamma_k^{\pm} = \begin{cases} \sup\limits_{K \in V_k^{\pm}} \min\limits_{x \in K} \pm g(x), \text{ if } V_k^{\pm} \neq \emptyset \\ 0, \text{ if } V_k^{\pm} = \emptyset. \end{cases} \tag{3.1}$$

The fundamental theorem of the Lusternik–Schnirelmann theory states that there exists a sequence of critical points $x_k^{\pm}$ of g such that $g(x_k^{\pm}) = \gamma_k^{\pm}$, $\pm\gamma_k^{\pm} \searrow 0$ and $x_k^{\pm} \to 0$ weakly.

Let the assumptions of Theorem 2.1 hold for a functional g. Furthermore, assume that g is even on S, i.e. $g(-x) = g(x)$ when $\|x\| = 1$. We assume that g has positive critical values. Let γ_1^+ and γ_2^+ be the first and second Lusternik–Schnirelmann critical values of $g|_S$, $\gamma_1^+ > \gamma_2^+$. Let there exist a positive constant ε such that there are no critical values in the interval $]\gamma_2^+ - \varepsilon, \gamma_2^+[$. Let K_1 be a compact symmetric subset of S such that $\operatorname{ord} K_1 \geq 2$, $g(x) > 0$ on K_1 and

$$\gamma_2^+ - \varepsilon < \min_{x \in K_1} g(x) < \gamma_2^+. \tag{3.2}$$

We denote by φ the function used to define the iteration in (2.4), i.e.

$$\varphi(x) = \frac{x + \theta g'(x)}{\|x + \theta g'(x)\|}. \tag{3.3}$$

Then φ is a well-defined, odd continuous map $S \to S$. Choose $x_1 \in K_1$, and put $x_{n+1} = \varphi(x_n) = \varphi^n(x_1)$, where φ^n denotes the n-fold composition $\varphi \circ \cdots \circ \varphi$.

Let $x_n^{(0)}$ be a vector from K_1 such that

$$\min_{x \in K_1} g(\varphi^n(x)) = g(\varphi^n(x_n^{(0)})) \tag{3.4}$$

for any integer n.

Theorem 3.1. *Let the above assumptions be fulfilled. Then the following assertions hold:*

(1) $\lim_{n \to \infty} g(\varphi^n(x_n^{(0)})) = \gamma_2^+$;

(2) *there exists* $x^{(0)} \in K_1$ *such that*

$$\lim_{n \to \infty} g(\varphi^n(x^{(0)})) = \gamma_2^+;$$

(3) *there exists a subsequence of* $(x_n^{(0)})_{n=1}^{\infty}$ *converging to* $x^{(0)}$

(4) *for each* $x^{(0)}$ *satisfying* (2) *there exists a subsequence of* $(\varphi^n(x^{(0)}))_{n=1}^{\infty}$ *converging to some* x_0 *such that* $g'(x_0) = \mu x_0$.

Corollary 3.2. *Let the assumptions of Theorem 3.1 be fulfilled. Let* $\gamma_1^+ \geq \cdots \geq \gamma_k^+ > \gamma_{k+1}^+ = \cdots = \gamma_{k+l}^+ > \gamma_{k+l}^+$ *be the positive Lusternik–Schnirelmann values of* $g|_S$. *Let there exist a constant* $\varepsilon > 0$ *such that there are no critical values in the interval* $]\gamma_{k+l}^+ - \varepsilon, \gamma_{k+l}^+[$. *Let* K_1 *be a compact symmetric subset of* S *such that* $\operatorname{ord} K_1 \geq k+1$ *and* $\gamma_{k+l}^+ - \varepsilon < \min_{x \in K_1} g(x) < \gamma_{k+l}^+$.

For $x \in K_1$ *let the sequence* $(x_n^{(0)})_{n=1}^{\infty}$ *be defined by (3.4). Then* $\lim_{n\to\infty} g(\varphi^n(x_n^{(0)})) = \gamma_{k+l}^+$, *and there exists a point* $x^{(0)} \in K_1$ *such that* $\lim_{n\to\infty} g(\varphi^n(x^{(0)})) = \gamma_{k+l}^+$. *Moreover, the assertions* (3) *and* (4) *of Theorem 3.1 hold.*

4. Bifurcation of solutions; examples

Let $g : H \to \mathbf{R}$ be a given function and for $r > 0$ define $g_r : H \to \mathbf{R}$, $g_r(x) = g(rx)$. Then $g_r'(x) = \mu x$ if and only if $g'(x_r) = \mu_r x_r$, where $x_r = rx$ and $\mu_r = \mu r^{-2}$. If the second derivative $g''(0)$ exists and $(\mu_0, 0)$ is a bifurcation point of (1.2), then μ_0 is an eigenvalue of $g''(0)$. To calculate solutions for equation (1.2) we will use the following method: Let $\mu_0 \neq 0$ be an eigenvalue and x_1 an eigenvector of $g''(0)$. Using Theorems 2.1 and 3.1 and Corollary 3.2 solve the equation $g_r'(x) = \mu x, x \in S$. Then (μ_r, x_r) gives a bifurcation branch of (1.2).

Theorem 4.1. *Let* $g : H \to \mathbf{R}$ *be a twice continuously differentiable function on a neighbourhood of the origin. Assume that* g' *is strongly continuously and that* $g''(0)$ *has an eigenvalue* $\mu_0 \neq 0$. *Then the point* $(\mu_0, 0)$ *is a bifurcation point of (1.2).*

The previous results can be applied to problem (1.1), where $G : \mathbf{R} \to \mathbf{R}$ is a given twice continuously differentiable function and $G'(u)$ denotes the function $t \mapsto G'(u(t))$. It is easily verified that the derivative g' is strongly continuous and that conditions (2.1), (2.2) and (2.3) are satisfied, if we assume that $G(0) = G'(0) = 0$, $G'(r)$ does not vanish on any interval and $G''(r) \geq -c$.

The linear eigenvalue problem $g''(0)u = \mu_0 u$ is equivalent to the boundary value problem

$$\begin{cases} -\mu_0 u'' = G''(0)\, u \\ u(a) = u(b) = 0. \end{cases} \tag{4.1}$$

If $G''(0) \neq 0$ then problem (4.1) has nonzero eigenvalues and we can apply Theorem 4.1 to prove that they give bifurcation branches for problem (1.1).

The algorithms have been tested numerically using FEM with piecewise linear elements to solve equation (2.4) for x_{n+1}. As an initial guess x_1 we have used the solution of the linearized problem (4.1).

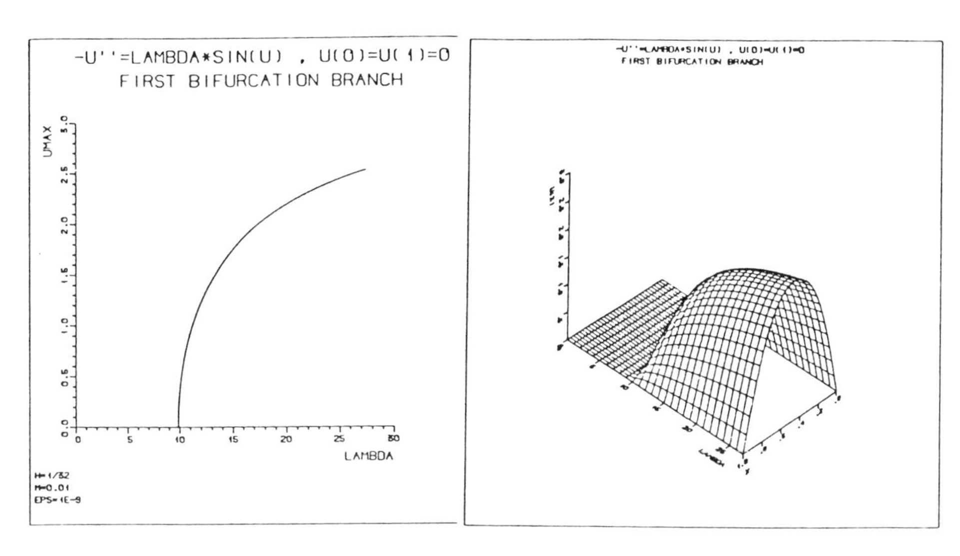

Fig. 1. First bifurcation branch for $-u''(t) = \lambda \sin u(t)$.

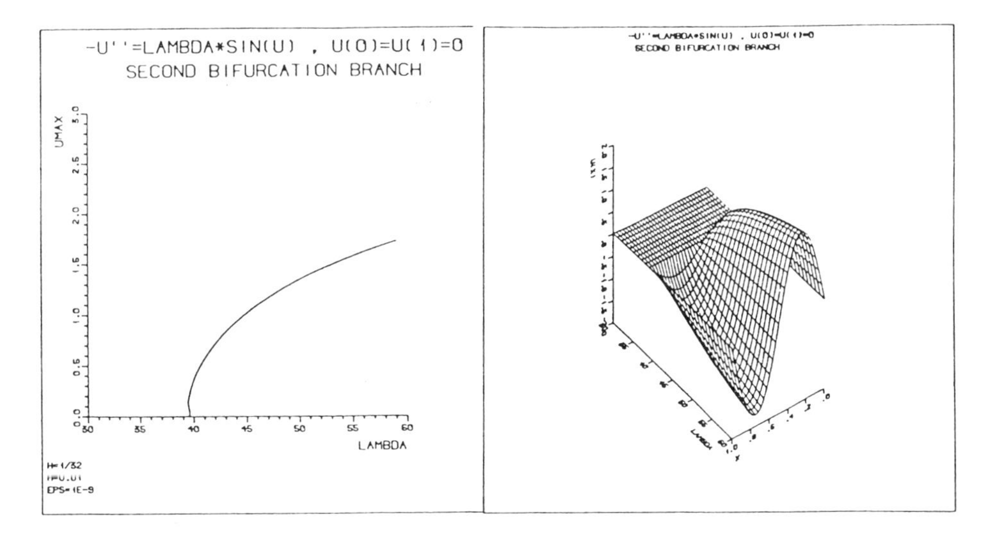

Fig. 2. Second bifurcation branch for $-u''(t) = \lambda \sin u(t)$.

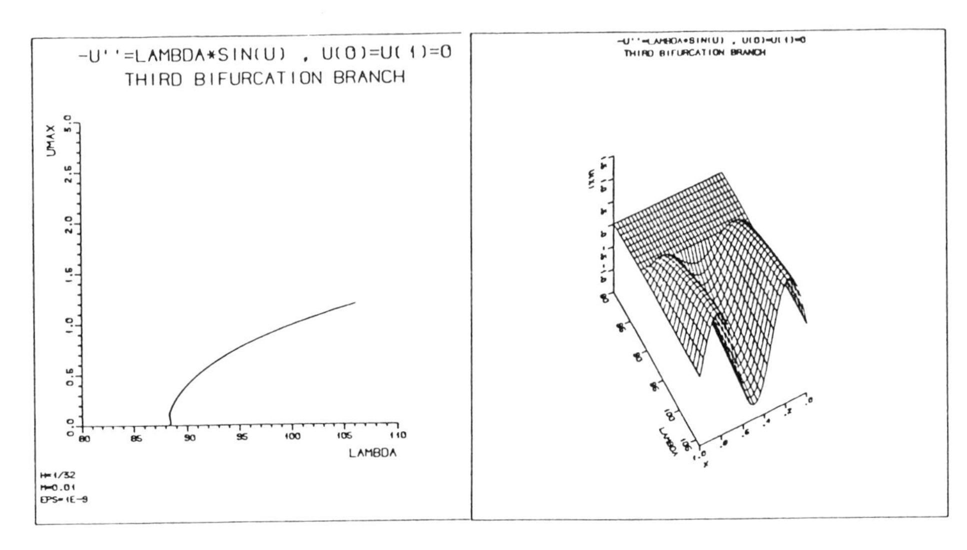

Fig. 3. Third bifurcation branch for $-u''(t) = \lambda \sin u(t)$.

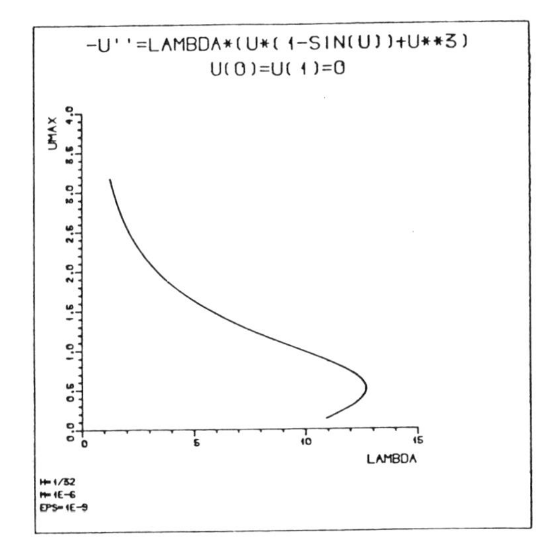

Fig. 4. First bifurcation branch for $-u''(t) = \lambda\big(u(t)(1 - \sin u(t)) + (u(t))^3\big)$.

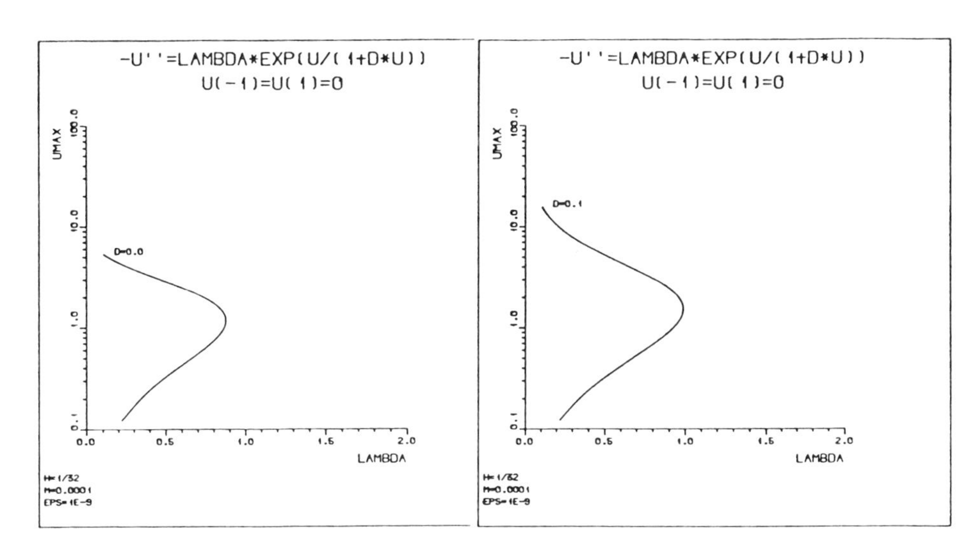

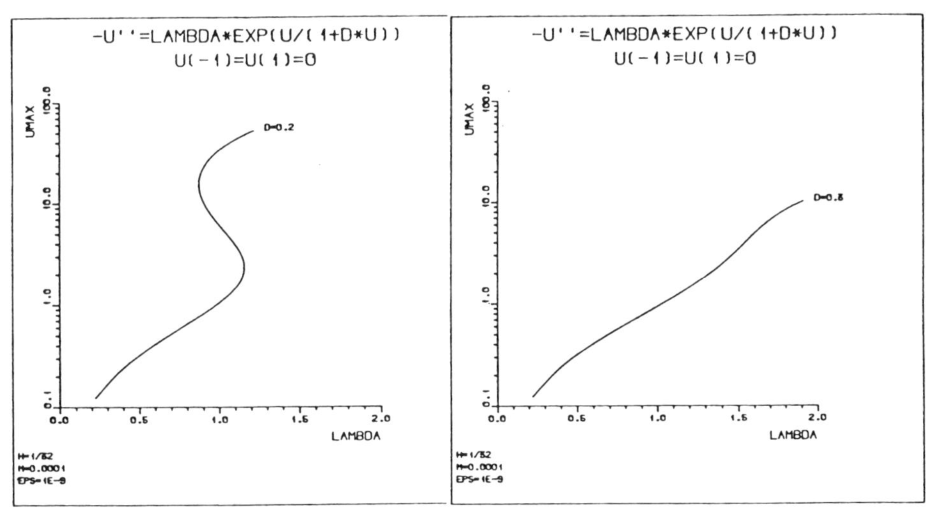

Fig. 5. Bifurcation branches for $-u''(t) = \exp\big(u(t)/(1 + Du(t))\big)$. Parameter values $D = 0$, $D = 0.1$, $D = 0.2$ and $D = 0.3$.

By the method presented here we can find the bifurcation point or turning point with a relatively good accuracy already with a rough discretization parameter. These examples are often used in literature to test numerical methods for nonlinear problems. Our results agree well with those given in literature, cf. [3], [6].

References

1. H. Amann: Lusternik-Schnirelmann Theory and Non-Linear Eigenvalue Problems. Math. Ann. **199**, 55–72 (1972)
2. S. Fučik, J. Nečas, J. Souček and V. Souček: Spectral Analysis of Nonlinear Operators. Berlin–Heidelberg–New York. Lecture Notes in Mathematics 346, Springer–Verlag 1973
3. R.Glowinski, H.Keller and L.Reinhart: Continuation-conjugate gradient methods for least square solution of nonlinear boundary value problems. INRIA, Rapports de Recherche **141** (1982)
4. A.Kratochvíl and J. Nečas: Secant Modulus Method for the Construction of a Solution of Nonlinear Eigenvalue Problems. Boll. Un. Mat. Ital. B(5) **16**, 694–710 (1979)
5. __________: Gradient Methods for the Construction of Lusternik–Schnirelmann Critical Values. RAIRO Anal. Numér. **14**, 43–54 (1980)
6. M.D.Mittelman and H.Weber (ed.): Bifurcation Problems and Their Numerical Solution. Basel. ISNM 54, Birkhäuser 1980
7. J. Nečas: An Approximate Method of Finding Critical Poins of Even Functionals. Proc. Steklov Inst. Math. **134**, 267–272 (1975)
8. J. Nečas, A. Lehtonen and P. Neittaanmäki: On the construction of Lusternik–Schnirelmann critical values with application to bifurcation problems. To appear
9. A. Lehtonen: On the construction of Lusternik–Schnirelmann critical values in Banach spaces. To appear
10. P.Neittaanmäki and K.Ruotsalainen: On the numerical solution of the bifurcation problem for the sine–Gordon equation. Arab. J. Math. (1986). To appear
11. E. Zeidler: Nonlinear Functional Analysis and its Applications III. Variational Methods and Optimization. New York–Berlin–Heidelberg–Tokyo. Springer–Verlag 1985

Ari Lehtonen, University of Jyväskylä, Department of Mathematics, SF-40100 Jyväskylä, FINLAND

International Series of
Numerical Mathematics, Vol. 79

ADAPTIVE NUMERICAL LYAPUNOV-SCHMIDT-REDUCTION

Wolfgang Mackens

RWTH Aachen, Federal Republic of Germany

I. Introduction

1. In this paper we develop an iteration to compute nondegenerate singular points of the solution set $\Phi^{-1}(0)$ of an underdetermined set of equations

(1.1) $\quad \Phi(y) = 0 \;, \quad \Phi \in C^3(\mathbb{R}^{n+1},\mathbb{R}^n) \;.$

$\hat{y}$ is a <u>regular point</u> of $\Phi^{-1}(0)$ if Φ's Jacobian has full rank n there. At <u>singular points</u> of $\Phi^{-1}(0)$ this condition is violated. Specifically if for $y^* \in \Phi^{-1}(0)$ we have

(1.2) $\quad \operatorname{rank} \Phi'(y^*) = n-m < n$

then y^* is called an <u>m-fold singular point</u> of $\Phi^{-1}(0)$. (For the relevance of singular points see e.g. [14]).
We call an m-fold singular point $y^* \in \Phi^{-1}(0)$ a <u>nondegenerate singular point</u> iff for some Lyapunov-Schmidt reduction

$$f(\beta) = 0 \;, \quad f\colon \mathbb{R}^{m+1} \longrightarrow \mathbb{R}^m$$

of (1.1) at y^* (see Section II for the notion of a Lyapunov-Schmidt reduction) the second derivative $D^2 f$ has full rank $m+1$ at the parameter value β^* corresponding to y^*.
Because of the equivalence of all Lyapunov-Schmidt reductions of (1.1) near y^* ([1], [13]) this definition is independent of the specific reduction. For simple singular points (m=1) it is equivalent to the usual notion that the quadratic form $P\,\Phi''(y^*)_{|N\times N}$ is nonsingular. Here P is the orthogonal projector onto Corange($\Phi'(y^*)$) and N denotes the kernel of $\Phi'(y^*)$. Usually more restrictive nondegeneracy conditions are met (cf. e.g. [2], [5], [17], [21]).

2. Most methods to compute singular points set up a system of "defining" [2] or "characterizing" [13] equations for the singular point. These are solved by a Newton-type iteration. Usually this set is an enlargement of

system (1.1) by some locally smooth version $S(y)=0$ of the singularity condition (1.2). Besides the unknown y these systems usually contain additional "unfolding-" or "imperfection"-parameters u_1, $u_2,\ldots$ which make the whole system square and regular near y^*:

(1.3.a) $\Phi(y, u_1, u_2,\ldots) = 0$,

(1.3.b) $S(y, u_1, u_2,\ldots) = 0$;

(cf. eg. [6] for a brief survey of defining systems (1.3) for simple bifurcation points, $m=1$).

3. The iteration to be given in Section III below has been designed to be used with the condensation method CNSP (cf. [11], [10], [12]). In this connection the system (1.1) arises from a reduction of a similar but very large system. Typically then the size of n will be moderate ($n \sim 10$) but

(1.4) the evaluation of Φ and specifically that of Φ' is extremely expensive compared with, say, linear algebra operations for $(n\times n)$-systems.

Whereas normally the linear algebra overhead of an iteration is tried to be kept small (e.g. by keeping the number of variables in (1.5) small, [20]) property (1.4) draws attention to saving of Φ- and Φ'-evaluations. Since a singularity condition (1.3.b) will always involve Φ', a Newton iteration for (1.3) will need second derivative information of Φ. Normally this is computed by differencing of Φ'-values. To reduce the corresponding effort, we want to use as few as possible second order directional derivatives of Φ per step retaining quadratic convergence. Actually our iteration will need $m+1$ of these directional derivatives per step. The characterization of m-fold singular points in Section II suggests that this number is minimal.
At least for $m>1$ the introduction of unfolding parameters u_1, $u_2,\ldots$ into system (1.1) appears to be a rather delicate undertaking depending on the specific nature of the singularity ([2], [13]). The elimination of unfolding parameters by use of symmetry considerations ([3], [22]) is not easily done in condensation algorithms. So it would be best if one could get rid of all unfolding parameters from the beginning. This is the second design aim.
The iteration is based on a variational characterization of nondegenerate singular points from the following section. It is defined and analysed in Section III, which contains an illustrative example, as well.

II. Variational characterization of nondegenerate singular points

1. Let y^* be an m-fold singular point of $\Phi^{-1}(0)$ such that rank $\Phi'(y^*) = n-m$. Then there are orthogonal matrices

$$H = (\underset{n-m}{H_1}, \underset{m}{H_2}) \in \mathbb{R}^{n\times n} \quad \text{and} \quad K = (\underset{n-m}{K_1}, \underset{m+1}{K_2}) \in \mathbb{R}^{(n+1)\times(n+1)}$$

such that within

$$(2.1) \quad H^T\Phi'(y)K = \begin{pmatrix} H_1^T\Phi'K_1 & H_1^T\Phi'K_2 \\ H_2^T\Phi'K_1 & H_2^T\Phi'K_2 \end{pmatrix}(y) =: \begin{pmatrix} \Gamma_{11}(y) & \Gamma_{12}(y) \\ \Gamma_{21}(y) & \Gamma_{22}(y) \end{pmatrix}$$

the leading $(n-m)\times(n-m)$-matrix $\Gamma_{11}(y)$ is nonsingular near y^*.

The matrices H and K induce a transformation of variables and equations :

$$\left.\begin{array}{ll} (2.1.a) & H_1^T\Phi(K_1\alpha + K_2\beta) = 0 \\ (2.1.b) & H_2^T\Phi(K_1\alpha + K_2\beta) = 0 \end{array}\right\} \quad \Longleftrightarrow \quad \Phi(y) = 0$$

with $\alpha = K_1^Ty \in \mathbb{R}^{n-m}$, $\beta = K_2^Ty \in \mathbb{R}^{m+1}$, $y = K_1\alpha + K_2\beta$.

Since $d/d\alpha\, H_1^T\Phi(K_1\alpha+K_2\beta) = \Gamma_{11}(y)$ is nonsingular near $y^*=K_1\alpha^*+K_2\beta^*$, equation (2.1a) can be solved for $\alpha(\beta)$ by the implicit function theorem:

$$(2.2) \quad H_1^T\Phi(K_1\alpha(\beta)+K_2\beta) = 0 \qquad \text{for } \beta \text{ near } \beta^*, \ \alpha(\beta^*)=\alpha^*.$$

Insertion of $\alpha(\beta)$ into (2.1.b) leads to the <u>Lyapunov-Schmidt-reduction</u>

$$f(\beta) := H_2^T\Phi(K_1\alpha(\beta)+K_2\beta) = 0$$

of problem (1.1).

Differentiation of f with respect to β (including implicit differentiation of $\alpha(\beta)$ by use of (2.2)) gives

$$(2.3) \quad f'(\beta) = (\Gamma_{22} - \Gamma_{21}\Gamma_{11}^{-1}\Gamma_{12})\,(K_1\alpha(\beta) + K_2\beta).$$

$\Gamma_{22} - \Gamma_{21}\Gamma_{11}^{-1}\Gamma_{12}$ is the Schur complement (cf. e.g. [8]) of Γ_{11} in Φ' such that

$$f'(\beta) = 0 \quad \Longleftrightarrow \quad \text{rank } \Phi'(K_1\alpha(\beta)+K_2\beta) = n-m \qquad \text{near } \beta^*$$

and β^* is a solution of this equation.

If y^* is nondegenerate then the second derivative $f''(\beta^*)$ has rank m+1 and we have

<u>Lemma 2.1 ([15]):</u> Let y^* be a nondegenerate m-fold singular point of $\Phi^{-1}(0)$. Then $\beta^*=K_2^Ty^*$ is

<u>case m = 1</u> : an isolated solution of the equation (cf. [4])

$$f'(\beta) = 0; \tag{2.4}$$

<u>case m ≥ 1</u> : a strong local solution of the unconstrained optimization problem

$$\Psi(\beta) := \| f'(\beta) \|_F \overset{!}{=} \min \quad (\| \ \|_F = \text{Frobenius-norm}) \tag{2.5}$$

with positive definite Hessian $D^2\Psi(\beta^*)$ and zero residual.

<u>Remark :</u> Notice that a solution of (2.4) or a strong minimum of (2.5), respectively, will survive under small smooth perturbations of f; these points may be used as "perturbed singular points" (cf. [4] for the case m=1).

2. Basically our iteration will be a kind of Newton-iteration for (2.4) or Gauss-Newton-iteration ([7]) to minimize (2.5), respectively. But, alas, carrying out this idea is not straightforward, since the Lyapunov-Schmidt-reduction involves the implicit function $\alpha(\beta)$, the computation of which is an infinite process in general (cf. [3]). However, we can get rid of $\alpha(\beta)$ simply by freeing α from its dependence on β in (2.4) and (2.5) and simultaneously complementing these problems by the equation (2.1.a) by which $\alpha(\beta)$ has been defined. In fact we have the following theorem ([15]).

<u>Theorem 2.2 :</u> Let y^* be a nondegenerate m-fold singular point of $\Phi^{-1}(0)$. Let the matrices $H=(H_1,H_2)$, $K=(K_1,K_2)$, $r_{11},\dots,r_{22}$ be defined as in paragraph 1. Let $S(\alpha,\beta) := (r_{22} - r_{21}r_{11}^{-1}r_{12})(K_1\alpha+K_2\beta)$ denote the Schur complement of r_{11} in Φ'.

Then $(\alpha^*, \beta^*) := (K_1^Ty^*, K_2^Ty^*)$ is

<u>case m = 1:</u> an isolated solution of the square system of equations

$$H_1^T\Phi(K_1\alpha+K_2\beta) = 0, \tag{2.7.a}$$

$$S(\alpha,\beta) = 0; \tag{2.7.b}$$

<u>case m ≥ 1:</u> a strong local solution of the constrained optimization problem

$$\| S(\alpha,\beta) \|_F^2 \overset{!}{=} \min \tag{2.8.a}$$

under the constraints

$$H_1^T\Phi(K_1\alpha+K_2\beta) = 0 . \tag{2.8.b}$$

The solution has zero residual and fulfills second order sufficiency conditions.

Remarks : (i) The "defining system" (2.7) for a nondegenerate simple singular point is related to the GRIEWANK-REDDIEN system for a "generalized turning point" ([9]).
(ii) It is a minimally extended system ([20]), whose number of variables can no more be reduced.
(iii) If (2.7) is used for the computation of a perturbed simple singular point, then $H_2^T\Phi(y^*)\neq 0$ may be interpreted as an unfolding parameter.
(iv) Lyapunov-Schmidt reduction has also be used by BEYN [3] and by JEPSON and SPENCE [13] to derive defining equations with unfolding parameters.

III. The iteration and its convergence

1. By Theorem 2.2 we can fulfill our second design aim: (2.7) and (2.8) do not contain unfolding parameters. We could apply Newton type iterations to calculate y^* with quadratic convergence. But this would in general defeat our primary design aim, to save derivatives of $S(\alpha,\beta)$.
In order to see that we can do with the number m+1 of derivatives of S assume for the moment that m=1 and that for the decomposition (2.1) we have the special form

$$(3.1)\qquad H_1^T\Phi'(y^*)K = \begin{pmatrix} r_{11}(y^*) & r_{12}(y^*) \\ r_{21}(y^*) & r_{22}(y^*) \end{pmatrix} = \begin{pmatrix} r_{11}^* & 0 \\ 0 & 0 \end{pmatrix},$$

such that the (nonsingular) Jacobian of (2.7) at y^* has the form

$$(3.2)\qquad \begin{pmatrix} r_{11} & 0 \\ S_\alpha & S_\beta \end{pmatrix}.$$

Then Chapter 10 of [19] tells us that the block Gauss-Seidel Newton iteration

$$(3.3.a)\qquad \alpha_{k+1} = \alpha_k - [r_{11}(K_1\alpha_k+K_2\beta_k)]^{-1}\, H_1^T\Phi(K_1\alpha_k+K_2\beta_k)$$

$$(3.3.b)\qquad \beta_{k+1} = \beta_k - [\partial/\partial\beta\; S(\alpha_{k+1},\beta_k)]^{-1}\, S(\alpha_{k+1},\beta_k)$$

converges locally and superlinearly to (α^*,β^*).

Remarks : (i) Notice that (3.3) uses exactly m+1 derivatives of $S(\alpha,\beta)$.
(ii) In (3.3.b) we freely interpret the matrix $S(\alpha,\beta)$ as a column vector.
(iii) The generalization of iteration (3.3) to the case "$m \geq 1$" is easily done by replacing the inverse of $\partial/\partial\beta\; S(\alpha_{k+1},\beta_k)$ in (3.3.b) by its Moore-Penrose generalized inverse ([8]).

Of course we can not construct H and K from (3.1), since we do not have y^* at hand. However, if y_d is sufficiently close to y^* then the SVD of $\Phi'(y^*)$ should deliver these matrices approximately. Thus using these approximations at least local linear convergence can be expected. This is the contents of the following main theorem ([15]).

Theorem 3.1 : Let y^* be a nondegenerate m-fold singular point of $\Phi^{-1}(0)$, $\Phi \in C^4(\mathbb{R}^{n+1},\mathbb{R}^n)$. Let the matrices

$H = (H_1,H_2) \in \mathbb{R}^{n\times(n-m)} \times \mathbb{R}^{n\times m}$ and $K = (K_1,K_2) \in \mathbb{R}^{(n+1)\times(n-m)} \times \mathbb{R}^{(n+1)\times(m+1)}$

be constructed by singular value decomposition [8] of $\Phi'(y_d)$ for y_d near y^*:

$$H^T\Phi'(y_d)K = [\mathrm{diag}(\sigma_1,\ldots,\sigma_n)|0] \in \mathbb{R}^{n\times(n+1)}, \quad \sigma_1 \geq \sigma_2 \geq \ldots \geq \sigma_n \geq 0 .$$

Then there is a neighbourhood U of y^* such that for all y_d and y_o from U

(i) the iteration with "$y_k \longrightarrow y_{k+1}$" consisting of the partial steps

a) $\alpha_k := K_1^T y_k$, $\beta_k := K_2^T y_k$

b) Execute a Newton step for $H_1^T\Phi(K_1\alpha+K_2\beta_k) = 0$ with respect to α at α_k giving α_{k+1}.

c) Execute a Gauss-Newton-Step for $\| S(K_1\alpha_{k+1}+K_2\beta) \|_F^2 = \min$ with respect to β at β_k giving β_{k+1} , where $S(y) := r_{22}(y) - r_{21}(y)r_{11}^{-1}(y)r_{12}(y)$ with $r_{ij} := H_i^T\Phi'()K_j$.

d) $y_{k+1} := K_1\alpha_{k+1} + K_2\beta_{k+1}$.

is well defined and stays in U.

(ii) There exist constants A and B (independent of $y_d \in U$) such that

$$\|y_{k+1} - y^*\| \leq A \|y_k - y^*\| \|y_d - y^*\| + B \|y_k - y^*\|^2 . \qquad (3.4)$$

From the independence of the constants in (3.4) of y_d we have at once

Corollary 3.2 : Under the assumptions of Theorem 3.1 the iteration sequence (y_k) converges quadratically to y^* if the matrices H and K are recalculated at $y_d := y_k$ after every step.

A numerical example : We treat the simple 4-point FD discretization

$$4u_{ij} - u_{3-i,j} - u_{i,3-j} - \tfrac{4}{9}\lambda\, u_{ij}\,(1 - \sin(u_{ij}) + u_{ij}^2) = 0 \quad , \quad i,j=1,2$$

of the elliptic problem $-\Delta u = \lambda\, u\,(1-\sin(u)+u^2)$ on $\Omega:=[-1,1]^2$, $u_{|\partial\Omega} = 0$, the main solution branches of which look as follows

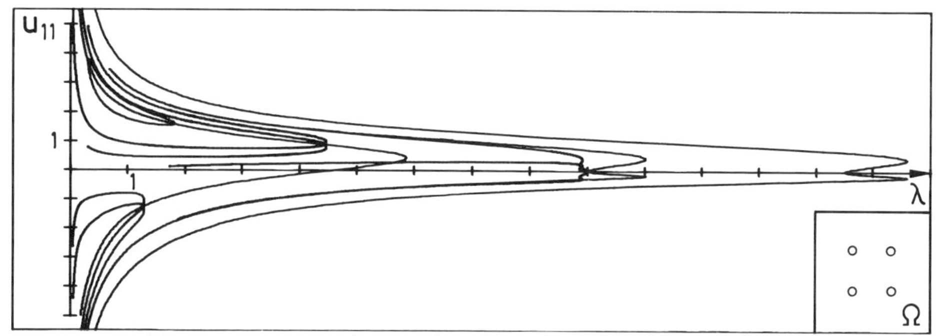

The quadratic convergence of the iteration from Corollary 3.2 is seen from a convergence history for the computation of the double bifurcation point ($u_{ij}=0, i,j=1,2,\ \lambda=9.0$):

u_{11}	u_{12}	u_{21}	u_{22}	λ
0.200 E 00	0.300 E 00	0.100 E 00	-0.200 E 00	9.100000000000
-0.946 E-01	-0.783 E-01	-0.258 E-01	-0.112 E-01	9.7328...
-0.118 E-01	-0.223 E-01	-0.991 E-02	-0.224 E-02	8.870267...
-0.169 E-02	-0.385 E-02	0.363 E-02	0.105 E-02	8.99789..
-0.866 E-04	-0.121 E-03	0.113 E-03	0.343 E-04	8.9999310...
-0.305 E-07	-0.131 E-07	-0.423 E-08	-0.243 E-07	8.999999909..
-0.203 E-15	-0.154 E-14	-0.150 E-14	-0.178 E-15	9.000000000000

The calculations have been done in Double Precision FORTRAN on an IBM AT-PC.

Concluding remarks : (i) The iteration has succesfully been implemented in the CNSP condensation method. Numerical results can be found in [11].
(ii) It can be extended to calculate certain multiparameter bifurcation points [15] and perturbed bifurcation points [16]. When used to compute perturbed nondegenerate simple singular points the iteration with permanent H-K-updating (Corollary 3.2) converges linearly (rate proportional to the perturbation) to the locally unique singular point which minimizes $\|\Phi\|$. This is the unfolded point delivered also by the defining system of MOORE [18].

IV. References :

[1] Beyn, W.-J. (1982) A note on the Lyapunov-Schmidt reduction. Manuscript, University of Konstanz, West Germany

[2] Beyn, W.-J. (1984) Defining equations for singular solutions and numerical applications. pp. 42-56 in [16].

[3] Beyn, W.-J. (1984) Zur numerischen Berechnung mehrfacher Verzweigungspunkte. ZAMM 65, T 370 - T 371

[4] Brezzi, F. , Rappaz, J. and P.A. Raviart (1981) Finite dimensional approximation of nonlinear problems. Part III. Simple bifurcation points. Numer. Math 38, 1-30

[5] Descloux,J. and J.Rappaz (1982) Approximation of solution branches of nonlinear equations. RAIRO, Anal. Numer. 16, 319-349

[6] Deuflhard, P., Fiedler, B. and P.Kunkel (1984) Efficient numerical path following beyond critical points. Preprint Nr. 278, University of Heidelberg, SFB 123

[7] Fletcher, R. (1981) Practical methods of optimization. Vol.2: Constrained Optimization (John Wiley, Chichester-New York-Brisbane-Toronto)

[8] Golub, G.H. and C.F. van Loan (1984) Matrix computations (The John Hopkins University Press, Baltimore)

[9] Griewank, A. (1984) Quadratically appended linear models for locating generalized turning points. pp. 162-170 in [16]

[10] Jarausch, H. and W. Mackens (1984) Numerical treatment of bifurcation by adaptive condensation. pp. 296-309 in [14].

[11] Jarausch, H. and W. Mackens (1985) Computing bifurcation diagrams for large nonlinear variational problems. Report No. 35 of the Inst. f. Geom. & Prakt. Math., RWTH Aachen (to appear in: Progress in large scale scientific computing, P.Deuflhard and B.Engquist (eds),Birkhäuser 1986)

[12] Jarausch, H. and W. Mackens (1986) Solving large nonlinear systems of equations by an adaptive condensation process. Report No. 29 of the Inst. f. Geom. & Prakt. Math., RWTH Aachen

[13] Jepson, A.D. and A. Spence (1984) Singular points and their computation. pp. 195-209 in [14]

[14] Küpper, T., Mittelmann, H.D. and H. Weber (eds) (1984) Numerical methods for bifurcation problems. (ISNM 70, Birkhäuser, Basel)

[15] Mackens, W. (1985) On an adaptive numerical Lyapunov-Schmidt-reduction to compute nondegenerate bifurcation points. Part I: The unperturbed case. Report No. 36 of the Inst. f. Geom. & Prakt. Math., RWTH Aachen.

[16] Mackens, W. On an adaptive numerical Lyapunov-Schmidt-reduction. Part II: The perturbed case. In preparation

[17] McLeod, J.B. and D. Sattinger (1973) Loss of stability and bifurcation at a double eigenvalue. J. Funct. Anal. 14, 62-84

[18] Moore, G. (1980) The numerical treatment of non-trivial bifurcation points. Numer. Func. Anal. and Optimization 6, 441-472

[19] Ortega, J.M. and W.C. Rheinboldt (1970) Iterative solution of nonlinear equations in several variables (Academic Press, New York and London)

[20] Pönisch,G. (1985) Computing simple bifurcation points using a minimally extended system of nonlinear equations. Computing 35, 277-294

[21] Rappaz,J. and G. Raugel (1981) Finite dimensional approximation of bifurcation problems at a multiple eigenvalue. Report no. 71, Centre de Mathematiques appliquees, Ecole Polytechnique, Paleseau

[22] Werner,B. (1984) Regular systems for bifurcation points with underlying symmetries. pp. 562-574 in [14]

Wolfgang Mackens, Institut für Geometrie und Praktische Mathematik der RWTH Aachen, Templergraben 55, 5100 Aachen, West Germany

International Series of
Numerical Mathematics, Vol. 79

FORMATION OF PERIODIC AND APERIODIC WAVES IN REACTION-DIFFUSION SYSTEMS

Miloš Marek and Igor Schreiber
Prague Institute of Chemical Technology, Dept. of Chemical Engineering,
Prague, Czechoslovakia

1. Introduction

It is well known that spatially distributed excitable media can conduct excitations (e.g., pulses) initiated by a sufficiently large perturbations. Neurones, Purkinje fibers and other excitable tissues in biological systems belong to the most often discussed examples of excitable media [4 ,8]. However, more simple chemical systems with reaction and diffusion can be also excitable and may thus serve as model systems [15]. Let us consider an S-component, spatially one-dimensional system

$$\frac{\partial X_i}{\partial t} = D_i \frac{\partial^2 X_i}{\partial z^2} + f_i (X_1,\dots,X_S) , \quad i = 1,\dots,S , \tag{1}$$

where D_i denote diffusion coefficients and f_i kinetic functions of a suitable form. The spatial coordinate z is from the interval $[0,L]$ and proper boundary conditions (for example, describing zero flux of the components X_i at the boundaries) are also defined. We shall assume such conditions (kinetic and diffusion parameter values) that if the system is unperturbed, it stays in a stationary state without spatial gradients.

If the characteristic length of the system L is not too small, then a sufficiently large perturbation at the left hand boundary can initiate a wave which travels along the system and is annihilated at the right hand boundary. Let as assume that the perturbation is defined by a single parameter - an amplitude A . If the perturbation is periodically repeated with a period T , a system of travelling waves arises. The magnitude of the perturbation amplitude and the period of the initiation at the boundary will determine whether the wave will be formed and whether the created wave pattern will be periodic or aperiodic.

The periodic regime is characterized by a repeating concentration profile after q perturbations; a "firing number" $\nu = p/q$ can be defined, here p is a number of travelling waves generated in the course of q external perturbations.

Provided the length L of the system is sufficiently large, the wave is well developed and physically well defined. A correct mathematical definition is not so evident, but the "firing number" ν can be in principle defined asymptotically also for aperiodic regimes.

The dependence of the "firing number" ν on the forcing amplitude was studied in an experimental spatially quasi-onedimensional reaction-diffusion excitable system, using a modification of the Belousov-Zhabotinski reaction mixture periodically perturbed by voltage pulses [7]. Very similar behaviour was also recorded in a lumped parameter system, involving an oscillatory Belousov-Zhabotinski reaction in a continuous stirred tank reactor [2 ,9 ,10]. Periodic regimes consisting of a combination of low-amplitude oscillations with high-amplitude oscillations were observed in these experiments. The "firing number" was then defined as a ratio of number of oscillations with the low amplitude to the overall number of oscillations in one period. The "firing number" was followed in dependence on a flow-rate.

In these experiments the dependence of the "firing number" on a parameter is a step-wise function. A rational value $\nu = p/q$ corresponds to every step and the associated dynamic regime is a q-periodic pattern. The arrangement of steps corresponds to the Farey sequences [5]. Such a function resembles to a devil´s staircase, i.e., the function which is constant on the intervals (where it takes on rational values) and increasing at points, (where it takes on irrational values). Devil´s staircases are common in dynamical systems with two competing frequencies, as are, for example, periodically forced oscillators (with their own intrinsic frequency at zero forcing). The periodically perturbed excitable systems do not belong directly into this class. However, they can be viewed as systems with a latent (but excitable) internal frequency.

Several authors suggested that the parametric dependence of the firing number can be understood on the basis of the study of iterates of certain one-dimensional maps. In particular, coupled neurons [1] and complex oscillations in the Belousov-Zhabotinski reaction [3 ,12 ,14] were modelled in this way. Properly chosen one-dimensional mappings are certainly able to produce devil´s staircases, but the relation between the original continuous dynamical system and one-dimensional mappings is not clear.

Hence our aim here is to derive as far as possible simple system which still has a direct relation to a periodically perturbed spatially one-dimensional excitable medium and study an arrangement of periodic and/or aperiodic regimes of pulse waves in dependence on the perturbation amplitude.

2. Model equations and reduction to low-dimensional approximations

Let us consider a periodically perturbed spatially one-dimensional two component reaction-diffusion system $(X_1 \equiv X \, , \, X_2 \equiv Y)$

$$\frac{\partial X}{\partial t} = D_X \frac{\partial^2 X}{\partial z^2} + f\,(X,Y)$$
$$\frac{\partial Y}{\partial t} = D_Y \frac{\partial^2 Y}{\partial z^2} + g\,(X,Y) \quad , \qquad z \in (0,L) \tag{2}$$

The excitable kinetics is modelled by the SH model [13]

$$f(X,Y) = \alpha \frac{\nu_0 + X^{\gamma}}{1 + X^{\gamma}} - X\,(1 + Y)$$
$$g(X,Y) = X\,(\beta + Y) - \delta\, Y \quad , \tag{3}$$

where α , δ , $\nu_0 > 0$ and γ, $\beta > 1$ are kinetic parameters. We suppose zero flux boundary conditions for the component Y at both ends of the system and for the component X at $z = L$:

$$\left.\frac{\partial Y}{\partial z}\right|_{z=0} = \left.\frac{\partial Y}{\partial z}\right|_{z=L} = \left.\frac{\partial X}{\partial z}\right|_{z=L} = 0 \quad . \tag{4}$$

The periodic perturbation is introduced via mixed boundary conditions for X at $z = 0$:

$$(1 - p(t)) \left.\frac{\partial X}{\partial z}\right|_{z=0} + p(t)(X|_{z=0} - A) = 0 \quad . \tag{5}$$

Here $p(t)$ is a T-periodic function consisting of rectangular pulses

$$p(t) = \begin{cases} 0 & \text{for} \quad t \in [\, t_k \,,\, t_k + T_1 \,) \\ 1 & \text{for} \quad t \in [\, t_k + T_1 \,,\, t_k + T \,) \,, \quad t_k = kT,\ k=0,1,\ldots, \,. \end{cases} \tag{6}$$

The function $p(t)$ switches between two types of boundary conditions for X at $x = 0$ and thus the periodic perturbation of the system with the period T and the amplitude A is defined.

The system (2) - (6) is nonautonomous with an infinite-dimensional functional phase space. A Poincare mapping can be defined by observing the solutions of the system (2) - (6) only at discrete times t_k , $k = 0,1,\ldots$. Hence we have an infinite-dimensional dynamical system with discrete time. This system yields a dependence of the firing number ν on A similar to that discussed above [7].

To simplify the analysis, we discretize the spatial coordinate and replace Eqs. (2) - (6) by a system of N coupled reaction cells described by a system of 2N ODE´s

$$\begin{aligned} \frac{dX_n}{dt} &= D_1 \,(X_{n+1} + X_{n-1} - 2X_n) + f\,(X_n, Y_n) \\ \frac{dY_n}{dt} &= D_2 \,(Y_{n+1} + Y_{n-1} - 2Y_n) + g\,(X_n, Y_n) \quad , \quad n = 1,\ldots,N \end{aligned} \tag{7}$$

with boundary conditions

$$Y_0 = Y_1 \,,\quad Y_N = Y_{N+1} \,,\quad X_N = X_{N+1} \tag{8}$$

$$(1 - p(t))\,(X_1 - X_0) + p(t)\,(X_0 - A) = 0 \tag{9}$$

The system (7) - (9) is again nonautonomous and its Poincare map has a 2N-dimensional phase space. Further simplification can be made in two ways. First, we may assume that the pulse wave is formed in the nearest neighbourhood of the left boundary of the distributed system and reduce the N-member cascade to a single cell. The single cell excitable system is then described as

$$\frac{dX}{dt} = D_1\, p(t)\,(A - X) + f\,(X,Y)$$

$$\frac{dY}{dt} = g\,(X,Y) \quad . \tag{10}$$

The corresponding Poincare mapping $(X_k,Y_k) \to P\,(X_k,Y_k)$ is two-dimensional.

The second way of the dimension reduction gives the same result if applied to the distributed system (2) - (6) or to the N-cascade (7) - (9). It is based on an assumption of an infinitely fast diffusion of both components. In this case spatial gradients disappear and the distributed system behave like the single cell system (10) but with $D_1 \to \infty$. If $t \in [t_k, t_k+T_1)$ then $p(t) = 0$ and the dynamics of X and Y is controlled only by the kinetic functions f and g . Let us assume that an initial state at the time t_k is (X_k,Y_k) and the system approaches the state (X,Y) as $t \to t_k + T_1$. The dynamics of the system (10) in the time interval $[t_k+T_1,\ t_k+T)$ (here $p(t)=1$) can be decomposed into two phases. Because of the infinitely fast diffusion a jump in the phase space at the time t_k+T_1 occurs, which changes the state of the system into $(A,\ Y)$. If $t \in (t_k+T_1,\ t_k+T)$ then $X = A$ due to infinitely fast diffusion and the dynamics of the variable Y is controlled by the equation

$$\frac{dY}{dt} = g\,(A,Y) \tag{11}$$

Hence $X_k = A$ for $k=1,2,\ldots$ and the Poincare mapping $Y_k \to P(Y_k)$ is one-dimensional, generally noninvertible and its phase space is the straight line $X = A$.

We are interested in solutions of the system (10) with a finite diffusion rate (we shall call it 2D model) and its one-dimensional limit for $D_1 \to \infty$ (further on referred to as 1D model). To be able to follow the correspondence between solutions of Eqs. (10) and the original distributed system we have to define a firing number for the 1D and 2D models.

A necessary condition for pulse waves to occur is that the corresponding lumped parameter system without external forcing will posses an excitable cycle. It means that one stable steady state S_1 exists in the phase plane (X,Y) ; however, there are orbits which form a large loop in the phase plane before approaching S_1 . This occurs, for example, if three stationary points S_1 , S_2 , S_3 exist such that the first one is a sink, the second one is a saddle and the third one is a node. In the case of the SH model this situation is observed for the set of parameters $\alpha = 12$, $\beta = 3$, $\gamma = 15$, $\delta = 1$, $\nu_0 = 0.01$ [13] cf. Fig. 1. The excitable cycles may be triggered by an external periodic perturbation, e.g., according to Eq. (10).

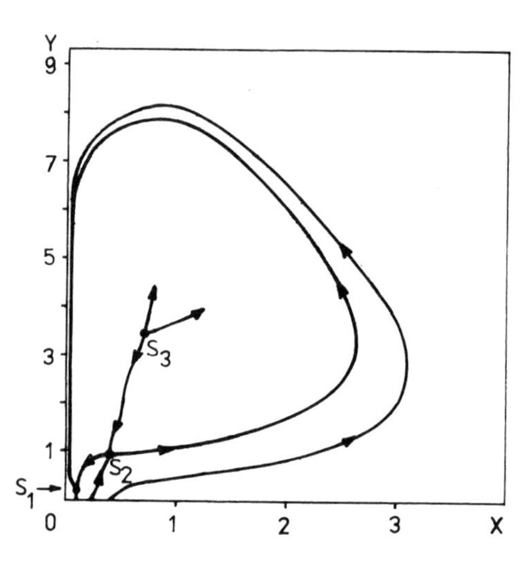

Fig. 1. Phase portrait of the unperturbed SH kinetics in single reaction cell.

Let us assume that "sufficiently large" excitable cycles correspond to the formation of travelling pulse waves, while "small" excitable cycles and the orbits which do not form loops do not lead to the formation of developed pulse waves. A "sufficiently large" excitable cycle for numerical computations was defined by the condition that an orbit must intersect the straight line $Y = 5$ from below in the time interval $[\, t_k, t_k + T_1)$. This condition is not very precise but it is sufficient for our purposes here. The firing number ν for a periodic orbit of period q is then defined as $\nu = p/q$, where

p is the number of "large" excitable cycles within the period q. The dependences of ν on A for the fixed period of the perturbation $T = 2.75$ and the length of the perturbation $\Delta T = T - T_1 = 0.6$ for several values of diffusion coefficient D_1 are shown in Fig. 2a - d. Only the largest plateaus are depicted in the Figure, but the similarity to the devil´s staircase function is apparent. An important observation which can be made from the Figure is that the behaviour is in principle not dependent on the magnitude of diffusion coefficient.

Dynamical behaviour for low values of A is characterized by the firing number $\nu = 0/1$. When A is increased, then a sequence of multiperiodic regimes evolves and, finally, a one-periodic regime with the firing number $\nu = 1/1$ sets in. This is in agreement both with the behaviour of the distributed system (2) - (6) and with experiments [7]. It confirms the idea that the essence of the inherent nonlinear dynamics is preserved in 2D and 1D approximations. In the next section we use the continuation of periodic orbits to elucidate instabilities leading to firing numbers different from 0 and 1.

3. Dependence of periodic orbits on A

Choosing initial conditions (X_k, Y_k) at time t_k, a solution (X,Y) of Eq. (10) at time $t_k + T_1$ can be written as

$$X = u_1 (X_k, Y_k)$$
$$Y = u_2 (X_k, Y_k) \tag{12}$$

Similarly the solution (X_{k+1}, Y_{k+1}) at time $t_k + T$ is

$$X_{k+1} = v_1 (X,Y,A)$$
$$Y_{k+1} = v_2 (X,Y,A) \tag{13}$$

or

$$Z_{k+1} = v (u (Z_k), A) \equiv w (Z_k, A) \quad , \tag{14}$$

where $Z = (X,Y)$, $v = (v_1,v_2)$, $u = (u_1,u_2)$. The 2D mapping $w : R^2 \times R \rightarrow R^2$ parametrized by A will have a periodic point of period q

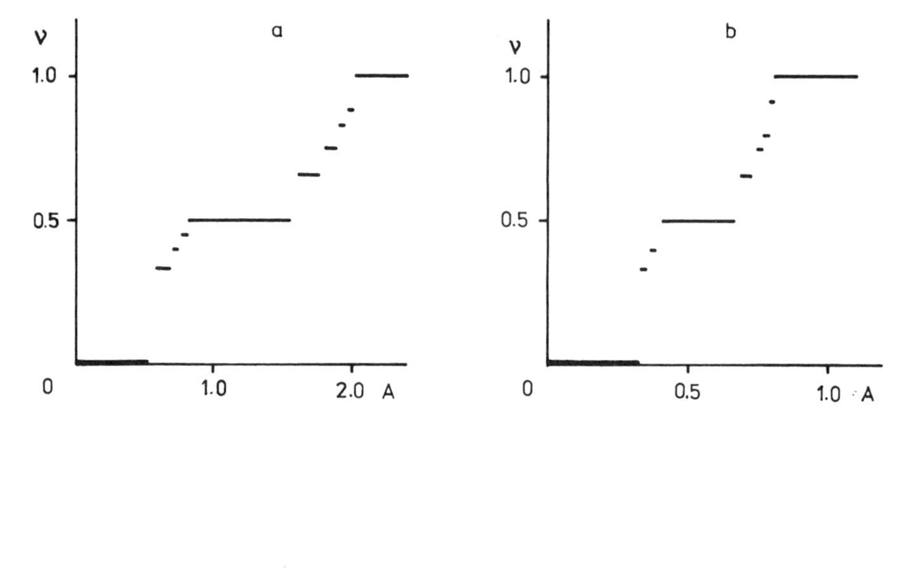

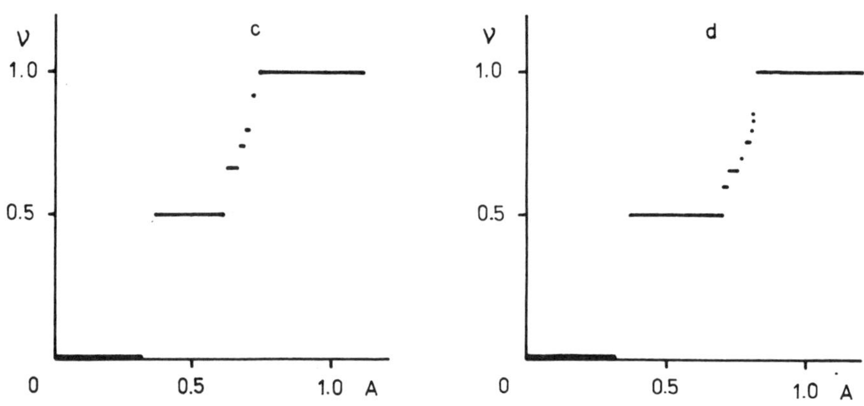

Fig. 2. Dependence of ν on A generated by Eg. (10) ; only the largest plateaus are shown ; a) $D_1 = 1$ b) $D_1 = 5$ c) $D_1 = 10$ d) $D_1 = \infty$

if

$$H\,(Z,\,A) \;\equiv\; w^q\,(Z,\,A) - Z \;=\; 0 \quad . \tag{15}$$

A solution $\overline{Z}$ to Eq. (15) parametrized by A can be found numerically. An algorithm for the continuation of the solution curve $(\overline{Z}\,(\varphi), A\,(\varphi))$ in $R^2 \times R$ parametrized by the arc-legth φ [6] was used. The method requires the knowledge of derivatives of H with respect to Z and A. They can be obtained by solving variational equations corresponding to Eq. (10) taking into consideration rules of differentiation of composed functions. The eigenvalues of the matrix $D_Z\, w^q$ which determine a stability of the orbit can be easily obtained as a by-product of the continuation.

An integration method with a constant step-size in the independent variable φ (the Adams-Bashforth method) is used in the original algorithm for the continuation of the curve $(\overline{Z}\,(\varphi), A\,(\varphi))$ [6]. Solution curves in our case have very sharp bends (it seems to be a general property of the systems which posses the devil´s staircase like behaviour) and the use of the original continuation algorithm becomes impractical. Hence we have used an integration method with an automatic control of the step.size (Runge-Kutta-Merson method, but other methods can be also used).

The 1D mapping can be written as

$$X_{k+1} = X_k \equiv A$$

$$Y_{k+1} = v\,(Y,A) = v\,(u_2\,(A,Y_k),\,A) \quad , \tag{16}$$

where $v\,(Y,A)$ is the solution of Eq. (11). To determine a periodic point of Eq. (16) we have to solve variational equations associated with Eqs. (10) and (11).

Only the most simple periodic orbits of period one were followed in this preliminary study. The examples of the results of the continuation corresponding to Figs. 2a - d are presented in Figs. 3a - d . Each solution curve has two stable parts, reflecting firing numbers $\nu = 0/1$ and $\nu - 1/1$. The loss of the stbility of periodic orbits with $\nu = 0/1$ at $A = A_1$ is associated with the bifurcation of a torus in the case of Fig. 3a and with the period-doubling in Figs. 3b - d . The direct integration of Eq. (10) shows that these bifurcations are subcritical. A sequence of periodic and aperiodic regimes with non-zero firing numbers arises to the right of the critical point

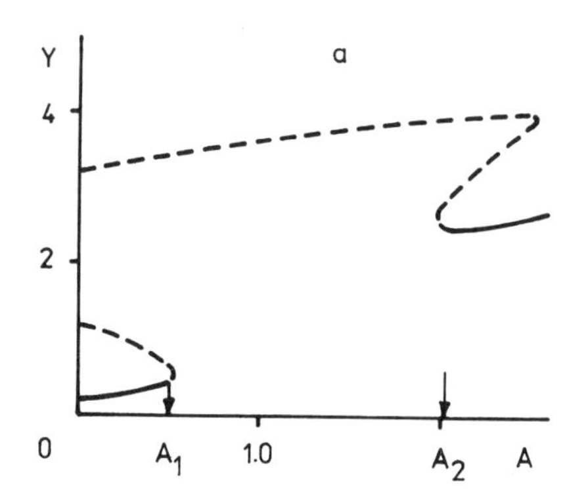

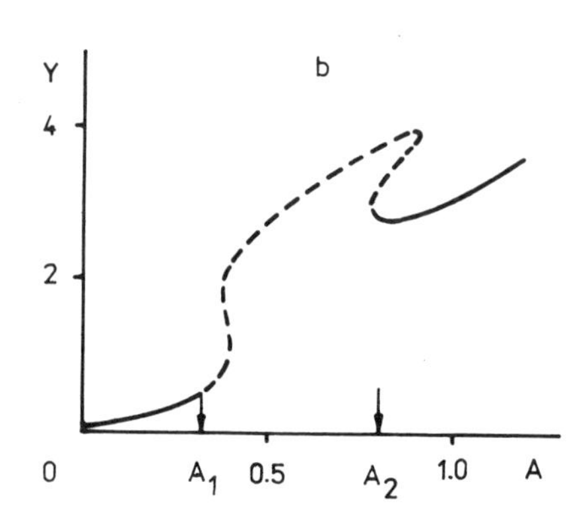

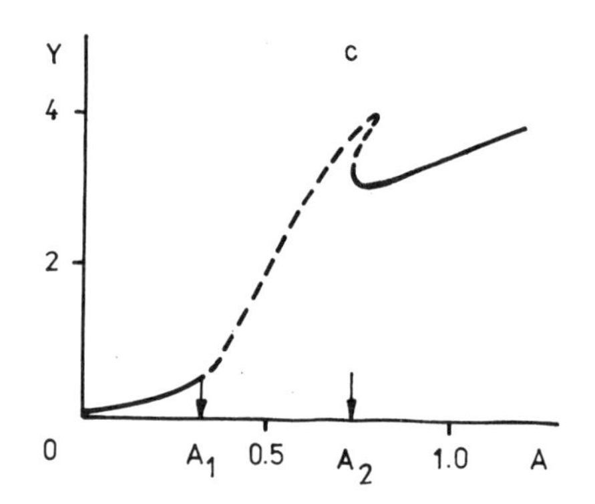

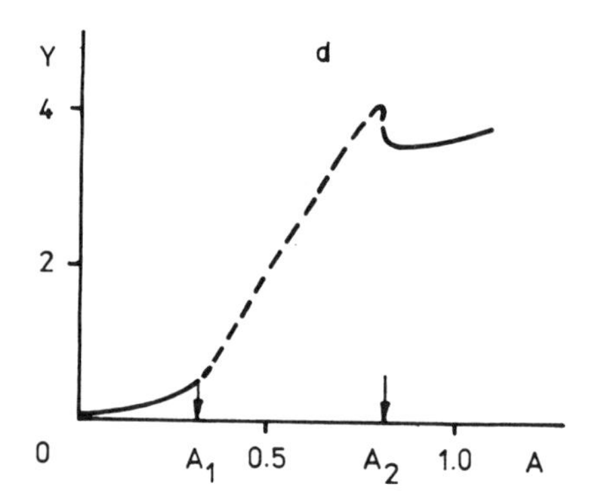

Fig. 3. Dependence of one-periodic orbit of Eg. (10) on A, the loss of stability at A_1 and A_2 corresponds to the first occurence of noninteger values of ν, cf. Fig. 2 ; a) $D_1 = 1$
b) $D_1 = 5$
c) $D_1 = 10$
d) $D_1 = \infty$

A_1 . The overall dynamics in the neighbourhood of the critical point is controlled by the type II (D_1 = 1) and the type III (D_1 = 5, 10,) intermittency according to the classification proposed by Pomeau and Mannevile [11]. On the other hand, the loss of stability of periodic orbits with $\nu = 1/1$ is associated with the limit point at $A = A_2$. The dynamics to the left of the critical point A_2 is controlled by the type I intermittency. Hence the devil's staircase-like behaviour in this problem provides a textbook example of various types of intermittencies in dynamical systems.

Let us remark that the transition of the critical point at $A = A_1$ from the period-doubling to a torus bifurcation when the value of D_1 is decreased is a codimension two bifurcation point where both eigenvalues of the matrix $D_Z w$ are equal to -1 . In addition, much more common cusp point exists in the neighbourhood of this point. However, despite the presence of more degenerate bifurcations, the devil's staircase-like dependence of ν on A is preserved in the system.

4. Conclusions

Devil's staircase-like dependence of the nonlinear resonance frequencies of the distributed reaction-diffusion excitable systems on the forcing amplitude reflects the dynamics of the pulse wave formation in the system and can be analyzed by an application of the path-following (continuation) techniques to appropriate low-dimensional models. In distributed excitable systems the internal frequencies are latent, but can be excited by the external forcing. In general, more complicated phenomena resulting, for example, from the excitation of multiple frequencies and their interactions can also be expected. Similar approach can be useful in various problems arising in other physical situations, e.g., in nonlinear mechanics.

5. References

[1] Allen, T. (1983) On the arithemic of phase locking: coupled neurones as lattice on R^2. Physica 6D, 305-320.

[2] Argoul, F., Arneodo, A., Richetti, P., Roux, J.C. (1986) From quasiperiodicity to chaos in the Belousov-Zhabotinskii reaction I., II. Preprint, Université de Bordeaux.

[3] Bagley, R.J., Mayer-Kress, G., Farmer, J. D. (1986) Mode locking, the Belousov-Zhabotinsky reaction, and one dimensional mappings. Phys. Lett. 114A, 419.

[4] Cooke, I., Lipkin, M., Eds. (1972) Cellular neurophysiology. Holt, Rinehart and Winston, New York.

[5] Hardy, G.H., Wright, E.M. (1965) An introduction of the theory of numbers, (Clarendon Press, Oxford).

[6] Kubíček, M., Marek, M. (1983) Computational methods in bifurcation theory and dissipative structures, (Springer, New York).

[7] Marek, M., Schreiber, I., Vroblová, L. (1986) Complex and chaotic waves in reaction-diffusion systems and on the effects of electric field on them. Proc. of the Midit (Lyngby) 1986 Workshop : "Structure, Coherence and Chaos in Dynamical Systems", to apper in series "Nonlinear Science, Theory and Applications, Manchester Univ. Press.

[8] Markin, V.S., Pastushenko, V.F., Chizmadzkev, Ju.A. (1981) Theory of excitable media. (in Russian), Nauka, Moscow.

[9] Maselko, J., Swinney, H.L. (1985) A complex transition sequence in the Belousov-Zhabotinskii reaction. Physica Scripta T9, 35.

[10] Maselko, J., Swinney, H.L. (1986) Complex periodic oscillations and Farey arithmetic in the Belousov-Zhabotinskii reaction. Preprint, University of Texas at Austin.

[11] Pomeau, Y., Manneville, P. (1980) Intermittent transition to turbulence in dissipative dynamical systems. Commun. Math. Phys. 74, 189.

[12] Rinzel, J., Schwartz, I.B. (1984) One variable map prediction of Belousov-Zhabotinskii mixed mode oscillations. J. Chem. Phys. 80, 5610-5615.

[13] Ševčíková, H., Marek, M. (1986) Chemical waves in electric field - modelling. Physica 21D, 61-77.

[14] Tsuda, I.-(1981) Self-similarity in the Belousov-Zhabotinsky reaction. Phys. Lett. 85A, 4-8.

[15] Zykov, V.S. (1984) Modelling of wave processes in excitable media. Nauka, Moscow.

Prof. Miloš Marek
Department of Chemical Engineering
Prague Institute of Chemical Technology
166 28 Praha 6, Czechoslovakia

International Series of
Numerical Mathematics, Vol. 79

STEADY AXISYMMETRIC TAYLOR VORTEX FLOWS WITH FREE STAGNATION POINTS OF THE POLOIDAL FLOW

RITA MEYER-SPASCHE and MATTHIAS WAGNER

The very reliable and efficient numerical methods of [8, 11] were used for investigating axisymmetric steady Taylor vortex flows between concentric cylinders with periodic boundary conditions. The Reynolds number Re and period λ were continuously varied for fixed radius ratio $\eta = 0.727$. The basic $(n, 2n)$ fold was found to exist up to at least $Re = 4.6Re_{cr}(T = 21T_{cr})$. A large multiplicity of solutions (≥ 21 for $\lambda = 2.5$) and of fold points (≥ 25) and period doubling and tripling bifurcations were detected for $Re = 3.65Re_{cr}, \eta = 0.727$, λ varying. Solutions with unusual flow patterns are displayed.

1. Formulation of the Problem

We consider steady axisymmetric Taylor vortex flows between two concentric cylinders. The inner cylinder rotates with angular speed ω_1 at $\hat{r} = R_1$, and the outer one is at rest, at $\hat{r} = R_2$. End effects are neglected (periodic boundary conditions). The Navier-Stokes equations used to compute these flows are, in dimensionless form,

$$u_r + \frac{1}{r}u + w_z = 0, \tag{1a}$$

$$\Delta u - \frac{1}{r^2}u - p_r = Re(uu_r + wu_z - \frac{v^2}{r}), \tag{1b}$$

$$\Delta v - \frac{1}{r^2}v = Re(uv_r + wv_z + \frac{uv}{r}), \tag{1c}$$

$$\Delta w - p_z = Re(uw_r + ww_z). \tag{1d}$$

Here lengths have been scaled by R_1, velocities by $\omega_1 R_1$, the Reynolds number is given by

$$Re := \omega_1 R_1^2/\nu ,$$

ν being the kinematic viscosity of the fluid, and Δ is the Laplacian,

$$\Delta := \frac{\partial^2}{\partial r^2} + \frac{1}{r}\frac{\partial}{\partial r} + \frac{\partial^2}{\partial z^2} .$$

Besides the Reynolds number, or Taylor number, $T = Re^2 \cdot$ const, there are two other independent parameters in the problem. These are usually taken to be $\eta := R_1/R_2$, the ratio of the radii, and k, the wave number. The dimensionless gap width $\delta := (R_2 - R_1)/R_1$ is related to η by $\delta = \eta^{-1} - 1$, and the period λ of the flow is related to k by $\lambda = 2\pi/k\delta$. With these definitions we can formulate the boundary conditions as

$$\begin{aligned} &u(1,z) = u(1+\delta,z) = 0\,, \\ &w(1,z) = w(1+\delta,z) = 0\,, \qquad -\frac{\pi}{k} \le z \le \frac{\pi}{k}\,, \\ &v(1,z) = 1, v(1+\delta,z) = 0\,, \end{aligned} \tag{2a}$$

$$\begin{aligned} &u(r,-\frac{\pi}{k}) = u(r,\frac{\pi}{k})\,, \\ &v(r,-\frac{\pi}{k}) = v(r,\frac{\pi}{k})\,, \qquad 1 \le r \le 1+\delta\,, \\ &w(r,-\frac{\pi}{k}) = w(r,\frac{\pi}{k}) = 0\,. \end{aligned} \tag{2b}$$

Note that these boundary conditions induce $\mathbf{v} \cdot \mathbf{n} = 0$, i.e. no fluid can leave the domain of computation. We solve these equations numerically, using an extended version of the TAYPERIO code described earlier, [6, 10, 11]. Comparisons of computed values with measurements have shown that eqs.(1), (2) are well suited to describing axisymmetric steady Taylor vortex flows in the middle portion of a long apparatus (comparison of torques for $Re \le 2.2Re_{cr}$ in [10], Figs.1 and 3; comparison of stability limits for the basic Taylor vortex flow for $Re \le 2Re_{cr}$ in [12, Figs.2 and 3]). In the present investigations we fixed η at a value which is known to allow axisymmetric steady flows up to $Re \ge 9Re_{cr}$ [1, 7]. Nevertheless it is possible that some of the solutions displayed here are absolutely unstable, i.e. that there are no experimental side conditions which could allow such flows to exist in the material world. We think that the investigation of such flows is nevertheless of some interest: 1) The computation of a branch of unstable flows might lead to the discovery of a branch of stable flows connected to it. 2) It might be of interest to see that flows do not exist or change substantially beyond their limit of stability [12]. 3) The Navier-Stokes equations are an important example of a nonlinear system of partial differential equations. Their mathematical investigation has given important impulses to mathematics (concept of weak solutions, development of nonlinear functional analysis ...) [9, 14]. Bifurcation patterns observed for eqs.(1), (2) [11, Figs.1, 3] have also been observed for other equations [2, 5, 13]. We are interested in questions such as: What types of flows are allowed by eqs.(1), (2) in the highly nonlinear regime ($Re \gg Re_{cr}$)? How many solutions exist for a given set of parameter values? What is the structure of the solution branches as Re and k vary? What folds and bifurcations do they undergo? At present, the proof of theorems answering these questions for the parameter values used in our computations seems to be out of range, [9, 14, §§9, 10]. Our "experimental" investigation of the solution set of eqs.(1), (2) might thus be a small contribution towards obtaining a better understanding of the mathematical properties of the Navier-Stokes equations. It might help a little bit to formulate realistic conjectures about these equations.

2. Numerical Methods

We give here a short review of the numerical methods used in the computations. More details are to be found in [10] and [11]. To discretize eqs.(1), (2), we use a Fourier decomposition in the axial direction and equidistant centered finite differences in the radial direction. This gives us an algebraic system of nonlinear equations

$$\mathbf{F}(\mathbf{x}, k, Re) = 0. \tag{3}$$

Here the vector $\mathbf{x}$ denotes the unknowns, i.e. the values of the Fourier components in the radial grid points, while k and Re are the parameters introduced earlier. Equations (1), (2) allow solutions which are antimetric with respect to the plane $z \equiv 0$. Our Fourier decomposition is such that all computed solutions exhibit this sort of symmetry. This is a restriction; the existence of asymmetric solutions has to be expected [2]. But the number of solutions left is still so large that this restriction might even be considered to be advantageous.

We investigate the solution set of (3) with the method of continuation [8] both with respect to Re and with respect to k, using pseudo arc length s to parameterize the curves

$$\mathbf{y}(s)^t = (\mathbf{x}(s)^t, Re(s)) \text{ and } \mathbf{y}(s)^t = (\mathbf{x}(s)^t, k(s)),$$

respectively. The enlarged system

$$G(\mathbf{y}(s), s) = 0 \tag{4}$$

is then solved by the (simplified) Gauss-Newton method. When (4) has been solved and $\mathbf{y}$ is known, the pressure p, the velocity components u, v, w and the stream function ψ are computed. ψ is defined by

$$u = \frac{1}{r}\frac{\partial\psi}{\partial z}, \quad w = -\frac{1}{r}\frac{\partial\psi}{\partial r}. \tag{5}$$

We thank Prof. D. Lortz for teaching us how to compute ψ by a numerically stable algorithm. All computations were performed with M2 = 33 radial grid points and N = 12 Fourier components on the Cray 1 of the ZIB, Freie Universität, Berlin.

3. Taylor Vortex Flows for $Re = 800 \approx 3.65 Re_{cr}$

In [11] one of the authors and Keller studied the structure of families of solutions or flows with respect to continuous variation of both parameters λ and Re, η being fixed at the same value $\eta = 0.727$ as in the Burkhalter-Koschmieder experiments [1] and in the present investigations. Varying λ with fixed $Re = \sqrt{1.5}Re_{cr}$ (or $T = 1.5T_{cr}$), they found a

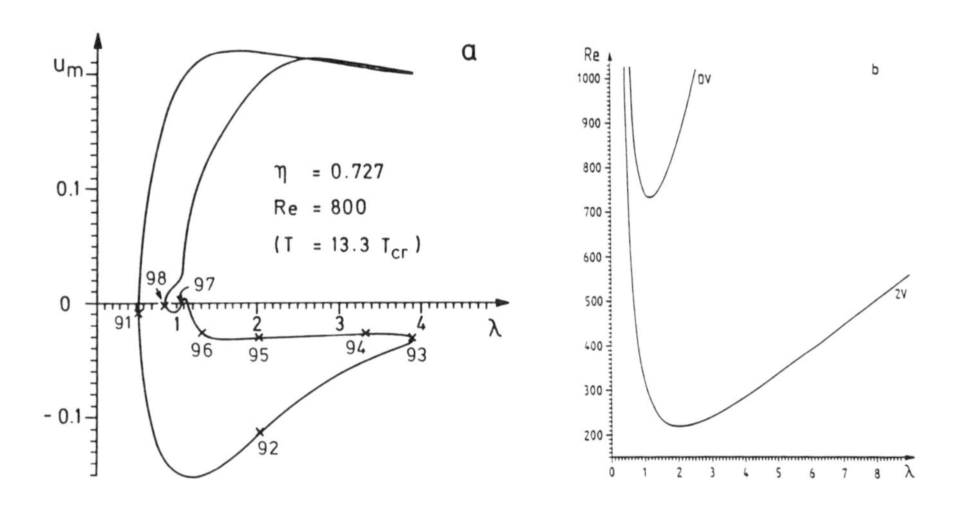

Fig.1: a) Basic 2-vortex branch. Variation with the period λ of the radial velocity at the midpoint, $u_m := u(1 + \frac{\delta}{2}, 0)$. Bifurcation of 2-vortex flow from Couette flow at $\lambda \approx 0.52$, fold point at $\lambda \approx 3.9$, bifurcation of double vortex flow to Couette flow at $\lambda \approx 0.85$. b) Curves of bifurcation from Couette flow. 2V: vortex flow. DV: double vortex flow.

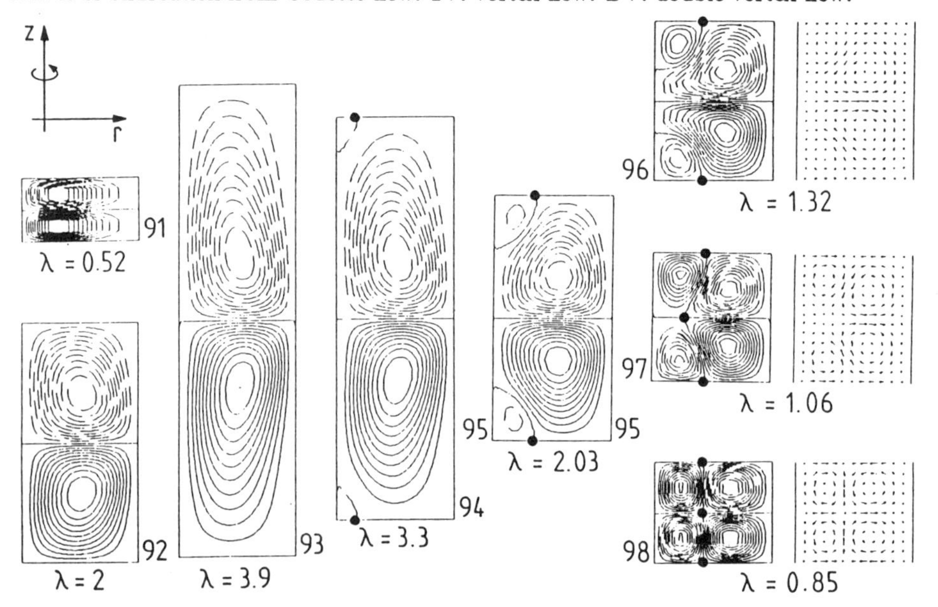

Fig.2: Streamline patterns of some of the flows on the basic branch, as marked in Fig.1. • free stagnation points.

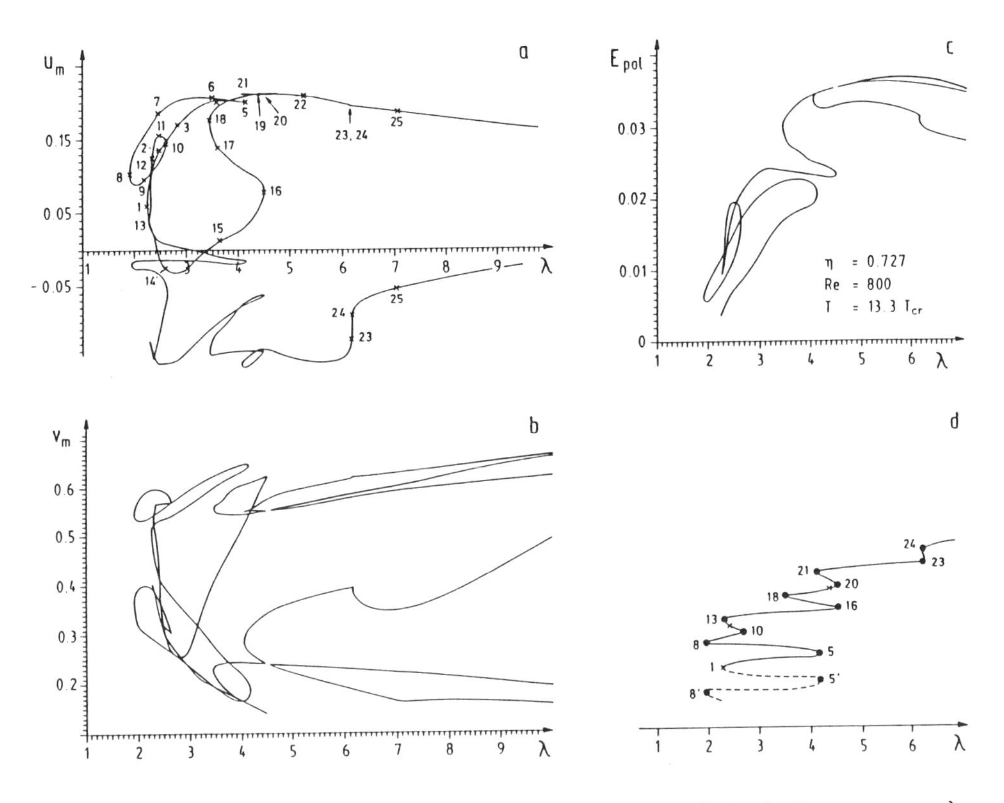

Fig.3: Second Re = 800-branch in different projections a) radial velocity u_m versus λ, b) angular velocity $v_m := v(1 + \frac{\delta}{2}, 0)$ versus λ, c) poloidal kinetic energy E_{pol} versus λ, d) sketch. This branch has 21 fold points and 3 bifurcation points. A third solution branch is encluded in parts b) and c), but not in parts a) and d) of this figure.

fold point near $\lambda = 2.68$ and a period-halving (or period-doubling) bifurcation near $\lambda = 2.59$. Varying the Taylor number T or Re^2 with fixed $\lambda = 2.65$, they also found a fold point and a period-halving bifurcation. Meanwhile, both fold points have been shown [4] to belong to the same fold. It is one of the various folds emanating from the double bifurcation point which is the intersection point of the curve of neutral stability of Couette flow with respect to n Taylor vortices and its periodic repetition. For reasons to be discussed elsewhere and following [4], we call this fold the basic $(n, 2n)$ fold. In [4] it was computed with a fold-following algorithm up to $Re^\delta \approx 290$ or $Re \approx 3.53 Re_{cr} (T \approx 12.3 T_{cr})$, and then numerical difficulties were reported. Using arc length continuation with respect

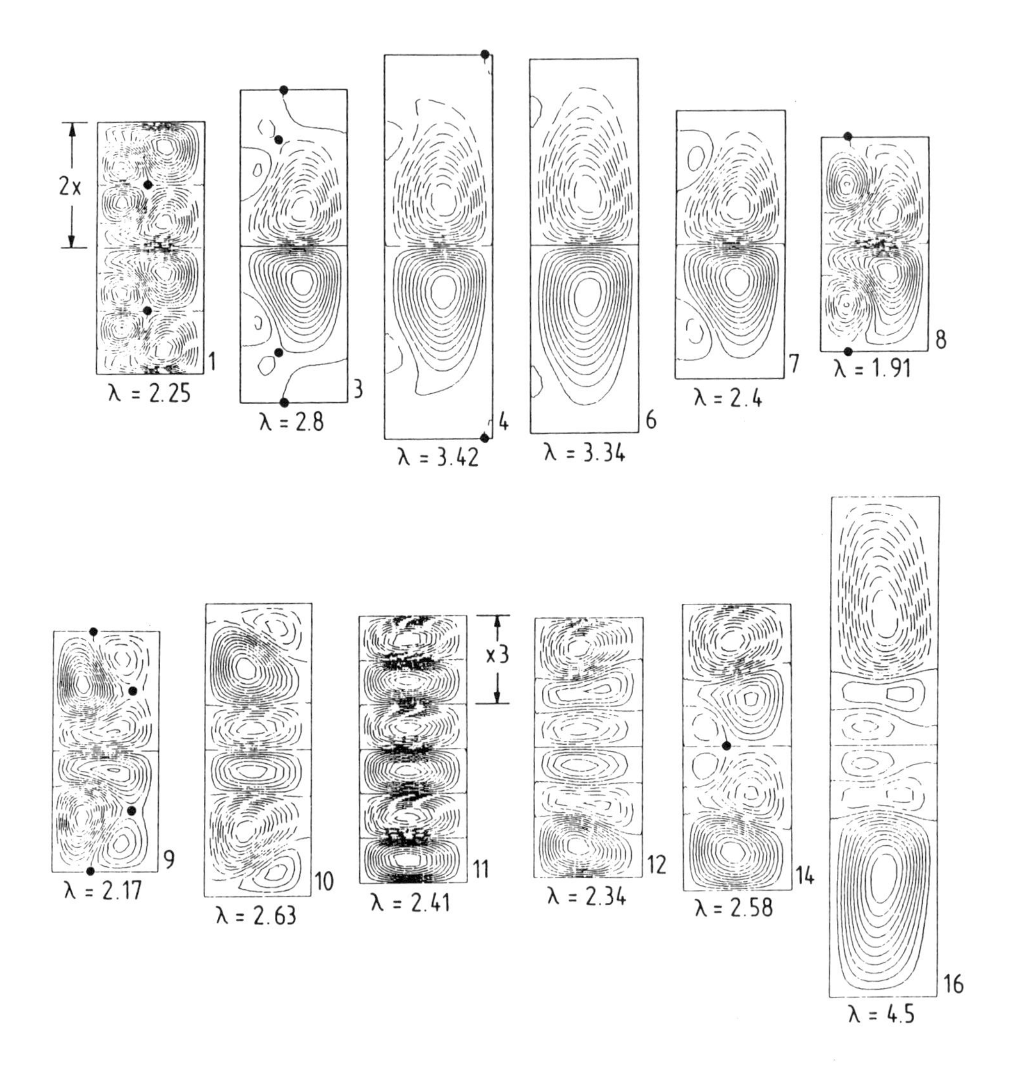

Fig.4: Change of streamline patterns with λ on the second $Re = 800$ branch. Numbers to the right coincide with numbers on Fig.3. Factors are given for the change of period in the 3 bifurcation points. • free stagnation points.

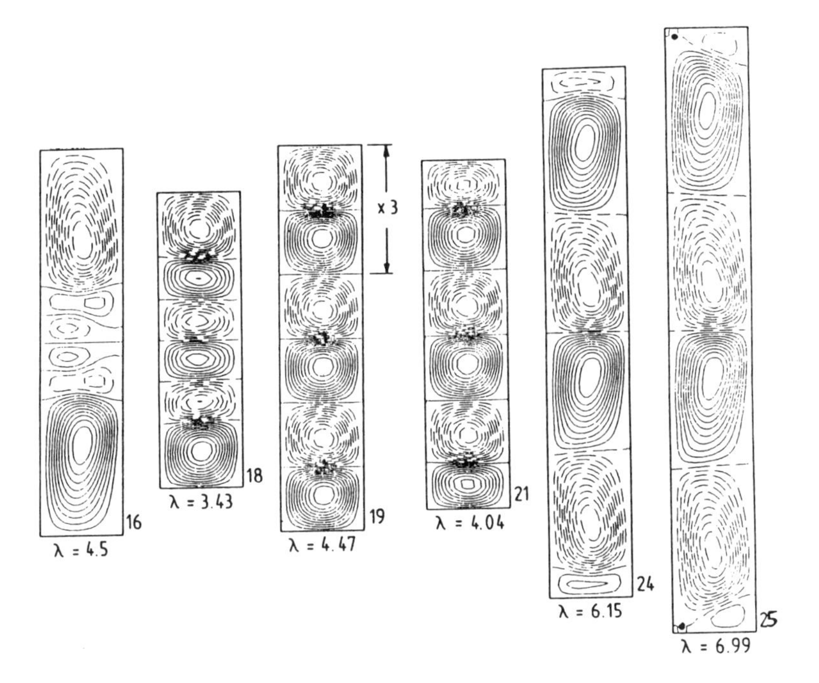

Fig.4, continued

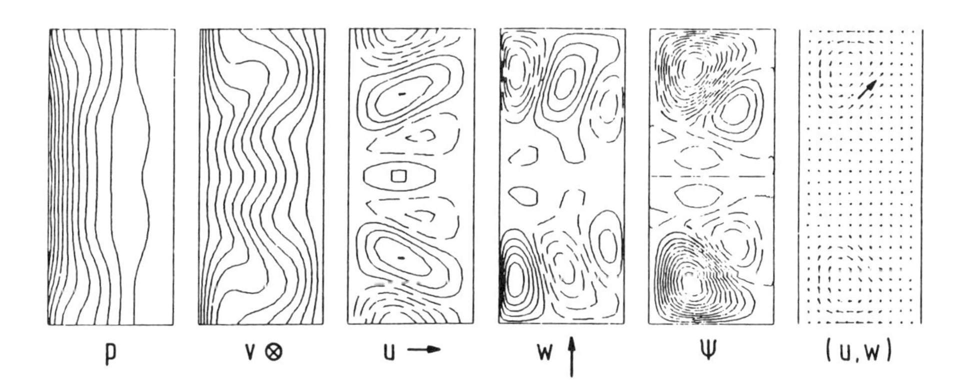

Fig.5: Level lines of pressure p, velocity components v, u, w and stream function ψ for $Re = 800$, $\eta = .727$, $\lambda = 2.27$. For this flow, the main outward movement is found to be in diagonal directions in (r, z).

to λ for a number of Re values, and arc length continuation with respect to Re for two fixed λ we found that this fold still exists at $T = 21T_{cr}$. It will be discussed elsewhere. Here we want to display Taylor vortex flows which we found when investigating the fold. Some of these flows have quite unusual structures, and we think that they might be of some interest. Fig.1a shows the branch which we get by continuation in λ, starting with the usual 2-vortex solution for $Re = 800$, $\lambda = 2$. Fig.2 shows the streamline patterns of some flows on this branch. Flows with 2 vortices in the radial direction have not been observed before. Theoretically, however, it is obvious that such flows also exist in the case where the outer cylinder is at rest [3]: They correspond to the second eigenvalue of the linear operator related to Couette Flow. We display the dependence of this eigenvalue on Re and λ in Fig.1b, together with the common neutral curve (first eigenvalue). In Fig.3 we display a second branch of solutions in different projections, and also a sketch of it. The picture given by Figs.1, 3 is not complete: no periodic repetitions of branches [11] are plotted, and there is at least one more original branch, separated from the main branch of Fig.3 by a perturbed bifurcation. For the values $\lambda = 2.5$, $Re = 800$, $\eta = 0.727$ Figs.1 and 3 tell us that there are at least 21 solutions, probably more.

In Fig.4 we display how the structure of the flow changes when the wave length or period λ is varied continuously. More details are given in the caption. The strangest flow is "# 13", which seems to contradict the common belief that the vortex movement is caused by centrifugal forces. We thus give more details about this flow in Fig.5.

4. Acknowledgement

Most of this work was done while the first author was visiting Freie Universität Berlin. Both authors would like to express their thanks to Prof. R. Gorenflo for valuable discussions and for providing financial support from FU-FPS "Modellierung und Diskretisierung". Special thanks are due to Prof. D. Lortz for very stimulating discussions. We also thank H. Specht, Dr. J. Bolstad, Prof. F. Busse, Dr. M. Henderson, and Prof. H.B. Keller.

5. References

[1] Burkhalter, J.E. and Koschmieder, E.L. (1974) Steady supercritical Taylor vortices after sudden starts. Phys. Fluids **17**, 1929 – 1935

[2] Busse, F.H. and Or, A.C. (1986) Subharmonic and asymmetric convection rolls. ZAMP **37**, 608–623; and Busse, F.H. Transition to asymmetric convection rolls. These proceedings

[3] Busse, F.H. (1978) private communication

[4] Dinar, N. and Keller, H.B., Computations of Taylor vortex flows using multigrid continuation methods. To be published

[5] Eilbeck, J.C. Numerical study of bifuraction in a reaction diffusion model using pseudospectral and path following methods. These proceedings

[6] Frank, G. and Meyer-Spasche, R. (1981) Computation of transitions in Taylor vortex flows. ZAMP **32**, 710 – 720

[7] Jones, C.A. (1981) Nonlinear Taylor vortices and their stability. JFM **102**, 249 – 261

[8] Keller, H.B. (1977) Numerical solution of bifurcation and nonlinear eigenvalue problems. In: Applications of Bifurcation Theory (ed. P. Rabinowitz), 359 – 384

[9] Ladyženskaya, O.A. (1969) The mathematical theory of viscous incompressible flow. (Russian) Akademie Verlag, Gordon and Breach

[10] Meyer-Spasche, R. and Keller, H.B. (1980) Computations of the axisymmetric flow between rotating cylinders. J. Comp. Phys. **35**, 100 – 109

[11] Meyer-Spasche, R. and Keller, H.B. (1985) Some bifurcation diagrams for Taylor vortex flows. Phys. Fluids **28**, 1248 – 1252

[12] Riecke, H. and Paap, H.-G. (1986) Stability and wave vector restriction of axisymmetric Taylor vortex flow. Phys. Rev. A **33**, 547

[13] Schreiber, I. and Kubiček, M. (1984) private communication

[14] Temam, R. (1983) Navier-Stokes equations and nonlinear functional analysis. SIAM Philadelphia

Dr. R. Meyer-Spasche
Max-Planck-Institut
für Plasmaphysik
Boltzmannstraße
8046 Garching

Matthias Wagner
Institut für Mathematik III
Freie Universität
Arminallee 2 – 6
1000 Berlin 33

International Series of
Numerical Mathematics, Vol. 79

SOME REMARKS ON THE DEFLATED BLOCK ELIMINATION METHOD

Gerald Moore

Department of Mathematics
Imperial College of Science & Technology
London, U.K.

1. Introduction

In many algorithms for computing singular points (or solution curves near such points) it is necessary to solve linear systems of the form

$$M \begin{pmatrix} \underset{\sim}{x} \\ y \end{pmatrix} \equiv \begin{pmatrix} A & \underset{\sim}{b} \\ \underset{\sim}{c}^T & d \end{pmatrix} \begin{pmatrix} \underset{\sim}{x} \\ y \end{pmatrix} = \begin{pmatrix} \underset{\sim}{f} \\ g \end{pmatrix} . \tag{1.1}$$

Here A is an N×N matrix; $\underset{\sim}{b}$, $\underset{\sim}{c}$, $\underset{\sim}{x}$ and $\underset{\sim}{f}$ N-vectors; and d, y and g scalars. The complete (N+1)×(N+1) coefficient matrix M is assumed to be well-conditioned, but A is expected to be singular with a one-dimensional null-space, or to have one relatively small singular value, for problems of interest. A, however, is also assumed to possess useful properties which may be lost if the augmented matrix M is dealt with directly; e.g. A might have a band structure which could be destroyed by pivoting on certain elements of M. Thus, for efficiency's sake, we only wish to solve linear systems with coefficient matrix A, but its possible ill-condition demands special care.

Various suggestions have been made to cope with this problem, but in our opinion the most attractive is the Deflated Block Elimination algorithm of CHAN (1984). This only requires a "black box" capable of solving

$$Az = p \tag{1.2}$$

and implicit deflation techniques are used to deal with an ill-conditioned matrix A. In the present paper we look at the algorithm in a somewhat different framework, which suggests new possibilities, and also include an error analysis which clarifies some of the difficulties.

2. The Deflated Block Elimination Algorithm

We introduce two pairs of unit vectors in R^N, $\{\vartheta,\gamma\}$ and $\{\eta,\xi\}$, with each pair being non-orthogonal. (The Euclidean norm is used throughout this paper.) This defines two projections

$$\text{(i)}\quad P_1 z \equiv (\gamma^T z)/(\gamma^T\vartheta)\vartheta \qquad \text{(ii)}\quad P_2 z \equiv (\xi^T z)/(\xi^T \eta)\eta, \tag{2.1}$$

and two decompositions of R^N,

$$\text{(i)}\quad z = P_1 z + (I-P_1)z \qquad \text{(ii)}\quad z = P_2 z + (I-P_2)z \tag{2.2}$$

or

$$\text{(i)}\quad R^N = \{\vartheta\} \oplus \{\gamma\}^\perp \qquad \text{(ii)}\quad R^N = \{\eta\} \oplus \{\xi\}^\perp. \tag{2.3}$$

To relate these decompositions to A we assume, in addition, that the pairs $\{\eta,\vartheta\}$ and $\{\gamma,\xi\}$ are linked by the formulae

$$\text{(i)}\quad \vartheta = A^{-1}\eta/\|A^{-1}\eta\| \qquad \text{(ii)}\quad \xi = A^{-T}\gamma/\|A^{-T}\gamma\|, \tag{2.4}$$

and so A must map $R(P_1)$ to $R(P_2)$ and $N(P_1)$ to $N(P_2)$, i.e.

$$\text{(i)}\quad A:\{\vartheta\} \longrightarrow \{\eta\} \qquad \text{(ii)}\quad A: \{\gamma\}^\perp \longrightarrow \{\xi\}^\perp. \tag{2.5}$$

Hence, using the notation $Q_i \equiv I-P_i$ $i=1,2$, the solution of (1.2) can be written

$$z = \alpha^{-1}(\xi^T p)/(\xi^T\eta)\vartheta + A^\dagger Q_2 p \quad , \tag{2.6}$$

where $\alpha \equiv (\|A^{-1}\eta\|)^{-1}$ and $A^\dagger$ is the inverse of A as the mapping

(2.5ii). Thus in order to solve (1.1) we can use the decomposition (2.2i) for $\underset{\sim}{x}$ and (2.2ii) for $\underset{\sim}{f}$ and $\underset{\sim}{b}$ to obtain

$$\underset{\sim}{x} = \mu\underset{\sim}{\vartheta} + \underset{\sim}{w}_f - y\,\underset{\sim}{w}_b \ , \tag{2.7}$$

where $\underset{\sim}{w}_f = A^{\dagger}Q_2\underset{\sim}{f}$ and $\underset{\sim}{w}_b = A^{\dagger}Q_2\underset{\sim}{b}$. This leaves the scalars μ and y to be determined by the 2×2 system formed by the equation for $P_1\underset{\sim}{x}$ and the last row of M, i.e.

$$D\begin{pmatrix}\mu\\ y\end{pmatrix} \equiv \begin{pmatrix}\alpha & (\underset{\sim}{\xi}^T\underset{\sim}{b})/(\underset{\sim}{\xi}^T\underset{\sim}{\eta})\\ \underset{\sim}{c}^T\underset{\sim}{\vartheta} & d-\underset{\sim}{c}^T\underset{\sim}{w}_b\end{pmatrix}\begin{pmatrix}\mu\\ y\end{pmatrix} = \begin{pmatrix}(\underset{\sim}{\xi}^T\underset{\sim}{f})/(\underset{\sim}{\xi}^T\underset{\sim}{\eta})\\ g-\underset{\sim}{c}^T\underset{\sim}{w}_f\end{pmatrix}. \tag{2.8}$$

The non-singularity of M implies that D is non-singular, and in fact it is easy to show that

$$\|D^{-1}\| \leqslant \|M^{-1}\|/\{1-(1-(\underset{\sim}{\gamma}^T\underset{\sim}{\vartheta})^2)^{\frac{1}{2}}\}. \tag{2.9}$$

Now we wish to consider the stability of the above algorithm for obtaining $(\underset{\sim}{x},y)$ when A becomes increasingly ill-conditioned. We shall use the standard notation of σ_j $j=1,\ldots,N$ for the singular values of A and $\underset{\sim}{u}_j,\underset{\sim}{v}_j$ for the corresponding left and right singular vectors. The only difficulty is that $A^{\dagger}$ may become unbounded and this will be avoided if $\underset{\sim}{\gamma}$ is not deficient in $\underset{\sim}{v}_N$ since then

$$\begin{aligned}(\|A^{\dagger}\|)^{-1} &= \min\{\|A\underset{\sim}{z}\| : \|\underset{\sim}{z}\| = 1,\ \underset{\sim}{\gamma}^T\underset{\sim}{z} = 0\}\\ &\geqslant \sigma_{N-1}\min\{(\sum_{j=1}^{N-1}(\underset{\sim}{v}_j^T\underset{\sim}{z})^2)^{\frac{1}{2}} : \|\underset{\sim}{z}\|=1,\ \underset{\sim}{\gamma}^T\underset{\sim}{z}=0\}\\ &\geqslant \sigma_{N-1}|\underset{\sim}{v}_N{}^T\underset{\sim}{\gamma}|.\end{aligned} \tag{2.10}$$

Hence we shall always have

$$\|A^{\dagger}\| \leqslant \min\{\sigma_N{}^{-1},\sigma_{N-1}^{-1}\ |\underset{\sim}{v}_N{}^T\underset{\sim}{\gamma}|^{-1}\} \tag{2.11}$$

and $\|\underset{\sim}{\xi}-\underset{\sim}{u}_N\| \to \underset{\sim}{0}$ as A becomes singular. Of course there are many vectors $\underset{\sim}{\gamma}$ and $\underset{\sim}{\eta}$ which satisfy our conditions and the choice must be made on practical grounds. Some possibilities are listed below.

(i) $\underset{\sim}{\gamma}=\underset{\sim}{v}_N$ and $\underset{\sim}{\eta}=\underset{\sim}{u}_N$, so that $\underset{\sim}{\xi}=\underset{\sim}{u}_N$ and $\underset{\sim}{\vartheta}=\underset{\sim}{v}_N$. Thus the projections P_i are orthogonal and $\alpha=\sigma_N$. Usually $\underset{\sim}{u}_N$ and $\underset{\sim}{v}_N$ would be computed by a variant of inverse iteration, cf. STEWART (1981).

(ii) $\underset{\sim}{\gamma}=\underset{\sim}{c}/\|\underset{\sim}{c}\|$ and $\underset{\sim}{\eta}=\underset{\sim}{\xi}$, so that P_2 is orthogonal. As $\sigma_N/\sigma_1 \to 0$ we cannot have $\underset{\sim}{c}$ deficient in $\underset{\sim}{v}_N$, else M would be ill-conditioned, cf. KELLER (1983). Thus

$$|\underset{\sim}{\gamma}^T\underset{\sim}{\vartheta}| \geq \max\{\sigma_N{}^2/\sigma_1{}^2, (\underset{\sim}{v}_N{}^T\underset{\sim}{c})^2/(\|\underset{\sim}{c}\|)^2\} \tag{2.13}$$

means that P_1 is well-defined.

(iii) $\underset{\sim}{\eta}=\underset{\sim}{b}/\|\underset{\sim}{b}\|$ and $\underset{\sim}{\gamma}=\underset{\sim}{\vartheta}$, so that P_1 is orthogonal. Again the assumption that M is well-conditioned means that $\underset{\sim}{b}$ cannot be deficient in $\underset{\sim}{u}_N$ as $\sigma_N/\sigma_1 \to 0$. Hence

$$|\underset{\sim}{\gamma}^T\underset{\sim}{\eta}| \geq \max\{\sigma_N{}^2/\sigma_1{}^2, (\underset{\sim}{u}_n{}^T\underset{\sim}{b})^2/(\|\underset{\sim}{b}\|)^2\} \tag{2.14}$$

implies that P_2 is well-defined and also $A^\dagger$ remains bounded because

$$|\underset{\sim}{v}_N{}^T\underset{\sim}{\gamma}| \geq |\underset{\sim}{u}_N{}^T\underset{\sim}{b}|/\|\underset{\sim}{b}\| . \tag{2.15}$$

The particular advantage of this choice is that $Q_2\underset{\sim}{b}=\underset{\sim}{0}$ and so $\underset{\sim}{w}_b=\underset{\sim}{0}$. Thus some gain in efficiency is achieved and also a possible loss of accuracy when forming (2.7), as noted in KELLER (1983), is avoided.

The practical problem that remains is, how to compute $\underset{\sim}{w}_f$ and $\underset{\sim}{w}_b$ in (2.7) if only systems of the form (1.2) can be solved. In CHAN (1984) it was proposed that the algorithm

(i) $A\underset{\sim}{z}=Q_2\underset{\sim}{f}$

(ii) $\underset{\sim}{w}_f = Q_1 \underset{\sim}{z}$ (2.16)

should be used, together with a similar calculation for $\underset{\sim}{w}_b$. The error analysis therein, however, could only be carried out in the absence of (2.16ii), although it was stated that the numerical results were satisfactory in either case. (Note that with exact arithmetic (2.16ii) should not be necessary). In the next section we present an error analysis which indicates the difference between including or omitting this final projection.

3. Error Analysis of the Algorithm

To avoid distracting complications we shall follow CHAN (1984) and assume that all calculations are performed exactly except for the key equations of the form (1.2) (with coefficient matrix A or A^T) which produce $\underset{\sim}{\vartheta}$, $\underset{\sim}{\xi}$, $\underset{\sim}{w}_f$ and $\underset{\sim}{w}_b$. In these cases we assume that an approximation $\hat{\underset{\sim}{z}}$ is obtained which satisfies

$$(A+E)\hat{\underset{\sim}{z}} = \underset{\sim}{p} \tag{3.1}$$

where $\|E\| \leqslant \varepsilon \|A\|$ and $\varepsilon \leqslant \text{const}(N)\varepsilon_M$. (Here ε_M is the machine precision and const(N) is an acceptable constant which usually depends on the pivoting strategy used.) Thus our computed quantities will be:-

(i) $(A+E)\hat{\underset{\sim}{z}}_1 = \underset{\sim}{\eta}$ $\quad \hat{\underset{\sim}{\vartheta}} = \hat{\underset{\sim}{z}}_1/\|\hat{\underset{\sim}{z}}_1\|$ $\quad \hat{\alpha} = (\|\hat{\underset{\sim}{z}}_1\|)^{-1}$

(ii) $(A^T+E)\hat{\underset{\sim}{z}}_1 = \underset{\sim}{\gamma}$ $\quad \hat{\underset{\sim}{\xi}} = \hat{\underset{\sim}{z}}_2/\|\hat{z}_2\|$

(iii) $(A+E)\hat{\underset{\sim}{w}}_f^* = Q_2\underset{\sim}{f}$ $\quad \hat{\underset{\sim}{w}}_f = Q_1\hat{\underset{\sim}{w}}_f^*$

(iv) $(A+E)\hat{\underset{\sim}{w}}_b^* = Q_2\underset{\sim}{b}$ $\quad \hat{\underset{\sim}{w}}_b = Q_1\hat{\underset{\sim}{w}}_b^*$

(iv) $\hat{\underset{\sim}{x}} = \hat{\mu}\hat{\underset{\sim}{\vartheta}} + \hat{\underset{\sim}{w}}_f - \hat{y}\hat{\underset{\sim}{w}}_b$ where

$$\hat{D}\begin{pmatrix}\hat{\mu}\\ \hat{y}\end{pmatrix} \equiv \begin{pmatrix}\hat{\alpha} & (\hat{\underset{\sim}{\xi}}^T\underset{\sim}{b})/(\hat{\underset{\sim}{\xi}}^T\underset{\sim}{\eta}) \\ \underset{\sim}{c}^T\hat{\underset{\sim}{\vartheta}} & d-\underset{\sim}{c}^T\hat{\underset{\sim}{w}}_b\end{pmatrix}\begin{pmatrix}\hat{\mu}\\ \hat{y}\end{pmatrix} \quad \begin{pmatrix}(\hat{\underset{\sim}{\xi}}^T\underset{\sim}{f})/(\hat{\underset{\sim}{\xi}}^T\underset{\sim}{\eta}) \\ g-\underset{\sim}{c}^T\hat{\underset{\sim}{w}}_f\end{pmatrix}$$

<u>or</u> $\hat{\underset{\sim}{x}}^* = \hat{\mu}^* \hat{\underset{\sim}{\vartheta}} + \hat{\underset{\sim}{w}}_f^* - \hat{y}^* \hat{\underset{\sim}{w}}_b^*$ where

$$\hat{D}^* \begin{pmatrix} \hat{\mu}^* \\ \hat{y}^* \end{pmatrix} \equiv \begin{pmatrix} \hat{\alpha} & (\hat{\underset{\sim}{\xi}}^T \underset{\sim}{b})/(\hat{\underset{\sim}{\xi}}^T \underset{\sim}{\eta}) \\ \underset{\sim}{c}^T \hat{\underset{\sim}{\vartheta}} & d - \underset{\sim}{c}^T \hat{\underset{\sim}{w}}_b^* \end{pmatrix} \begin{pmatrix} \hat{\mu}^* \\ \hat{y}^* \end{pmatrix} \begin{pmatrix} (\hat{\underset{\sim}{\xi}}^T \underset{\sim}{f})/(\hat{\underset{\sim}{\xi}}^T \underset{\sim}{\eta}) \\ g - \underset{\sim}{c}^T \hat{\underset{\sim}{w}}_f^* \end{pmatrix}$$

and the starred version means that the projection (2.16ii) is omitted.

First let us consider the error in $\hat{\underset{\sim}{\vartheta}}$, $\hat{\underset{\sim}{\xi}}$ and $\hat{\alpha}$. Standard error analysis gives

$$\begin{aligned} &\text{(i)} && \|\underset{\sim}{\vartheta} - \hat{\underset{\sim}{\vartheta}}\| \leqslant \text{const}(\sigma_1/\sigma_N)\varepsilon \\ &\text{(ii)} && \|\underset{\sim}{\xi} - \hat{\underset{\sim}{\xi}}\| \leqslant \text{const}(\sigma_1/\sigma_N)\varepsilon \\ &\text{(iii)} && |\alpha - \hat{\alpha}| \leqslant \text{const}(\sigma_1/\sigma_N)\sigma_1\varepsilon \end{aligned} \tag{3.2}$$

and so we have to proceed more carefully in order to obtain realistic bounds as $\sigma_1/\sigma_N \to 0$. In this case we shall insist that $\sigma_{N-1} - \sigma_N \geqslant 2(1+\sqrt{2})\sigma_1\varepsilon$, which implies that the smallest singular value remains apart from the other singular values when A is perterbed to A+E. We shall also require the following perturbation results (cf. GOLUB & VAN LOAN (1983))

$$\begin{aligned} &\text{(i)} && |\sigma_N - \hat{\sigma}_N| \leqslant \sigma_1\varepsilon \\ &\text{(ii)} && \max\{\|\underset{\sim}{u}_N - \underset{\sim}{u}_N\|, \|\underset{\sim}{v}_N - \underset{\sim}{v}_N\|\} \leqslant 2\sqrt{2}\,\sigma_1\varepsilon/(\sigma_{N-1} - \sigma_N - 2\sigma_1\varepsilon) \\ &\text{(iii)} && \|A^+ - (A+E)^+\| \leqslant (1+\sqrt{5})\sqrt{2}\sigma_1\varepsilon/\{(\sigma_{N-1} - \sigma_N - 2\sigma_1\varepsilon) \\ &&& \qquad (\sigma_{N-1} - \sigma_1\varepsilon)\}\ , \end{aligned} \tag{3.3}$$

where A^+ and $(A+E)^{|}$ are the pseudo-inverses with largest singular values (σ_N^{-1} and $\hat{\sigma}_N^{-1}$ respectively) set to zero. Now our conditions on the choice of $\underset{\sim}{\gamma}$ mean that $\underset{\sim}{\eta}$ cannot be deficient in $\underset{\sim}{u}_N$ and therefore

$$\underset{\sim}{\vartheta}=\{A^{+}\underset{\sim}{\eta}+\sigma_N{}^{-1}(\underset{\sim}{u}_N{}^T\underset{\sim}{\eta})\underset{\sim}{v}_N\}\{\|A^{+}\underset{\sim}{\eta}\|^2+\sigma_N{}^{-2}(\underset{\sim}{u}_N{}^T\underset{\sim}{\eta})^2\}^{-\frac{1}{2}}$$

$$= \{\underset{\sim}{v}_N+\sigma_N(\underset{\sim}{u}_N{}^T\underset{\sim}{\eta}\}^{-1}A^{+}\underset{\sim}{\eta}\}\{1+\sigma_N{}^2\|A^{+}\underset{\sim}{\eta}\|^2(\underset{\sim}{u}_N{}^T\underset{\sim}{\eta})^{-2}\}^{-\frac{1}{2}}. \qquad (3.4)$$

Similarly, if $|\underset{\sim}{u}_N{}^T\underset{\sim}{\eta}| \geqslant 2\ \|\underset{\sim}{u}_N-\hat{\underset{\sim}{u}}_N\|$ so that $\underset{\sim}{\eta}$ is not deficient in $\hat{\underset{\sim}{u}}_N$, we have

$$\hat{\underset{\sim}{\vartheta}}=\{\hat{\underset{\sim}{v}}_N+\hat{\sigma}_N(\hat{\underset{\sim}{u}}_N{}^T\underset{\sim}{\eta})^{-1}(A+E)^{+}\underset{\sim}{\eta}\}\{1+\hat{\sigma}_N{}^2\|(A+E)^{+}\underset{\sim}{\eta}\|^2(\hat{\underset{\sim}{u}}_N{}^T\underset{\sim}{\eta})^{-2}\}^{-\frac{1}{2}} \qquad (3.5)$$

and hence, using (3.3) repeatedly,

$$\|\underset{\sim}{\vartheta}-\hat{\underset{\sim}{\vartheta}}\| \leqslant \text{const}(\sigma_1/\sigma_{N-1},\sigma_1/(\sigma_{N-1}-\sigma_N),|\underset{\sim}{u}_N{}^T\underset{\sim}{\eta}|^{-1})\varepsilon. \qquad (3.6)$$

Thus, not surprisingly, as A becomes ill-conditioned the (relative) error in $\hat{\underset{\sim}{\vartheta}}$ depends on the condition of the restriction of A to $\{\underset{\sim}{v}_N\}^{\perp}$, the separation of σ_N from the other singular values of A, and the non-deficiency of the $\underset{\sim}{u}_N$-component of $\underset{\sim}{\eta}$. The error in $\hat{\underset{\sim}{\xi}}$ satisfies a similar bound with $\underset{\sim}{v}_N{}^T\underset{\sim}{\gamma}$ replacing $\underset{\sim}{u}_N{}^T\underset{\sim}{\eta}$ while

$$|\alpha-\hat{\alpha}| \leqslant \text{const}(\sigma_{N-1}/(\sigma_{N-1}-\sigma_N),\ |\underset{\sim}{u}_N{}^T\underset{\sim}{\eta}|^{-1})\sigma_1\varepsilon\ . \qquad (3.7)$$

(Note that the relative error in $\hat{\alpha}$ is $O(\sigma_1\varepsilon/\sigma_N)$ and hence is unbounded as A becomes ill-conditioned). Finally the errors in $\hat{\underset{\sim}{\vartheta}}$ and $\hat{\underset{\sim}{\xi}}$ immediately imply

$$\text{(i)}\qquad \|P_1-\hat{P}_1\| \leqslant \text{const}(|\underset{\sim}{\gamma}^T\underset{\sim}{\vartheta}|^{-1})\|\underset{\sim}{\vartheta}-\hat{\underset{\sim}{\vartheta}}\|$$

$$\text{(ii)}\qquad \|P_2-\hat{P}_2\| \leqslant \text{const}(|\underset{\sim}{\xi}^T\underset{\sim}{\eta}|^{-1})\|\underset{\sim}{\xi}-\hat{\underset{\sim}{\xi}}\|\ , \qquad (3.8)$$

provided $|\underset{\sim}{\gamma}^T\underset{\sim}{\vartheta}| \geqslant 2\ \|\underset{\sim}{\vartheta}-\hat{\underset{\sim}{\vartheta}}\|$ and $|\underset{\sim}{\xi}^T\underset{\sim}{\eta}| \geqslant 2\ \|\underset{\sim}{\xi}-\hat{\underset{\sim}{\xi}}\|$ so that $\hat{P}_1$ and $\hat{P}_2$ are well-defined.

Now we shall consider the error in $\hat{\underset{\sim}{w}}_f$ and $\hat{\underset{\sim}{w}}_b$. It is easy to show that

$$\text{(i)}\qquad \|\underset{\sim}{w}_f-\hat{\underset{\sim}{w}}{}_f^{*}\|\leqslant\text{const}(\sigma_1/\sigma_N)\varepsilon\|A^{-1}Q_2\underset{\sim}{f}\|+\text{const}\|P_2-\hat{P}_2\|\|A^{-1}\|\|\underset{\sim}{f}\|$$

$$\text{(ii)} \qquad \|\underset{\sim}{w}_f-\hat{\underset{\sim}{w}}_f\| \leqslant \text{const}(\sigma_1/\sigma_N, |\underset{\sim}{\gamma}^T\underset{\sim}{\vartheta}|^{-1})\varepsilon\|A^{-1}Q_2\underset{\sim}{f}\| + \text{const}(|\underset{\sim}{\gamma}^T\underset{\sim}{\vartheta}|^{-1})\|P_2-\hat{P}_2\|\|A^{-1}\|\|\underset{\sim}{f}\| + \text{const}(|\underset{\sim}{\xi}^T\underset{\sim}{\eta}|^{-1})\|P_1-\hat{P}_1\|\|A^{-1}\|\|\underset{\sim}{f}\| \tag{3.9}$$

and a similar result for $\hat{\underset{\sim}{w}}_b^*$ and $\hat{\underset{\sim}{w}}_b$. The first term on the r.h.s in each case is the standard relative error, but this is spoiled by the additional terms since $\|Q_2\underset{\sim}{f}\|$ may be much smaller than $\|\underset{\sim}{f}\|$. To obtain a more useful bound when A is ill-conditioned we decompose $(A+E)\hat{\underset{\sim}{w}}_f^* = \hat{Q}_2\underset{\sim}{f}$ according to (2.3),

$$\begin{aligned} &\text{(i)} \qquad (A+Q_2E)Q_1\hat{\underset{\sim}{w}}_f^* = Q_2\hat{Q}_2\underset{\sim}{f}-\hat{t}_fQ_2E\underset{\sim}{\vartheta} \\ &\text{(ii)} \qquad (\alpha\underset{\sim}{\xi}^T\underset{\sim}{\eta}+\underset{\sim}{\xi}^TE\underset{\sim}{\vartheta})\hat{t}_f = \underset{\sim}{\xi}^T\hat{Q}_2\underset{\sim}{f}-\underset{\sim}{\xi}^TEQ_1\hat{\underset{\sim}{w}}_f \ , \end{aligned} \tag{3.10}$$

where $\hat{t}_f$ denotes the $\underset{\sim}{\vartheta}$-component of $\hat{\underset{\sim}{w}}_f^*$. If A is sufficiently well-conditioned in (2.5ii), i.e. $\|Q_2E\|\|A^\dagger\| \leqslant 1/2$, then $(A+Q_2E)^\dagger$ exists and we obtain

$$\begin{aligned} \underset{\sim}{w}_f-Q_1\hat{\underset{\sim}{w}}_f^* &= A^\dagger Q_2\underset{\sim}{f}-(A+Q_2E)^\dagger Q_2\hat{Q}_2\underset{\sim}{f}+\hat{t}_f(A+Q_2E)^\dagger Q_2E\underset{\sim}{\vartheta} \qquad (3.11) \\ &= (A^\dagger-(A+Q_2E)^\dagger)Q_2\underset{\sim}{f}-(A+Q_2E)^\dagger Q_2(P_2-\hat{P}_2)\underset{\sim}{f} \\ &\quad + t_f(A+Q_2E)^\dagger Q_2E\underset{\sim}{\vartheta} \ , \end{aligned}$$

and consequently

$$\begin{aligned} \|\underset{\sim}{w}_f-Q_1\hat{\underset{\sim}{w}}_f^*\| \leqslant\ & \text{const}(\|A\|\|A^\dagger\|, |\underset{\sim}{\xi}^T\underset{\sim}{\eta}|^{-1})\varepsilon\|A^\dagger Q_2\underset{\sim}{f}\| \\ &+ \text{const}(|\underset{\sim}{\xi}^T\underset{\sim}{\eta}|^{-1})\|P_2-\hat{P}_2\|\|A^\dagger\|\|\underset{\sim}{f}\| \qquad (3.12) \\ &+ \text{const}(\|A\|\|A^\dagger\|, |\underset{\sim}{\xi}^T\underset{\sim}{\eta}|^{-1})\varepsilon|\hat{t}_f| \ . \end{aligned}$$

The first two terms on the r.h.s. of (3.12) have analogies in (3.9) but the third term, involving $\hat{t}_f$, must be bounded by inserting $Q_1\hat{\underset{\sim}{w}}_f^*$ into (3.10ii) to obtain

$$\hat{t}_f=\{\underset{\sim}{\xi}^T\hat{Q}_2\underset{\sim}{f}-\underset{\sim}{\xi}^TE(A+Q_2E)^\dagger Q_2\hat{Q}_2\underset{\sim}{f}\}/\{\alpha\underset{\sim}{\xi}^T\underset{\sim}{\eta}+\underset{\sim}{\xi}^TE\underset{\sim}{\vartheta}-\underset{\sim}{\xi}^T(A+Q_2E)^\dagger Q_2E\underset{\sim}{\vartheta}\}. \tag{3.13}$$

Now as A becomes increasingly ill-conditioned it is possible for this quantity to be unbounded, but we would expect

$$|\hat{t}_f| \leqslant \text{const}(\|A\|\|A^\dagger\|, |\xi^T\eta|^{-1})\|A^\dagger\|\|f\| \qquad (3.14)$$

unless a fortuitous correlation in errors occurred. Thus $P_1\hat{w}_f^*$, which should be zero,may be O(1) although $Q_1\hat{w}_f^*$ will be within $O(\varepsilon)$ of w_f. It is here that the significance of the projection (2.16ii) is felt, since

$$w_f-\hat{w}_f = w_f-Q_1\hat{w}_f^*-\hat{t}_f\hat{Q}_1\vartheta \qquad (3.15)$$

and $\hat{Q}_1\vartheta=(P_1-\hat{P}_1)\vartheta$. Hence for an accurate approximation to w_f it is necessary, as $\sigma_N/\sigma_1 \to 0$, to use $\hat{w}_f$ rather than $\hat{w}_f^*$. If, however, we are only interested in accurately approximating x and y, we shall see later that working with $\hat{w}_f^*$ need not be disastrous. Finally, all the results of this paragraph apply equally well to the computation of w_b.

In order to bound $\|x-\hat{x}\|$ and $|y-\hat{y}|$, it only remains to consider the error in $\hat{\mu}$ and $\hat{y}$ obtained from the approximation to equation (2.8). Since $\|M\| \geqslant \max\{\|A\|,\|b\|,\|c\|\}$, we have

$$\|D-\hat{D}\| \leqslant 2\max\{|\alpha-\hat{\alpha}|, |c^T(\vartheta-\hat{\vartheta})|, |c^T(w_b-\hat{w}_b)|, |(\xi^Tb)/(\xi^T\eta)- (\hat{\xi}^Tb)/(\hat{\xi}^T\eta)|\} \leqslant \text{const}\|M\|\varepsilon , \qquad (3.16)$$

where the values upon which the constant depends, in particular $(\|M\|\|A^\dagger\|)$, have been suppressed. Hence if M is well enough conditioned for $\|D^{-1}\|\|D-\hat{D}\| \leqslant 1/2$, then $\hat{D}$ will be invertible with $\|\hat{D}^{-1}\| \leqslant 2\|D^{-1}\|$. Thus

$$\begin{pmatrix} \mu-\hat{\mu} \\ y-\hat{y} \end{pmatrix} = D^{-1}\begin{pmatrix} (\xi^Tf)/(\xi^T\eta) \\ g-c^Tw_f \end{pmatrix} - \hat{D}^{-1}\begin{pmatrix} (\hat{\xi}^Tf)/(\hat{\xi}^T\eta) \\ g-c^T\hat{w}_f \end{pmatrix}$$

$$=(D^{-1}-\hat{D}^{-1})\begin{pmatrix} (\xi^Tf)/(\xi^T\eta) \\ g-c^Tw_f \end{pmatrix}+\hat{D}^{-1}\begin{pmatrix} (\xi^Tf)/(\xi^T\eta)-(\hat{\xi}^Tf)/(\hat{\xi}^T\eta) \\ c^T(\hat{w}_f-w_f) \end{pmatrix} \qquad (3.17)$$

leads to

$$\|(\mu-\hat{\mu},y-\hat{y})\| \leqslant \text{const } \|D^{-1}\|\varepsilon\{\|M\|\|(\mu,y)\| + \|f\|\}, \qquad (3.18)$$

where we have again suppressed the values upon which the constant depends. As we have seen before, the r.h.s. of (3.18) consists of a standard relative error bound plus an additional term which, through $\|f\| \leqslant \|M\|\|(\underset{\sim}{x},y)\|$, bounds the error with respect to the complete solution $(\underset{\sim}{x},y)^T$. Consequently, writing

$$\|\underset{\sim}{x}-\hat{\underset{\sim}{x}}\|\leqslant|\mu|\,\|\underset{\sim}{\vartheta}-\hat{\underset{\sim}{\vartheta}}\|+|\mu-\hat{\mu}|+\|\underset{\sim}{w}_f-\hat{\underset{\sim}{w}}_f\|+|y-\hat{y}|\,\|\hat{\underset{\sim}{w}}_b\|+\|\underset{\sim}{w}_b-\hat{\underset{\sim}{w}}_b\|\,|y| \qquad (3.19)$$

and using all the results of this section, we arrive at

$$\|(\underset{\sim}{x}-\hat{\underset{\sim}{x}},y-\hat{y}\,\| \leqslant \text{const } \varepsilon\ \|(\underset{\sim}{x},y)\|\ , \qquad (3.20)$$

where the most important value upon which the constant depends is $\|M\|(\|D^{-1}\|+\|A^{\dagger}\|)$.

To conclude this section we indicate why $\hat{\underset{\sim}{x}}^*$ and $\hat{y}^*$ will also generally be acceptable, even though the errors in $\hat{\underset{\sim}{w}}_f^*$ and $\hat{\underset{\sim}{w}}_b^*$ may be large when A is ill-conditioned. Denoting the $\underset{\sim}{\vartheta}$-components of $\hat{\underset{\sim}{w}}_f^*$ and $\hat{\underset{\sim}{w}}_b^*$, which may be O(1), by $\hat{\tau}_f$ and $\hat{\tau}_b$ respectively. we have

$$\hat{D}^* \begin{pmatrix} \hat{\mu}^*-\hat{\mu}+\hat{\tau}_f-\hat{y}\hat{\tau}_b \\ \hat{y}^*-y \end{pmatrix} = \begin{pmatrix} \hat{\alpha}(\hat{\tau}_f-\hat{y}\hat{\tau}_b) \\ 0 \end{pmatrix} \qquad (3.21)$$

and the r.h.s. will be $O(\varepsilon)$ when $\sigma_N/\sigma_1 = O(\varepsilon)$, unless a catastrophic correlation of errors occurs. Additionally, although $\|D-\hat{D}^*\|$ need not be small in this case, we will have

$$|\det(D)-\det(\dot{D}^*)| = O(\varepsilon)\ . \qquad (3.22)$$

Thus

$$\|(\hat{\mu}^*-\hat{\mu}+\hat{\tau}_f-\hat{y}\hat{\tau}_b,\hat{y}^*-\hat{y})\| = O(\varepsilon) \qquad (3.23)$$

and

$$\|(\hat{x}^* - \hat{x}, \hat{y}^* - \hat{y})\| = O(\varepsilon) \quad , \tag{3.24}$$

since the large errors in $\hat{w}_f^*$ and $\hat{w}_b^*$ will cancel with those in $\hat{\mu}^*$. Therefore, in general, it is satisfactory to approximate $(x,y)^T$ by $(\hat{x}^*, \hat{y}^*)$. We would still recommend, however, that $(\hat{x}, \hat{\tilde{y}})$ be used, since the computation of $\hat{\tau}_f$ and $\hat{\tau}_b$ is trivial and their size will indicate if any instability has occurred. If it has, then it should only be necessary to repeat the computation with the data perturbed by $O(\varepsilon_M)$.

4. Extension of the Algorithm

In this final section we briefly describe the natural extension of the algorithm to the problem

$$M \begin{pmatrix} x \\ y \end{pmatrix} \equiv \begin{pmatrix} A & b \\ c^T & d \end{pmatrix} \begin{pmatrix} x \\ y \end{pmatrix} = \begin{pmatrix} f \\ g \end{pmatrix} , \tag{4.1}$$

where A is still an N×N matrix but now b and c are N×p matrices and d a p×p matrix. This type of system often arises in the determination of higher-order singularities and M can be non-singular with A having a p-dimensional null-space. The procedure below corresponds to the p=1 case of section 2, but we have no computational experience with it yet.

Let $\{S_i, T_i\}$ i=1,2 be two pairs of p-dimensional subspaces of R^N such that

$$S_i \cap \{T_i\}^{\perp} = \{0\} , \tag{4.2}$$

and hence S_i and $\{T_i\}^{\perp}$ are complementary subspaces, i.e.

$$R^N = S_i \oplus \{T_i\}^{\perp} \quad i=1,2. \tag{4.3}$$

The projections of R^N onto S_i are again denoted by P_i, with $Q_i \equiv I-P_i$. If ϑ_j j=1,...,p and γ_j j=1,...,p are orthonormal bases for S_1 and T_1 respectively then $P_1 z = \sum_{j=1}^{p} \mu_j \vartheta_j$ with

$$\underset{\sim}{\mu} = [\gamma^T\vartheta]^{-1}\ [\gamma^T\underset{\sim}{z}] \quad , \tag{4.4}$$

where γ and ϑ are the N×p matrices with columns $\underset{\sim}{\gamma}_j$ and $\underset{\sim}{\vartheta}_j$. Similarly, if $\underset{\sim}{\eta}_j$ and $\underset{\sim}{\xi}_j$, j=1,...,p, are orthonormal bases for S_2 and T_2 respectively then $P_2\underset{\sim}{z} = \sum_{j=1}^{p} \nu_j\underset{\sim}{\eta}_j$ with

$$\underset{\sim}{\nu} = [\xi^T\eta]^{-1}[\xi^T\underset{\sim}{z}]. \tag{4.5}$$

Now to relate the above projections to A we insist that the columns of $A^{-1}\eta$ span S_1 and the columns of $A^{-T}\gamma$ span T_2, and hence A will decompose so that

$$\text{(i)}\quad A:S_1\to S_2 \qquad\qquad \text{(ii)}\quad A:\{T_1\}^{\perp}\to\{T_2\}^{\perp}. \tag{4.6}$$

ϑ and ξ are obtained by the orthonormalisation of $A^{-1}\eta$ and $A^{-T}\gamma$ respectively, i.e. a QR factorisation as in orthogonal iteration, and if $\underset{\sim}{z} = \sum_{j=1}^{p} \mu_j\underset{\sim}{\vartheta}_j$ then $A\underset{\sim}{z} = \sum_{j=1}^{p} \nu_j\underset{\sim}{\eta}_j$, where $\underset{\sim}{\nu}=L\underset{\sim}{\mu}$ and L is a lower triangular p×p matrix. The solution $\underset{\sim}{x}$ of (4.1) can then be written in the form

$$\underset{\sim}{x} = \sum_{j=1}^{p} \mu_j\underset{\sim}{\vartheta}_j+\underset{\sim}{w}_f - [w_b\underset{\sim}{y}] \quad , \tag{4.7}$$

where $\underset{\sim}{w}_f = A^{\dagger}Q_2\underset{\sim}{f}$ and w_b is the N×p matrix $A^{\dagger}[Q_2b]$. $A^{\dagger}$ again denotes the inverse of A in (4.6ii) and in practice $\underset{\sim}{w}_f$ and w_b will be computed by the algorithm in (2.16). We are then left with the following 2p×2p system of equations for $\underset{\sim}{\mu}$ and $\underset{\sim}{y}$,

$$\begin{aligned} L\underset{\sim}{\mu} + [\xi^T\eta]^{-1}[\xi^Tb]\underset{\sim}{y} &= [\xi^T\eta]^{-1}[\xi^T\underset{\sim}{f}] \\ [c^T\vartheta]\underset{\sim}{\mu} + (d-[c^Tw_b])\underset{\sim}{y} &= \underset{\sim}{g}-[c^T\underset{\sim}{w}_f] \quad . \end{aligned} \tag{4.8}$$

As with the p=1 case, we have to be careful with the choice of T_1 when A is ill-conditioned. In particular if A has p small singular values then we require

$$T_1 \cap \{V_p\}^{\perp} = \{\underset{\sim}{0}\} \ , \qquad (4.9)$$

where V_p is the p-dimensional subspace spanned by $\underset{\sim}{v}_N, \ldots, \underset{\sim}{v}_{N-p+1}$. The possibilities for $\underset{\sim}{\gamma}$ and $\underset{\sim}{\eta}$ in section 2 have analogies here.

References

Chan, T.F. (1984) Deflation techniques and block-elimination algorithms for solving bordered singular systems. SIAM J. Sci. Stat.Comp. 5, 121-134.

Golub, G.H. & van Loan, C.F. (1983) Matrix computations, (North Oxford Academic, Oxford).

Keller, H.B. (1983) The bordering algorithm and path following near singular points of higher nullity. SIAM J.Sci.Stat.Comp. 4, 573-582.

Stewart, G.W. (1981) On the implicit deflation of nearly singular systems of linear equations. SIAM J.Sci.Stat.Comp. 2, 136-140.

Dr Gerald Moore, Department of Mathematics, Imperial College of Science & Technology, Huxley Building, Queen's Gate, London SW7 2BZ, U.K.

International Series of
Numerical Mathematics, Vol. 79

NONLINEAR DYNAMIC BEHAVIOR AND BUCKLING OF SHELLS OF REVOLUTION

H. Obrecht, W. Redanz, W. Wunderlich

Lehrstuhl IV, Institut für Konstruktiven Ingenieurbau,
Ruhr-Universität Bochum,
Bochum, West-Germany

1. Introduction

Most of the available nonlinear dynamic analyses of engineering structures are based on the standard displacement finite element method. It follows the Lagrangian formulation of dynamics and leads to a nonlinear algebraic system of second-order ordinary differential equations with respect to time, which may then be solved by a variety of single or multi-step methods. In the first part of this paper we briefly describe and discuss this approach and then use it to analyze the dynamic response of a clamped spherical cap subjected to a uniform step pressure of moderate magnitude.

Alternatively, an equivalent set of nonlinear algebraic first-order ordinary differential equations with respect to time may be derived and solved by suitable extrapolation methods. This approach has a firm theoretical foundation in the Hamiltonian formulation of dynamics and, moreover, offers some significant numerical advantages over the former. Nevertheless, so far it appears to have received little, if any, attention in the field of numerical structural analysis. We therefore give a brief description of the formulation and of the numerical solution procedure as well as a discussion of their relative merits. As an illustrative example the analysis of the dynamic snap-through behavior of a spherical cap is presented which, due to the formulation employed, automatically yields complete information both on the time-history of the displacements and on the orbit of the motion in state-space.

While nonlinear dynamic analyses as such are no longer an exception, there are relatively few general treatments of problems concerning the dynamic buckling and stability of structures under arbitrary loading histories and initial conditions. Moreover, most of the investigations which are available on this subject, approach it in either a rather specialized or in a somewhat heuristic manner. In particular, there is a conspicuous lack of stability criteria which are sufficiently general and, at the same time, suitable for application in numerical analyses. The quantitative criterion used here to assess the stability behavior of a spherical cap is believed to satisfy these requirements.

2. Lagrangian Formulation and Numerical Solution Procedure

The analyses presented in this paper are based on the geometrically nonlinear shell equations described, for example, in [1-4] and the references given there. They are valid for large displacements and moderate rotations of the tangential base vectors and also account for the influence of general initial imperfections. In the particular version of the theory used here the field equations are referred to a closed rotationally symmetric reference surface of arbitrary meridional shape which, in the case of a perfect initial configuration, is taken to coincide with the undeformed shell's middle surface.

As shown in some detail in [1,2], our spatial discretization approach starts from a mixed variational principle which contains both displacement quantities and stress resultants as independent variables. The associated Euler equations are a set of first-order partial differential equations in those field variables which appear in the line-integral along a circumferential boundary. Unlike in most two-dimensional finite element formulations the terms which result from the nonlinear contributions to the midsurface stretching strain tensor are treated in the form of pseudo-loads, and Fourier series are used to approximate the circumferential variation of all geometrical and physical quantities. Truncation of the series after N terms leads to N decoupled sets of implicit first-order ordinary differential equations with respect to the meridional coordinate involving the natural stress-resultants and displacement variables. They are integrated numerically along a finite section of the meridian to give the related transfer matrices from which, after eliminating the stress resultants, the linear stiffness matrices $\overset{(n)}{\underline{k}}_L$ of a circular ring element are obtained. The latter are associated with the displacement vectors $\overset{(n)}{\underline{v}}$ which contain the nodal displacement degrees of freedom of the elements and the superscripts n range from 0 to N. By an analogous procedure one finds the vectors of the nodal values of the applied external loads $\Lambda(t)\cdot\overset{(n)}{\underline{p}}$, where $\overset{(n)}{\underline{p}}$ denotes normalized load intensities and the time-dependent scalar function $\Lambda(t)$ characterizes the loading history. Furthermore, one obtains the consistent mass matrices $\overset{(n)}{\underline{m}}$, the damping matrices $\overset{(n)}{\underline{c}}$, and the nonlinear pseudo-load vectors $\overset{(n)}{\underline{p}}_N(\overset{(k)}{\underline{v}})$. The latter depend cubically on the components of $\overset{(k)}{\underline{v}}$ $(0 \leq k \leq N)$. From the element matrices and vectors a system of algebraic equations for the entire discretized shell structure may be derived in the usual manner. This leads to

$$\underline{M}_{(n)}\ddot{\underline{V}}_{(n)} + \underline{C}_{(n)}\dot{\underline{V}}_{(n)} + \underline{K}_{L(n)}\underline{V}_{(n)} = \Lambda(t)\underline{P}_{(n)} + \underline{P}_{N(n)}[\underline{V}_{(k)}]. \qquad (1)$$

In (1) $\underline{V}_{(n)}$ denotes the n-th harmonic of the global vector of nodal degrees of freedom, $\dot{\underline{V}}_{(n)}$ and $\ddot{\underline{V}}_{(n)}$ the corresponding vectors of the nodal velocities and accelerations, and $\underline{M}_{(n)}$, $\underline{C}_{(n)}$ and $\underline{K}_{L(n)}$ are the associated global mass, damping and linear elastic stiffness matrices. The vector $\underline{P}_{(n)}$ on the right-hand side of (1) denotes the normalized global load vector and $\underline{P}_{N(n)}[\underline{V}_{(k)}]$ is the global pseudo-load vector which accounts for the geometrically nonlinear terms. Now, since the N sets of equations (1) represent a discrete statement of force equilibrium, they must be satisfied at all times within the time range considered.

Taking, on the other hand, the deformed configuration of the shell at a given time t_o as a fixed reference state we may denote the associated global displacement vectors, velocities and accelerations by $\overset{\circ}{\underline{V}}_{(n)}$, $\overset{\circ}{\dot{\underline{V}}}_{(n)}$ and $\overset{\circ}{\ddot{\underline{V}}}_{(n)}$ and the respective finite increments by $\underline{v}_{(n)}$, $\dot{\underline{v}}_{(n)}$ and $\ddot{\underline{v}}_{(n)}$. The latter are then governed by the N nonlinear sets of incremental equations

$$\begin{aligned}\underline{M}_{(n)}\ddot{\underline{v}}_{(n)} + \underline{C}_{(n)}\dot{\underline{v}}_{(n)} + \underline{K}_{L(n)}\underline{v}_{(n)}\\ = \lambda(t_o,t)\ \underline{P}_{(n)} + \underline{p}_{N(n)}[\overset{\circ}{\underline{V}}_{(k)},\underline{v}_{(k)}]\end{aligned} \tag{2}$$

which hold for all $t \geq t_o$. In (2) the incremental function $\lambda(t_o,t)$ is referred to t_o and $\underline{p}_{N(n)}\ [\overset{\circ}{\underline{V}}_{(k)},\underline{v}_{(k)}]$ denotes the incremental pseudo-load vector which depends cubically on both $\overset{\circ}{\underline{V}}_{(k)}$ and $\underline{v}_{(k)}$ $(0 \leq k \leq N)$.

Finally, equations (1) may be differentiated with respect to time to give the rate equations

$$\begin{aligned}\underline{M}_{(n)}\dddot{\underline{V}}_{(n)} + \underline{C}_{(n)}\ddot{\underline{V}}_{(n)} + \underline{K}_{L(n)}\dot{\underline{V}}_{(n)}\\ = \dot{\lambda}\ (t_o,t)\ \underline{P}_{(n)} + \dot{\underline{p}}_{T(n)}[\overset{\circ}{\underline{V}}_{(k)},\dot{\underline{V}}_{(k)}]\end{aligned} \tag{3}$$

which are only valid at $t = t_o$. It should be noted that despite their similarity equations (2) and (3) differ significantly. In (3) the pseudo-load vector $\dot{\underline{p}}_{T(n)}\ [\overset{\circ}{\underline{V}}_{(k)},\dot{\underline{V}}_{(k)}]$, which takes the nonlinear contributions of $\mathring{\underline{V}}_{(k)}$ to the familiar tangential elastic stiffness matrix into account, is only formally equivalent to the incremental vector $\underline{p}_{N(n)}$ in (2). Both depend cubically on $\mathring{\underline{V}}_{(k)}$, but $\underline{p}_{N(n)}$ is also a cubic function of the finite increments $\underline{v}_{(k)}$, whereas the dependence of the rate vector $\dot{\underline{p}}_{T(n)}$ on $\dot{\underline{v}}_{(k)}$ is merely linear. Thus, (3) represents N sets of implicit linear equations.

Equations (1-3) may be integrated with respect to time by any of the single and multi-step methods described, for example, in [5]. When, for instance, the well-known Newmark algorithm is applied to the solution of (2) one obtains

$$\underline{K}_{LE(n)}\underline{v}_{(n)} = \lambda(t_o,t)\underline{P}_{(n)} + \underline{p}_{N(n)}[\mathring{\underline{V}}_{(k)},\underline{v}_{(k)}] + \underline{p}_{D(n)}, \tag{4}$$

where

$$\underline{K}_{LE(n)} = \underline{K}_{L(n)} + a_o\underline{M}_{(n)} + a_1\underline{C}_{(n)} \tag{4a}$$

denotes the so-called effective linear elastic stiffness matrix associated whith the n-th Fourier index, and the dynamic pseudo-load increment $\underline{p}_{D(n)}$ is given by

$$\underline{p}_{D(n)} = \underline{M}_{(n)}(a_2\dot{\mathring{\underline{V}}}_{(n)} + a_3\ddot{\mathring{\underline{V}}}_{(n)}) + \underline{C}_{(n)}(a_4\dot{\mathring{\underline{V}}}_{(n)} + a_5\ddot{\mathring{\underline{V}}}_{(n)}). \tag{4b}$$

The constants a_1 through a_5 appearing in (4a,b) are functions of the 'Newmark parameters' β and γ (see [5]) and of the chosen time interval Δt. Since we are dealing with non-linear dynamic problems the latter is, of course, not a constant but depends on the loading and on the shell's response.

As mentioned earlier, the pseudo-load vector $\underline{p}_{N(n)}[\mathring{\underline{V}}_{(k)},\underline{v}_{(k)}]$ depends implicitly on the unknown displacement increments $\underline{v}_{(k)}$ $(0 \leq k \leq N)$ so that the N sets of nonlinear algebraic equations (4) cannot be solved directly. Instead, suitable iterative methods must be employed. To do this we first use a Newton-type procedure to linearize $\underline{p}_{N(n)}$ whith respect to $\underline{v}_{(k)}$, and then solve the resulting sets of linear algebraic equations by a modified preconditioned conjugate gradient method [4].

3. Numerical Example

To illustrate the numerical formulation given in the previous chapter we consider the dynamic response of a spherical cap (see Fig. 1) to a uniform step pressure. A similar problem has earlier been treated in [6] and elsewhere. We assume homogeneous initial

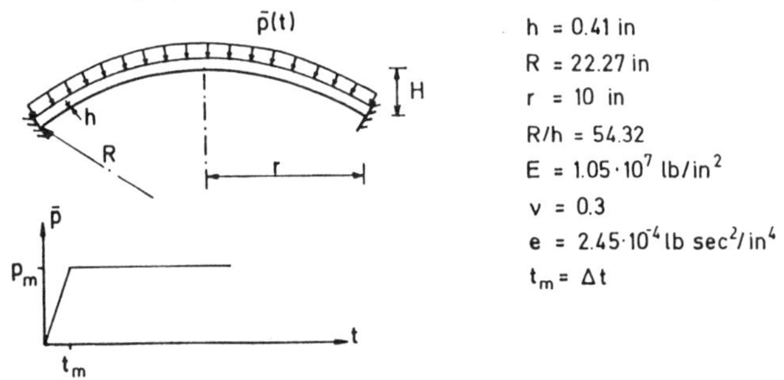

Fig. 1 Spherical cap under uniform step-pressure

conditions and a load amplitude of 4.7 percent of that pressure which causes the perfect elastic cap to bifurcate into a nonaxisymmetric mode [7]. Choosing a constant time-step of $\Delta t = 5 \cdot 10^{-5}$ sec we obtain the numerical results plotted in Fig. 2. The curves

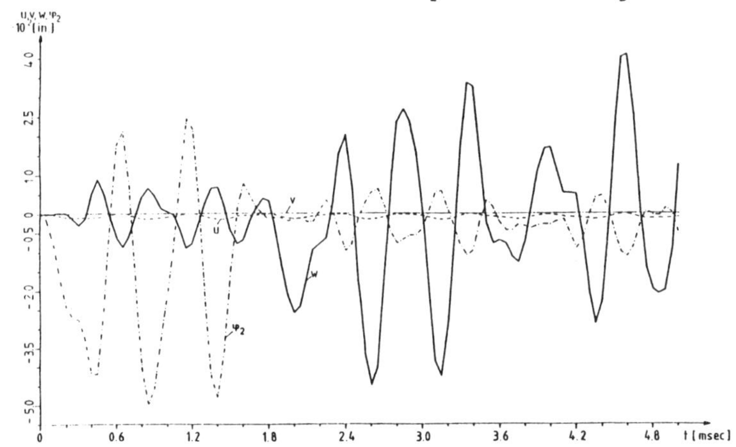

Fig. 2 Spherical cap - Time histories of apex deflections

shown there represent the time histories of the circumferential, meridional and normal displacement components u, v and w as well as that of the meridional angle of rotation ϕ_2 at the apex. It is seen that initially the amplitudes of w are comparatively small, whereas those of ϕ_2 are quite large. This indicates that the non-symmetric modes have a considerable influence. Then, however, at about $t = 1.8 \cdot 10^{-3}$ sec the situation reverses and the maximum amplitudes of w increase by about a factor of six while at the same time those of ϕ_2 are reduced by approximately a factor of one-seventh. The circumferential displacement component u, on the other hand, is unaffected by this interaction between w and ϕ_2. Its time-history is almost periodic and its amplitudes amount to only about ten to twenty percent of those of w. Finally it is noted that the meridional component v remains vanishingly small.

4. Hamiltonian Formulation and Numerical Solution Procedure

Results such as those in Fig. 2 are typical of displacement-based finite element analyses. They are very useful because, within the limits of numerical accuracy attainable, they yield detailed information about the time-dependence of the deflections and stress resultants at any given node. Unfortunately, however, local information of this kind does not usually permit more general conclusions to be drawn about the overall dynamic behavior of a structure and, in particular, about its stability. When the boundary conditions are holonomic, as they are assumed to be here, this wider class of problems can be dealt with more appropriately from the point-of-view of Hamiltonian dynamics [8]. As is well known, it differs from the Lagrangian approach employed above in that generalized displacements as well as generalized momenta are used as primary variables. As a result, the governing differential equations only contain first-order but no second-order time derivatives. As will be seen, this change of variables significantly simplifies the numerical solution procedure as well as the physical interpretation of the results.

Following [9] we use the spatial discretization procedure described briefly in chapter 2 and again denote the n-th Fourier components of the global displacement and velocity vectors by $\underline{V}_{(n)}$ and $\dot{\underline{V}}_{(n)}$. The corresponding generalized momenta, defined by

$$\underline{Q}_{(n)} = \underline{M}_{(n)} \dot{\underline{V}}_{(n)} , \tag{5}$$

are conjugate to $\dot{\underline{V}}_{(n)}$. Denoting their time-derivatives by $\dot{\underline{Q}}_{(n)}$ it is easy to show that equations (2) may also be expressed in the Hamiltonian form

$$\begin{bmatrix} \dot{\underline{V}}_{(n)} \\ \dot{\underline{Q}}_{(n)} \end{bmatrix} = \begin{bmatrix} \underline{0} & \underline{M}^{-1}_{(n)} \\ -\underline{K}_{L(n)} & -\underline{C}_{(n)} \end{bmatrix} \begin{bmatrix} \underline{V}_{(n)} \\ \underline{Q}_{(n)} \end{bmatrix} + \Lambda(t) \begin{bmatrix} \underline{0} \\ \underline{P}_{(n)} \end{bmatrix} + \begin{bmatrix} \underline{0} \\ \underline{P}_{N(n)}[\underline{V}_{(k)}] \end{bmatrix}, \tag{6}$$

or, more concisely, as

$$\dot{\underline{Y}}_{(n)} = \underline{A}_{L(n)} \underline{Y}_{(n)} + \Lambda(t) \underline{Y}_{p(n)} + \underline{Y}_{N(n)}[\underline{Y}_{(k)}], \tag{6a}$$

where the state vectors $\underline{Y}_{(n)}$ are given by

$$\underline{Y}^T_{(n)} = [\underline{V}^T_{(n)} \underline{Q}^T_{(n)}]. \tag{6b}$$

The definition of the matrices $\underline{A}_{L(n)}$ as well as of the load and pseudo-load vectors $\underline{Y}_{p(n)}$ and $\underline{Y}_{N(n)}$ appearing in (6a) should be obvious from (6). Clearly, since the dimensions of the vectors and matrices in (6a) are twice as large as those in (1), a numerical solution scheme based on (6a) requires more storage space. This can be a disadvantage when large problems are considered. On the other hand, the numerical time integration of nonlinear first-order ordinary differential equations, such as (6a), (see for instance [10]) can be done in a more direct and efficient manner than that of second-order equations such as (1). In particular, the methods available for this purpose may be applied directly to the full nonlinear equations (6a), rather than to their incremental or rate counterparts and, what is very attractive from a numerical point-of-view, they do not require any matrix inversions. Thus, since the matrices $\underline{A}_{L(n)}$ do not change with time, the solution of (6a) only involves vector operations.

For the purposes of this paper the particular algorithm described in some detail in [9] was used. It has been derived from the work summarized in [10] and essentially relies on a quadratic extrapolation of the vectors $\underline{Y}_{(n)}$ in the time interval Δt, with the nonlinear pseudo-load terms $\underline{Y}_{N(n)}$ being treated in an iterative fashion. During the analysis a time-step control algorithm was employed to ensure that iterations converge within at most 12 steps and that the absolute values of the errors of each row of (6a) remain smaller than $\varepsilon = 10^{-10}$ at all times.

5. Numerical Example

As a numerical example we again consider the problem of a clamped spherical cap subjected to a uniform step pressure. For simplicity the same geometrical parameters are used as before, but the pressure amplitude is now taken to be 0.935 times the bifurcation load of the perfect cap. For homogeneous initial conditions we obtain the time-history of the apex deflection shown in Fig. 3.

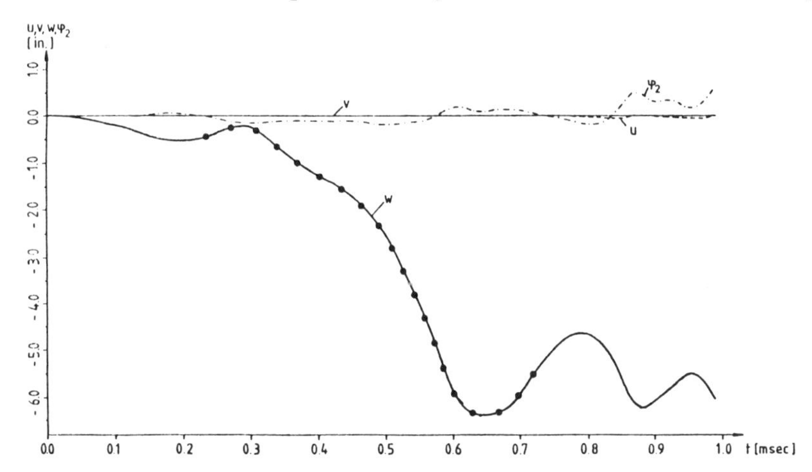

Fig. 3 Spherical cap - Time histories of apex deflections

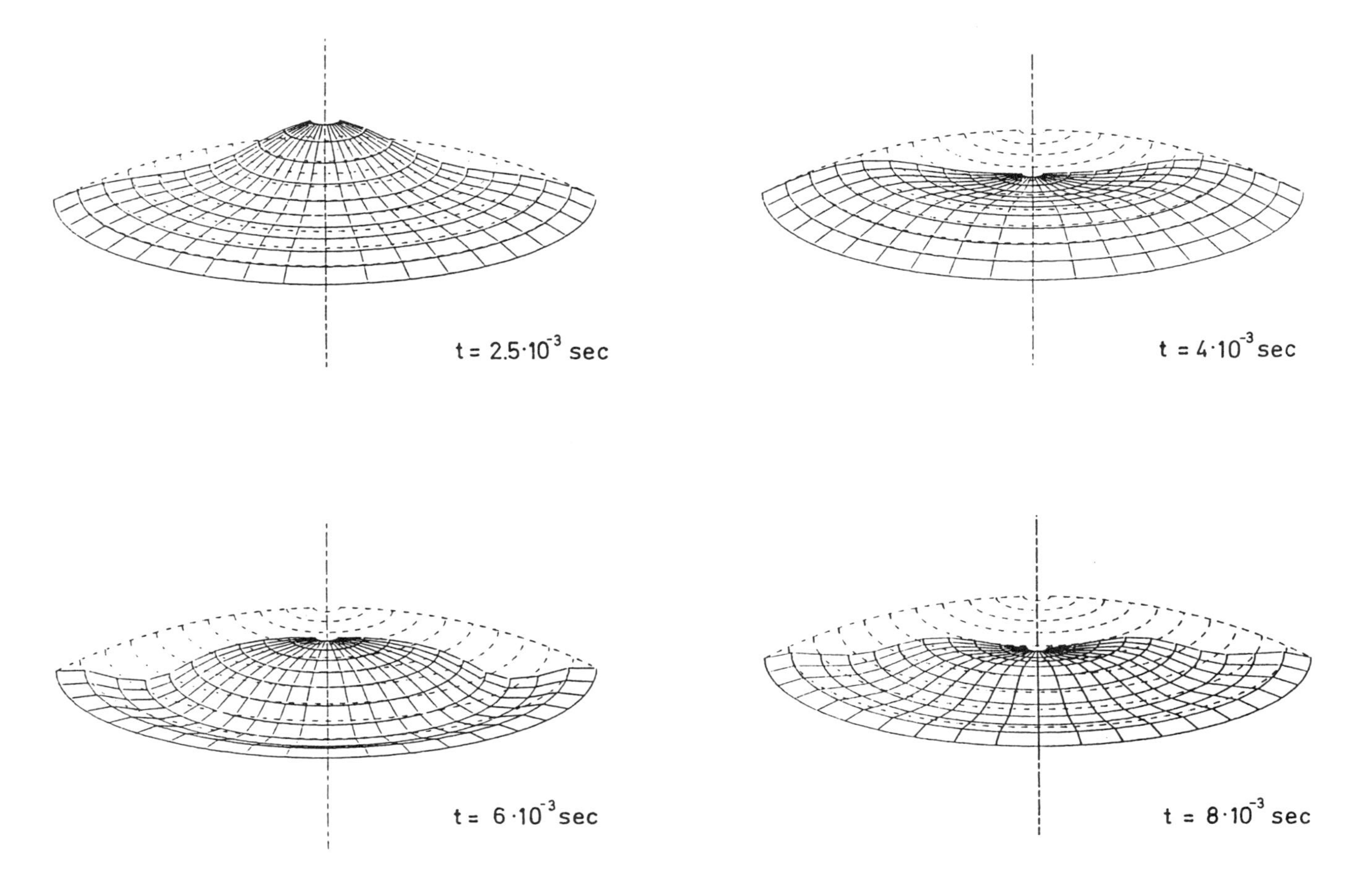

Fig. 4 Deflected shapes of spherical cap

It is seen that the shell first oscillates at relatively moderate amplitudes. Then, in the time interval 4.5 sec $< t \cdot 10^4 <$ 6.5 sec, there is a continuous increase in deflections to about 15 times the shell thickness followed by an oscillatory motion about a finitely deformed reference state. Interestingly, this snapping-through to an inverted position is not particularly sudden as might be thought. Instead, it has approximately the same duration as the periods of the initial and final oscillations. For various values of time the deformed shapes of the cap have been plotted in Fig. 4. They show that the response is essentially axisymmetric even though the load amplitude is quite close to the pressure at which, in the static case, unstable nonsymmetric buckling modes appear. This indicates that static and dynamic buckling behavior are not normally related in a straightforward manner. Furthermore, the plots in Fig. 4 show that the snap-through process is accompanied by a significant reduction in the characteristic wave-length of the deflection pattern.

Finally, Fig. 5 gives a phase-plane representation of the above numerical results in the form of a plot of the Euclidean norm of all velocity components versus that of all displacement components. It shows that from the start the shell moves away from the origin rather steadily, and that the motion does not become bounded until after the shell has reached its inverted position.

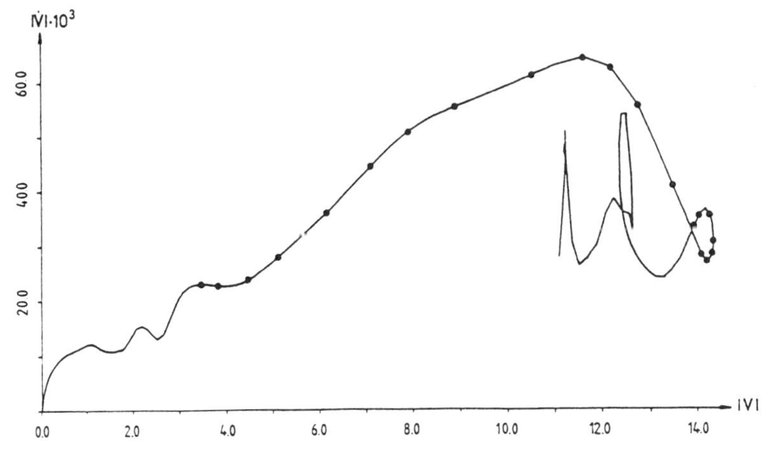

Fig. 5 Spherical cap - Orbit in phase plane

6. Stability Considerations

Comparison of the results in Figs. 2 and 3 shows that under the first, relatively small, pressure the shell performs small-amplitude oscillations in the vicinity of the unloaded state. Hence, the motion is stable according to Liapunov's general definition of dynamic stability. At the higher load level, on the other hand, the shell soon leaves the vicinity of the initial configuration and moves to the neighborhood of a distant reference state, or focus. Thus, the motion in the second example becomes unstable and the shell buckles dynamically.

Now, whereas the general concept of stability is quite well established, there still is a conspicuous lack of stability criteria which are both general and easy to apply in practical situations. Several criteria have been proposed in the literature but they mainly serve the purposes of the particular disciplines for which they have been developed, and most of them are not directly applicable to the kinds of problems we have discussed here. The absence of a unified and generally agreed-upon criterion is not very satisfactory, but this situation is also a reflection of the fact that stability and instability, being qualitative concepts, depend on defined measures which, unfortunately, are not intrinsic properties of a system. Therefore, any quantitative stability criterion necessarily involves a certain degree of arbitrariness.

Here we employ the criterion suggested in [9]. It essentially stipulates that a motion becomes unstable when it ceases to be oscillatory, that is, when at some time t_o and over a small time interval Δt there exists a positive acceleration away from the tangent to the orbit in state space. A concise formal expression of this statement may be obtained by combining all N Fourier harmonics $\dot{\underline{V}}_{(n)}$, $\dot{\underline{Q}}_{(n)}$ and $\dot{\underline{Y}}_{(n)}$ into the vectors $\dot{\underline{V}}^T = [\dot{\underline{V}}^T_{(1)}\ \dot{\underline{V}}^T_{(2)} \cdots \dot{\underline{V}}^T_{(N)}]$, $\dot{\underline{Q}}^T = [\dot{\underline{Q}}^T_{(1)}\ \dot{\underline{Q}}^T_{(2)} \cdots \dot{\underline{Q}}^T_{(N)}]$, $\dot{\underline{Y}}^T = [\dot{\underline{V}}^T\ \dot{\underline{Q}}^T]$, and denoting the individual components of $\dot{\underline{V}}$ and $\dot{\underline{Q}}$ by $\dot{V}_i$, $\dot{Q}_i$. The criterion then states that whenever the inequality

$$\sum_{i,j} [\dot{V}_i(t_o)\, \dot{Q}_j(t_o+\Delta t) - \dot{V}_j(t_o+\Delta t)\, \dot{Q}_i(t_o)] < 0 \qquad (7)$$

(where the sum is taken over all components of $\dot{\underline{V}}$ and $\dot{\underline{Q}}$) holds in $t_o \leq t \leq t_o+\Delta t$, it may be inferred that the motion is stable and continues to remain in the vicinity of the reference configuration associated with t_o. When, on the other hand, the left-hand side of (7) becomes zero or negative, it is concluded that the motion is unstable and thus begins to depart from the neighborhood of the respective reference state.

Since, now, inequality (7) is essentially based on the evolution of the velocity fields $\dot{\underline{V}}$ and $\dot{\underline{Q}}$ in the immediate vicinity of $t = t_o$, it is a local criterion which strongly depends on the details of a given motion. It must therefore be examined at all points along its orbit. In the numerical analyses presented above this has been done at all time steps and those points in Figs. 2, 3 and 4 at which (7) was not satisfied have been marked by a solid dot. As can be seen, there are no dots in Fig. 2 which indicates that in the time range considered that motion is stable in the sense just discussed. In Figs. 4 and 5, on the other hand, criterion (7) is violated along the dotted parts of the curves. Not surprisingly, they represent precisely that section of the orbit along which the cap moves to its inverted position without oscillating. Since the first dot separates the stable from the unstable regimes it denotes the starting point of the dynamic buckling process. Thus, it marks the limit of the structure's practical usefulness and in this sense it has the same significance as the unstable bifurcation and limit-points occurring under static loading conditions.

7. Concluding Remarks

The above discussion shows that the question of dynamic stability can only be answered conclusively with respect to a given motion, a given reference state and a given time interval. Finally, dynamic buckling may be defined to occur at that point of a structure's orbit at which criterion (7) is first violated.

References

[1] Wunderlich, W. (1982) Zur nichtlinearen Berechnung von Rotationsschalen (On the nonlinear analysis of shells of revolution). Wiss. Zeitschr. d. Hochschule f. Arch. u. Bauwesen, Weimar 28, 221-225

[2] Rensch, H.J. (1982) Elastoplastisches Beulen und Imperfektionsempfindlichkeit torisphärischer Schalen (Elastic-plastic buckling and imperfection-sensitivity of torispherical shells). Techn. Wiss. Mitteilungen Nr. 82-13, Institut für Konstruktiven Ingenieurbau, Ruhr-Universität Bochum

[3] Wunderlich, W., Rensch, H.J., Obrecht, H. (1982) Analysis of elastic-plastic buckling and imperfection-sensitivity of shells of revolution. In: Ramm, E., ed., Buckling of Shells (Springer, Berlin) 137-174

[4] Wunderlich, W., Cramer, H., Obrecht, H. (1985) Application of ring elements in the nonlinear analysis of shells of revolution under nonaxisymmetric loading. Comp. Meth. Appl. Mech. Eng. 51, 259-275

[5] Belytschko, T., Hughes, T.J.R., eds. (1983) Computational Methods for Transient Analysis (North-Holland, Amsterdam)

[6] Nagarajan, S. (1973) Nonlinear static and dynamic analysis of shells of revolution under axisymmetric loading. Report UCSESM 73-11, Civil Engineering Dept., University of California, Berkeley

[7] Huang, N.C. (1964) Unsymmetrical buckling of thin shallow spherical shells. J. Appl. Mech. 31, 447-457

[8] Goldstein, H. (1980) Classical Mechanics, 2nd edn. (Addison-Wesley, Reading)

[9] Redanz, W. (1987) Nichtlineares dynamisches Verhalten und Stabilität von Systemen der Strukturmechanik (Nonlinear dynamic behavior and stability of structural systems). Dissertation, Fakultät für Bauingenieurwesen, Ruhr-Universität Bochum

[10] Deuflhard, P. (1983) Recent progress in extrapolation methods for ordinary differential equations. SIAM 30th Anniversary Meeting, Stanford, CA, July 1982, Preprint Nr. 224

Hans Obrecht, Ph.D., Lehrstuhl KIB IV, Institut für Konstruktiven Ingenieurbau, Ruhr-Universität, D-4630 Bochum, W. Germany

International Series of
Numerical Mathematics, Vol. 79

THE CURIOUS LINK CHAIN

Christoph Pospiech

Mathematisches Institut
Universität Bayreuth
West Germany

1. Introduction

First some strange example will be presented that shows what curious things may occur, when you deal with bifurcation under the presence of some symmetry group. In a second part it will be shown that such behavior is impossible for analytic bifurcation equations. Indeed, in this case the solution set near the bifurcation point can be decomposed into finitely many cells of (possibly) different dimension but constant symmetry. As a consequence, every solution sufficiently near to the bifurcation point can be joined to this point by an analytic curve of solutions all having the same symmetry except for the bifurcation point.

2. The curious link chain

Consider the zeros of

$$F : \mathbb{R} \times \mathbb{R}^3 \longmapsto \mathbb{R}^3$$

defined by

$$F(\lambda, x, y, \rho) = (x\lambda - y^3 \ , \ y\lambda + x^5 \sin\frac{1}{x} - \rho^2 \ , \ \rho\lambda \ )$$

F commutes with the elements of the group $\{\pm 1\}$, which operates on $\mathbb{R}^3$ in the following way:

$$\sigma(x, y, \rho) := (x, y, \sigma\rho) \qquad \text{for all} \quad (x, y, \rho) \in \mathbb{R}^3 \ , \ \sigma \in \{\pm 1\}$$

Considering the cases $x = 0$ and $x \neq 0$ these zeros are easily computed. They are sketched in the follwing figure (Fig. 1), where we dropped the y-axis, because the $\lambda - y$ dependence is strictly monotone. The broken lines therein indicate zeros with trivial symmetry, whereas the solid lines represent solutions with high symmetry, i.e. those solutions that stay fixed

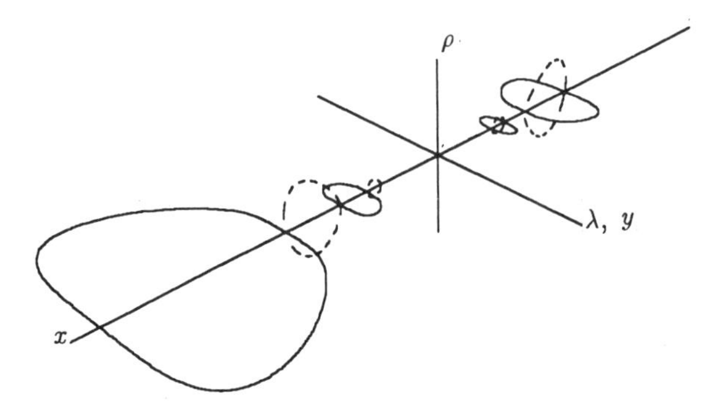

Fig. 1. The curious link chain

under the action of the group. For $x > 0$ the symmetric and nonsymmetric zeros together form links of a chain, in the following sense: The solution set for $x > 0$ is a countable set of loops such that the loops with high symmetry are perpendicular to the loops with trivial symmetry and meet those in the points $(0, 0, 0, 1/k\pi)$, $k \in \mathbb{Z}$. The links of this chain differ in size and accumulate at the bifurcation point $(0, 0, 0, 0)$, where they meet with the trivial solution $\{ (\lambda, 0, 0, 0) \mid \lambda \in \mathbb{R} \}$. For $x < 0$ the situation is different in so far as the links do not connect to a chain.

This example shows what strange things can bifurcate from a given (trivial) solution. Indeed, you cannot talk about the bifurcating branch of a given symmetry, because in this example the symmetry on the bifurcating branch changes even in arbitrarily small neighbourhoods of the bifurcation point. However, if you look only at solutions with some given symmetry, the chain looses half of its links and disconnects to an infinite series of loops accumulating at the bifurcation point.

3. The cell decomposition

Such weird things cannot occur if F happens to be analytic. To make things precise, let X and Y be Banach spaces and let G be a finite group operating on X and Y. That is, associated to every $g \in G$ there are bounded linear operators on X and Y, respectively, also denoted by g, such that $1 \in G$ is mapped to the identity and products of group elements correspond to products of linear operators. Speaking of the symmetry G_y of some element y we mean its isotropy subgroup,

$$G_y := \{ g \in G \mid gy = y \} .$$

Theorem 3.1. For every $F: X \mapsto Y$ being Fredholm (of any index) and equivariant — i.e. F commutes with every $g \in G$ — and any zero y of F the following holds: In a sufficiently small closed neighbourhood of y all zeros of F form a finite complex of cells with constant symmetry, in the following sense: Any two points in the cell stay fixed under precisely the same elements of the group.

Before we can proceed with the proof we first have to explain the notion of "cell":

Definition 3.2. A cell $a \subseteq X$ is a subset of X topologically equivalent to an n-ball $b \subseteq \mathbb{R}^n$ via some map φ, that extends to a homeomorphism of the closures $\bar{b}$ and $\bar{a}$. The number n is called the dimension of the cell. Cells of dimension 0 are points of X. $\partial a := \bar{a} - a$ is called the boundary of the cell.

Definition 3.3. A cell complex $(K, \mathcal{K})$ is is a subset $K \subseteq X$ equipped with a set $\mathcal{K}$ of cells satisfying the following:

1. K is the disjoint union of all cells in $\mathcal{K}$.
2. For every $a \in \mathcal{K}$, ∂a is the union of finitely many cells in $\mathcal{K}$.
3. For every point in K, some neighbourhood meets only a finite number of cells.

A pair of subsets $(K', \mathcal{K}')$ of $(K, \mathcal{K})$, that is a complex itself, is called a subcomplex.

The proof of theorem 3.1. makes use of the following theorem of Koopman and Brown(KOOPMAN-BROWN, [3], p. 242; GIESEKE, [2], p. 185):

Theorem 3.4. Let the functions $\Theta_i(x_1, \ldots, x_n)$, $i = 1, \ldots, m$ be real and analytic at the point $x = 0$, vanishing there but not vanishing identically. Then, after a suitable real change of coordinate axes which keeps the origin fixed, it is possible to enclose the origin in a closed region $A = \{ x \in \mathbb{R}^n \mid |x_i| \leq a_i, a_i > 0 \}$, which coincides with a cell complex $(A, \mathcal{A})$, such that each of the loci $\Theta_i(x) = 0$ in A coincides with a subcomplex $(K_i, \mathcal{K}_i)$ of $(A, \mathcal{A})$.

Proof of theorem 3.1. First we can assume, that $G = G_y$, because in a small neighbourhood U of y, the symmetry $G_{y'}$ of $y' \in U$ is a subgroup of G_y. This follows from the fact that the orbit $Gy = \{ gy \mid g \in G \}$ is a discrete set and we can choose some open set U that meets the orbit Gy only in the point y and does not intersect its images gU under $g \in G \setminus G_y$. Assuming now $G = G_y$ neither the equivariance of F nor the symmetry of its zeros are affected by a translation by y. So we can assume that $y = 0$. Moreover the linearization $DF(0)$ of F at this point is assumed to have a nontrivial kernel (being identified with $\mathbb{R}^n$ for $n = \dim \ker DF(0)$). For otherwise 0 would be an isolated zero of F and the cell complex would be just a single point.

Now we are ready to perform Liapunov-Schmidt reduction according to the detailed description in VANDERBAUWHEDE [5]. We arrive at analytic bifurcation equations $\Theta_i(x_1, \ldots, x_n)$, $i = 1, \ldots, k$ the zeros of which are in one to one correspondence to the zeros of F. This correspondence is given by some equivariant projection onto the kernel of $DF(0)$ and as a consequence any zero of the Θ_i has the same symmetry as the corresponding zero of F. To include the symmetry in our considerations we form for every subgroup $H \subseteq G$ of G the so called fixed space $(\mathbb{R}^n)^H$ given by

$$(\mathbb{R}^n)^H := \{ x \in \mathbb{R}^n \mid hx = x \quad \text{for all} \quad h \in H \},$$

where the group action on $\mathbb{R}^n$ is given by the above identification with the kernel of $DF(0)$. Each of these is a subspace of $\mathbb{R}^n$ and can thus be decribed as the common zeros of some linear equations Θ_i, $i > k$. Now we apply theorem 3.4 to all Θ_i constructed so far and immediately find that the cell complex given by the intersection of the $(K_i, \mathcal{K}_i)$ for $i \leq k$ describes the zeros of the bifurcation equations and subsequently the zeros of F. It remains to be shown that the symmetry is constant along a cell. To do this let us remark that the fixed spaces being common zeros of some Θ_i also form subcomplexes of $(A, \mathcal{A})$. Now let a be a cell of A, p and q be points thereof and let $H := G_p$ be the symmetry of p. Then p must lie in some cell b of the cell decomposition of $(\mathbb{R}^n)^H$, which also must be a cell of A. Since cells are disjoint, $a = b$ is a cell of $(\mathbb{R}^n)^H$. This implies $G_p \subseteq G_q$. Replacing p and q in this argument we get $G_q \subseteq G_p$ and equality follows. Thus any two elements of the cell a have the same symmetry and the proof is complete.

4. Curve selection

Given some zero y of F and the cell complex as in theorem 3.1 we can find a neighbourhood $U(y)$ that meets only those cells containing y in their closure. Then the following holds:

Theorem 4.1. Any zero of F in the above neigbourhood $U(y)$ can be joined to y by some analytic curve of zeros that stays in the same cell except possibly for the endpoint y that may be on the boundary of this cell. As a consequence the symmetry is constant along the curve except possibly for y.

Proof. The closure of every cell is homeomorphic to some closed n-ball and any two points in this n-ball can be joined by a straight line. The homeomorphic image of this straight line yields a curve meeting all requirements of 4.1 except for the smoothness of the curve. In order to obtain a curve that can be analytically parametrized we have to go into the details of the proof of the theorem of Koopman and Brown. To do this let $\mathcal{F}^n$ be the set of Θ_i mentioned in theorem 3.4 . Applying the Weierstrass preparation theorem to every Θ_i you obtain a product

$$\Theta = U \cdot \prod_j P_j ,$$

where U is a unit in the ring $\mathbb{R}\{x_1, \dots, x_n\}$ of convergent power series and P_j are irreducible elements of the subring $\mathbb{R}\{x_1, \dots, x_{n-1}\}[x_n]$ of power series that are polynomials in the variable x_n. Even more, the P_j are Weierstrass polynomials, i.e. they have the special form

$$P(x_1, \dots, x_n) = \sum_{\nu=0}^{k} a_\nu(x_1, \dots, x_{n-1}) x_n^{k-\nu} ,$$

where

i.) $a_0 \equiv 1$

ii.) $a_\nu(0) = 0$ for $\nu = 1, \dots, k$.

We take the discriminant of every element and the resultant of any two elements of the set $\mathcal{P}^n$ of Weierstrass polynomials obtained so far. These are elements of the ring

$\mathbb{R}\{x_1 \ldots, x_{n-1}\}$ and none is identically zero since the polynomials are irreducible. Those discriminants and resultants vanishing at 0 form a second set $\mathcal{F}^{n-1}$ and the above process is repeated ending at the set $\mathcal{P}^1$ that is either empty or consists of the single element x_1.

In the proof of theorem 3.4 now a series of cell complexes $(K^\mu, \mathcal{K}^\mu)$, $\mu = 1, \ldots, n$ is constructed inductively, such that $(K^\mu, \mathcal{K}^\mu)$ meets the requirements of 3.4 for the functions in $\mathcal{F}^\mu$. $\mathcal{K}^1$ consists of the points $-a_1, 0, a_1$ and the two open intervalls bounded by 0 and $\pm a_1$. In the induction step from μ to $\mu+1$ every polynomial $P \in \mathcal{P}^{\mu+1}$ is split into linear factors

$$P(\bar{x}, x_{\mu+1}) = \prod_{\nu=1}^{k} \left(x_{\mu+1} - Z_\nu(\bar{x})\right)$$

for every $\bar{x} \in K^\mu$. If the argument $\bar{x}$ is restricted to a cell $a \in \mathcal{K}^\mu$, the zeros $Z_\nu(\bar{x})$ form continuous complex functions in $\bar{x} \in a$. Moreover, if any two of these Z-functions (may be from different polynomials) coincide for some $\bar{x} \in a$, they coincide all along a.

Now the cells $\hat{a}$ of $\mathcal{K}^{\mu+1}$ are the graphs of the real Z-functions for every cell $a \in \mathcal{K}^\mu$,

$$\hat{a} = \left\{ \left(\bar{x}, Z(\bar{x})\right) \mid \bar{x} \in a \right\}$$

as well as the space in between two consecutive real Z-functions Z_1 and Z_2,

$$\hat{a} = \left\{ (\bar{x}, y) \mid Z_1(\bar{x}) < y < Z_2(\bar{x}), \bar{x} \in a \right\}.$$

For any details see KOOPMAN-BROWN [3] or GIESECKE [2].

In view of the proof of theorem 3.1 the following lemma establishes the proof of theorem 4.1 :

Lemma 4.2. Taking any point $(x_1, \ldots, x_\mu)$ in some cell $a \in \mathcal{K}^\mu$ containing $0 \in \mathbb{R}^\mu$ in its closure, we can find an analytic map $p: [0,1] \mapsto K^\mu$ such that

a. $p(0) = 0$ and $p(1) = (x_1, \ldots, x_\mu)$.

b. $p(t) \in a$ for every $t > 0$.

Proof. We proceed by induction on $\mu \in \{1, \ldots, n\}$. For $\mu = 1$ the function $p(t) := t \cdot x_1$ will do. Suppose now that the lemma is proved for μ and let $(x_1, \ldots, x_{\mu+1}) \in \hat{a}$ for some $\hat{a} \in \mathcal{K}^{\mu+1}$ that has 0 on its boundary. By the above construction the projection onto $\mathbb{R}^\mu$ leaving the first μ coordinates fixed maps $\hat{a}$ onto a cell $a \in \mathcal{K}^\mu$. By induction we already have an analytic curve p in a joining 0 with $(x_1, \ldots, x_\mu)$. Now for every $t_0 \in [0,1]$ and every polynomial $P \in \mathcal{P}^{\mu+1}$ we can decompose $P\left(p(t), x_{\mu+1}\right)$ into irreducible elements

$$P\left(p(t), x_{\mu+1}\right) = \coprod_{\kappa} \tilde{P}_\kappa(t, x_{\mu+1})$$

with respect to the ring $\mathbb{R}\{t - t_0\}[x_{\mu+1}]$ of convergent power series in $t - t_0$ and $x_{\mu+1}$ that are polynomials in the second variable. For every $t_0 \in [0,1]$ all irreducible factors obtained this way form a set $\mathcal{P}_{t_0}^{\mu+1} \subseteq \mathbb{R}\{t - t_0\}[x_{\mu+1}]$. Since every $\tilde{P} \in \mathcal{P}_{t_0}^{\mu+1}$ is irreducible, its discriminant does not vanish identically near t_0. Indeed, it does not vanish for every $t_0 > 0$. Otherwise two Z-functions $\tilde{Z}_1(t) := Z_1\left(p(t)\right)$ and $\tilde{Z}_2(t) := Z_2\left(p(t)\right)$ describing complex zeros of $\tilde{P} \in \mathcal{P}_{t_0}^{\mu+1}$ would coincide at the point $p(t_0) \in a$ and thus all along the curve $p(t)$ causing the discriminant of $\tilde{P} \in \mathcal{P}_{t_0}^{\mu+1}$ to vanish there identically. As a consequence, the zeros of $\tilde{P} \in \mathcal{P}_{t_0}^{\mu+1}$ are simple near t_0 and the implicit function theorem implies that the functions $\tilde{Z}(t) = Z(p(t))$ are analytic near $t_0 > 0$. This argument does not work for $t_0 = 0$, since the elements of $\mathcal{P}_0^{\mu+1}$ can be assumed to be Weierstrass polynomials and hence either their degree is 1 or their discriminant has an — necessarily isolated — zero at $t = 0$. This difficulty is overcome by considering complex arguments as has been done in §14 of VAN DER WAERDEN [6]: Since the zeros of any $\tilde{P} \in \mathcal{P}_0^{\mu+1}$ are simple for $t \neq 0$, we can follow any zero of $\tilde{P}$ along a small circle in complex plane centered at 0 by the process of analytic continuation. Having gone all round the circle once counterclockwise we may end up at a different zero of $\tilde{P}$ and so we may have to repeat it l' times before we reach the zero again we started with. Now we take l to be the least common multiple of all these l' — considering all $\tilde{P} \in \mathcal{P}_0^{\mu+1}$. Then, if t goes all round the circle once, t^l does it l times. So all functions $\tilde{Z}(t^l) = Z(p(t^l))$ are analytic for $t \neq 0$ and bounded for t small, since the moduli of the roots of any polynomial can be estimated by its coefficients. Thus all $\tilde{Z}(t^l)$ are analytic near $0 \in \mathbb{C}$.

Now, if $\hat{a} \in \mathcal{K}^{\mu+1}$ is the graph of a real Z-function, we define the desired curve $\hat{p}$ to be

$$\hat{p}(t) := \left(p(t^l), Z\left(p(t^l)\right)\right) .$$

However, if $\hat{a}$ fills the space between two consecutive real Z-functions Z_1 and Z_2, we can form two curves $\hat{p}_1$ and $\hat{p}_2$ according to the previous formula with Z being replaced by Z_1 and Z_2, respectively. Then the desired curve $\hat{p}$ is given by

$$\hat{p}(t) := s \cdot \hat{p}_1(t) + (1 - s) \cdot \hat{p}_2(t) ,$$

where

$$s := \frac{x_{\mu+1} - Z_2\left(p(1)\right)}{Z_1\left(p(1)\right) - Z_2\left(p(1)\right)} .$$

This completes the induction step and thus the proof of 4.2 .

Remark 4.3. Theorem 3.1 and 4.1 can be generalized to the situation, where in addition to F some inequalities

$$g_i(x) < 0, \quad i = 1, \ldots, m'$$

are given, where $g_i: X \mapsto \mathbb{R}$ is analytic for $i \in \{1, \ldots, m'\}$ and can be assumed to satisfy $g_i(y) = 0$. Indeed, we can find some cell complex meeting the requirements of theorem 3.1, such that for every cell a in this complex and every $i \in \{1, \ldots, m'\}$ one of the following is true:

i.) $g_i(x) < 0$ for all $x \in a$.

ii.) $g_i(x) = 0$ for all $x \in a$.

iii.) $g_i(x) > 0$ for all $x \in a$.

In particular, if $p: [0, 1] \mapsto X$ is the analytic curve given by theorem 4.1, $p(t)$ satisfies the inequalities $g_i\left(p(t)\right) < 0$ for all $t > 0$, if the endpoint $p(1)$ of this curve does.

Proof. As in the proof of 3.1 we first apply Liapunov-Schmidt reduction to F and arrive at a set $\{\Theta_i(x_1,\ldots,x_n) \mid i=1,\ldots,k\}$ of analytic bifurcation equations the zeros of which are in one to one correspondence to the zeros of F. In one direction this correspondence is given by some equivariant projection Π, the other direction is described by a nonlinear analytic map $\Phi\colon \mathbb{R}^n \mapsto X$, such that $\Pi\Phi = \mathrm{Id}$. Now we append the functions Θ_i, $i = k+1,\ldots,m$ describing the fixed spaces $(\mathbb{R}^n)^H$ as well as the functions $\Theta_{m+i} = g_i \circ \Phi$, $i = 1,\ldots,m'$ to the set of bifurcation equations and applying theorem 3.4 we get a cell complex $(A, \mathcal{A})$. Given now any cell $a \in \mathcal{A}$ and any $i \in \{1,\ldots,m'\}$, the equation $\Theta_{m+i}(x) = 0$ is satisfied either for all $x \in a$ or for no point in a. In the second case the sign of $\Theta_{m+i}(x)$ is constant along a. For if not we can join two points of different sign by a curve p lying in a and the intermediate value theorem applied to $\Theta_{m+i} \circ p$ yields a point $p(t_0) \in a$ such that $\Theta_{m+i}(p(t_0)) = 0$. This completes the proof of remark 4.3.

Remark 4.4. Theorem 4.1 — or rather remark 4.3 — generalizes "curve selection lemmas" given by MILNOR ([4], p. 25) and BÖHME ([1], p. 118) to the situation where some symmetry group is present. Even more, some curve is selected meeting a given end point.

This paper was written at the Sonderforschungsbereich 123, Universität Heidelberg, West-Germany and is part of the author's Ph.D. thesis.

References

1.) Böhme, R. (1972) *Die Lösung der Verzweigungsgleichungen für nichtlineare Eigenwertprobleme.* Math. Z. 127, 105–126

2.) Giesecke, B. (1964) *Simpliziale Zerlegung abzählbarer analytischer Räume.* Math. Z. 83, 177–213

3.) Koopman, B. O., Brown, A. B. (1932) *On the covering of analytic loci by complexes.* Trans. Am. Math. Soc. 34, 231–251

4.) Milnor, J. (1968) *Singular points of complex hypersurfaces.* (Priceton University Press, Princeton).

5.) Vanderbauwhede, A. (1982) *Local bifurcation and symmetry* (Pitman, Boston/London/Melbourne).

6.) van der Waerden, B. L. (1973) *Einführung in die algebraische Geometrie* 2nd edn. (Springer, Berlin/Heidelberg/New York).

Christoph Pospiech, Mathematisches Institut der Universität Bayreuth, Postfach 10 25 51, 8580 Bayreuth, West-Germany.

International Series of
Numerical Mathematics, Vol. 79

On a Moving-Frame Algorithm and the Triangulation of Equilibrium Manifolds

by

Werner C. Rheinboldt [1]

1. Introduction

Nonlinear, parametrized equations

$$F(z,\lambda) = 0, \tag{1.1}$$

represent models of equilibrium problems for many physical systems. If $F: R^n \to R^m$, $n=m+p$, $p \geq 1$, is continuously differentiable on R^n, then the regular solution manifold

$$M = \{ x \in R^n ; F(x) = 0 , \text{rank } DF(x) = m \} \tag{1.2}$$

is a p-dimensional, differentiable manifold in R^n without boundary. We shall assume always that F is at least of class C^r, $r \geq 2$.

The standard procedures for the computational analysis of such solution manifolds are the continuation methods. When the parameter dimension p exceeds unity, these methods require a restriction to some path on the manifold and then produce a sequence of points along that path. In general, it is not easy to develop a good picture of a multi-dimensional manifold from information along one-dimensional paths; thus there is growing interest in computational methods which generate multi-dimensional grids of solution points. Up to now, the only such method appears to be that of E.L. Allgower and P.H. Schmidt [1]. It utilizes a

---

1) University of Pittsburgh, Pittsburgh, PA 15260, USA
This work was in part supported by the National Science Foundation under grant DCR-8309926, the Office of Naval Research under contract N-00014-80-C-9455, and the Air Force Office of Scientific Research under grant 84-0131.

simplicial continuation algorithm to triangulate a p-dimensional manifold by means of p-simplices.

In [10] a new algorithm was developed for computing vertices of a triangulation (by p-simplices) of certain subsets of a p-dimensional solution manifold (1.2). It depends on an algorithm for constructing a moving frame on these subsets of M. We present here an overview of these two algorithms and illustrate their effectiveness with some numerical examples.

2. Local Coordinate Systems

At any point x of M the tangent space T_xM may be identified with the kernel of the Jacobian DF(x),

$$T_xM = \ker DF(x) = \{ u \in R^n ; DF(x)u = 0 \}, \qquad (2.1)$$

and then the corresponding normal space N_xM is specified as the orthogonal complement $N_xM = T_xM^{\perp} = \text{rge } DF(x)^T$.

A given p-dimensional subspace $T \subset R^n$ induces a local coordinate system of M at any point $x \in M$ where

$$T \cap N_xM = \{0\} \qquad (2.2)$$

As shown, for instance, in [4] or [9], at any $x \in M$ where (2.2) holds there exist neighborhoods $V_1 \subset T$ and $V_2 \in R^n$ of the origins of T and R^n, respectively, and a unique C^{r-1} function $w: V_1 \to T^{\perp}$, $w(0) = 0$, such that

$$M \cap V_2 = \{ y \in R^n; \ y = x + t + w(t), \ t \in V_1 \} \qquad (2.3)$$

A well-known procedure for computing tangent bases is provided by the QR-decomposition

$$DF(x)^T = Q \begin{bmatrix} R \\ 0 \end{bmatrix} , \quad Q = (Q_1, Q_2) , \tag{2.4}$$

where the $n \times n$ matrix Q is orthogonal, Q_1 has m columns, and the $m \times m$ matrix R is upper triangular and non-singular for $x \in M$. Then the p columns of Q_2 form an orthonormal basis of T_xM.

If $x \in M$ is a point where the QR-decomposition (2.4) has been computed, then with any starting-point $y = y^0$ sufficiently near x in $x+T_xM$ we may apply the chord-Gauss-Newton process:

For $k=0,1,\ldots$ until convergence
1) solve $R^T z = F(y)$ for $z \in R^p$ (2.5)
2) $y := y - Q(z,0)^T$

The convergence theory of these methods is well understood. In particular, a theorem of Deuflhard and Heindl [3] can be used to ensure that there exists for any $x \in M$ a neighborhood $V = V(x)$ of x in $x+TxM$ such that for any y in $V(x)$ the process (2.4) converges to some $y^* \in M$. Moreover, we can show readily that $y^*-y^0 \in N_xM$ and hence that, in the notation of (2.3), we have $y^* = x+t+w(t)$, $t=y^0-x$. In other words, the process (2.5) represents an implementation of the "corrector" mapping w of the local coordinate representation (2.3).

3. The Moving Frame Algorithm

Recall that a vector field of class C^s, $s \leq r$, on an open subset M_0 of M is a C^s function $u : M_0 \to TM$ into the tangent bundle TM such that $u(x)$ belongs to T_xM for x each M_0. A moving frame of class C^s on M_0 associates with each x of M_0 an ordered basis (frame) $\{u^1,\ldots,u^p\}$ of T_xM such that each coordinate map $u^i : M_0 \to TM$, $i=1,\ldots,p$ defines a vector field of class C^s on M_0. We shall consider only orthonormal moving frames.

In our setting, an algorithm for constructing a moving frame has to generate for each x of some open subset M_0 of M an $n \times p$ matrix $U(x)$ with orthonormal columns such that $DF(x)U(x) = 0$ and that the mapping $U : M_0 \to R^{p \times n}$ is of class C^s on M_0. As noted in [2], the QR-decomposition (2.4) does not produce continuously varying matrices $U(x)$. This observation extends to other algorithms of a similar nature. The three remedies proposed in [2] do not concern the generation of a moving frame.

For the moving frame algorithm developed in [10] we assume that some method is available for computing at the points x of M some $n \times p$ matrix $U_0(x)$ with orthonormal columns that span T_xM. Of course, $U_0(x)$ is not expected to depend continuously on x. For instance, we may use the QR-decomposition (2.4).

The algorithm is based on the selection of an $n \times p$ reference matrix T_r with orthonormal columns. Then for a point x of the manifold we proceed as follows:

(1) Compute the tangent basis matrix $U_0(x)$;
(2) form $U_0 := U_0(x)T_r$;
(3) compute the singular value decomposition $A^TU_0B = \Sigma$ and save A and B ; (3.1)
(4) with $Q = AB^T$ form the basis matrix $U_0(x)Q$.

The following result, proved in [10], guarantees the validity of this algorithm:

Theorem: Let M_0 be the open subset of M where the subspace of R^n spanned by the columns of the reference matrix T_r induces a local coordinate system. Then the mapping $x \in M \Rightarrow U_0(x)Q \in R^{n \times p}$ given by the algorithm (3.1) is of class C^{r-1} on M_0 and defines an orthonormal moving frame on M_0.

If the QR-decomposition is used in step (1) and the dimension of the manifold is small in comparison with the space dimension, then the principal cost of (3.1) derives from the approximately $(2/3)n^3$ flops needed for the decomposition of $DF(x)^T$.

In practice, it has turned out to be advantageous to construct the reference matrix T_r in the following manner. We select a reference point x^r on M. Then the Euclidean norms

$$\tau_i = \| U_0(x^r)^T e^i \|_2 \ , \ i=1,\dots,n$$

of the rows of $U_0(x^r)$ are the cosines of the principal angles between the tangent space of M at x^r and the i-th natural basis vector e^i of R^n. The τ_i are independent of the choice of the basis matrix $U_0(x)$. Let $i_1,\dots,i_p$ be the indices of the p largest of these τ_i (with ties broken, say, lexicographically). Then we form the desired reference matrix T_r as the matrix with the columns e^i , $i=i_1,\dots,i_p$. This construction is analogous to the local parameter selection in the continuation program PITCON, [11].

4. The Triangulation Algorithm

For the triangulation of a p-dimensional manifold we begin by constructing a reference triangulation on R^p. Let Σ be the collection of simplices of this triangulation. Except for considerations of computational efficiency and simplicity, no restrictions are placed on Σ. We refer, for example, to [12] for various algorithms for triangulating R^p. For our purposes, the well-known Kuhn-triangulations have been useful, and, in the case p = 2, triangulations of R^2 by equilateral triangles have been applied as well.

Let ξ denote a given vertex of this triangulation in R^p and $h > 0$ a fixed steplength. Then for any point $x \in M$ where a basis matrix U of T_xM is known, the mapping

$$A: R^p \rightarrow x + T_xM, \quad A\eta = x + hU(\eta-\xi), \ \eta \in R^p \qquad (4.1)$$

transfers Σ from R^p onto $x + T_xM$. As before, let $V(x) \subset x + T_xM$ denote the local convergence domain of the Gauss-Newton process (2.5) . If η is a vertex of Σ for which $A\eta \in V(x)$, then (2.5) can be used to map $A\eta$ into a point $y \in M$. The set $\Gamma(\xi,x,U)$ of vertices of Σ that can be mapped onto M in this way shall be called the "patch" corresponding to ξ,x,U . (The steplength h will be held fixed throughout).

An "idealized" form of our algorithm can now be phrased as follows:

(1) Select a reference vertex ξ^* of Σ ;
(2) Select a reference point $x^* \in M$ and let M_0 be the subset
    where, by the theorem, the moving frame algorithm applies ;
(3) Set $x = x^*$, $\xi = \xi^*$;
(4) Mark the vertex ξ as "used" ;
(5) While $x \in M_0$
    (5a) Mark ξ as a "center"
    (5b) Compute the frame $U(x)$ by algorithm (3.1) ;
    (5c) Select all vertices of the patch $\Gamma(\xi,x,U(x))$
        which have not yet been marked "used" ;
    (5d) Map these vertices onto M and mark them "used" ;
    (5e) Choose a "used" vertex ξ of Σ not marked a "center"
        and let x be its computed image on M ;

The points computed on M inherit the connectivity pattern of the original simplices of Σ which, in turn, induces a simplicial approximation M_Σ of M in R^n.

The algorithm is still "idealized" because, in practice, it is impossible to check the condition $x \in M_0$ and to identify the vertices of Σ that belong to $\Gamma(\xi,x,U(x))$. Thus, special provisions have to be added in order to overcome the possible failures due to these missing checks. We shall not go into details here. The principal approach is to select a "standardized" patch of Σ which is used in step (5c) in place of $\Gamma(\xi,x,U(x))$. Then, in step (5d), appropriate alternatives are introduced for all vertices where a failure of the corrector iteration is encountered.

As noted, for two-dimensional manifolds a reference triangulation of equilateral triangles can be used. Then the "standardized" patch is the hatched, star-shaped region in the center of Figure 1. At each vertex, the second of the two integers is a counter and the first one identifies the "center" ξ that is used in mapping that vertex onto M. Thus, after the reference vertex 0, the nodes 7,...,12 become centers which serve to map the nodes 13,...,42 onto M. Then the process continues with nodes 17,18,19,23,24,28,29,33,34,38,39,42 as centers. This is no longer shown in the figure, but, in practice, we always continued through this further stage. It results in a total of 114 triangles on M and involves 19 centers and hence as many Jacobian evaluations. This indicates the efficiency of the algorithm. In fact, in terms of computed points per Jacobian evaluation, the method performes better than most continuation processes.

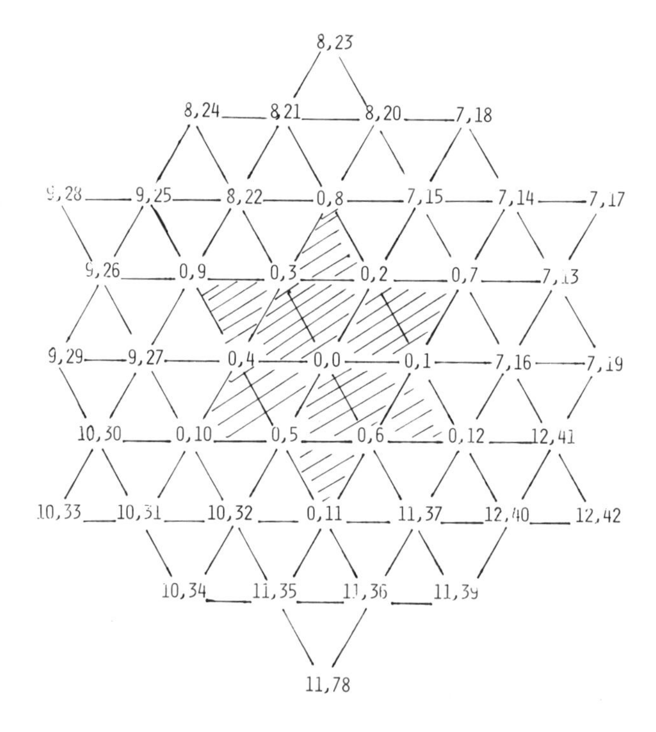

Figure 1

5. Examples

We present now a few numerical examples to indicate the performance of the methods. But space limitations force us to be brief. More extensive examples will be given elsewhere.

Our first example concerns the well-known Belousov-Zhabotinskii reaction [13]. As in [6] we write the mass balance equations in the form

$$\begin{aligned} (\mu - x_1)x_2 + x_1(1-x_1) - \varepsilon_1 \beta x_1 &= 0 \\ -(\mu + x_1)x_2 + x_3 + \varepsilon_2 \beta(\alpha - x_2) &= 0 \\ x_1 - x_3(1-\beta) &= 0 \end{aligned} \tag{5.1}$$

If $\varepsilon_1 = 1/1{,}500$, $\varepsilon_2 = 1/56{,}250$, and $\mu = 8.4\times10^{-6}$, then, as discussed in [6], there is an isola point approximately at the point with the coordinates

Figure 2

$x_1 = 0.249$, $x_2 = 0.750$, $x_3 = 0.125$, $\alpha = 3{,}508$, $\beta = 0.997$. This point was used as our reference point on M, and Figure 2 shows the computed simplicial triangulation (based on the reference triangulation of Figure 1). The printed page is the α,β-plane and x_2 is the third coordinate in the figure.

Our second example concerns the roll stability of maneuvering airplanes. Without going into details, we use the equations originally formulated in [7] and given in [8] and [5] in a simplified form $Ax+\Phi(x) = 0$, $x \in R^8$. Here A is a 5 x 8 matrix and $\Phi: R^8 \to R^5$ a quadratic function. The (dimensionless) control parameters x_6, x_7, x_8 denote the elevator, aileron, and rudder deflections, respectively. The bifurcation diagram for rudder deflections $x_8 = 0$ was given in [8] and again (with some extensions) in [5]. In the neighborhood of the origin of the x_6,x_7-plane it has the form shown in Figure 3.

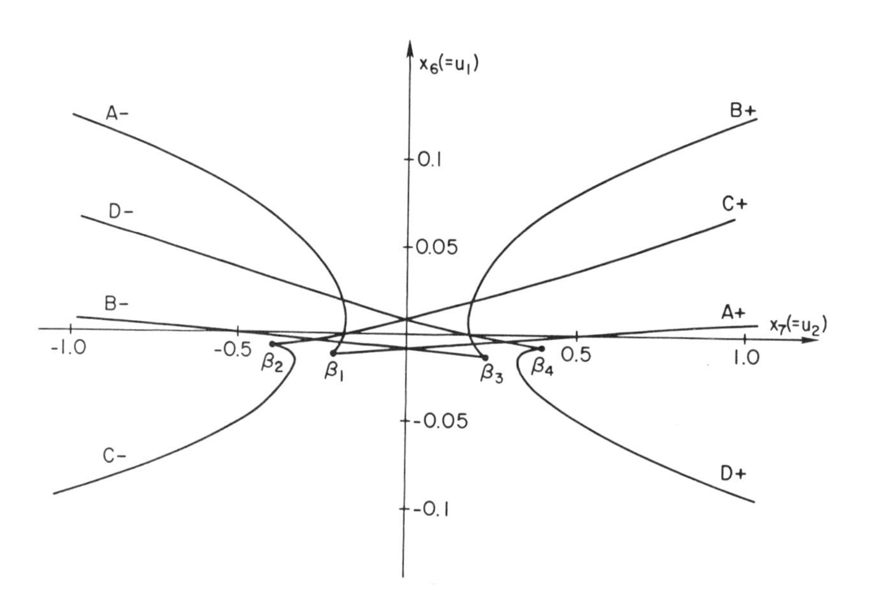

Figure 3

The process was applied with an approximation of the bifurcation point β_2 as center. The results are shown in Figure 4, where the printed page is the x_6,x_7-plane and the third coordinate the roll-rate x_1. The triangulation contains also the bifurcation point β_1, and the foldlines emanating from these two bifurcation points are clearly visible.

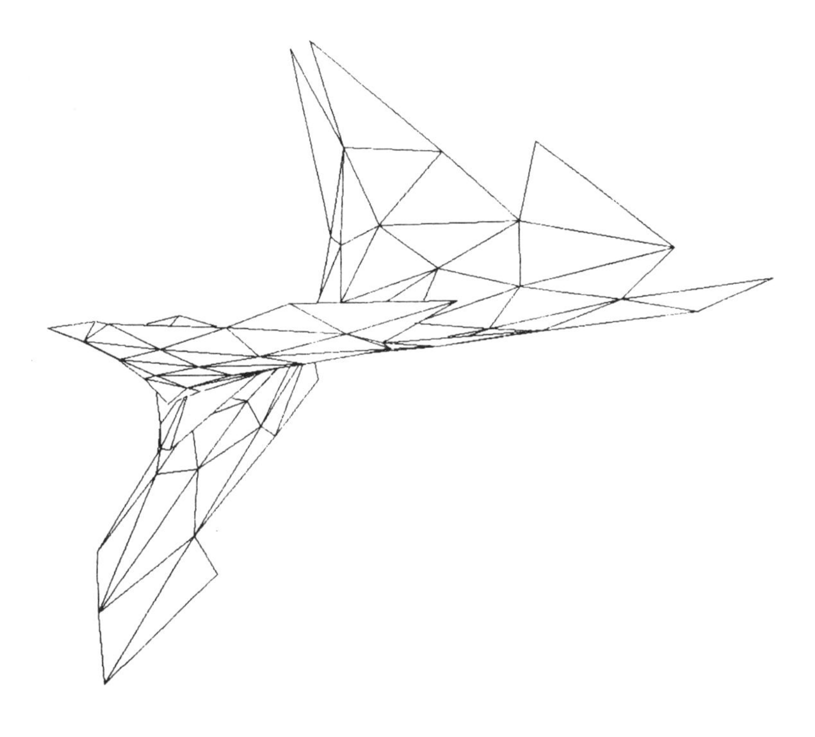

Figure 4

The examples indicate that the algorithms work very efficiently, even around singularities. Thus, as intended, they do indeed provide a new tool for deriving information about the shape and features of the manifold. Of course, besides any graphical representation, the extensive numerical output of the process contains a wealth of further information. For instance, linear interpolation between the computed points defines the earlier mentioned simplicial approximation M_Σ of M. The corrector process can be started from any point of M_Σ to produce additional points of M. In addition, for any given functional it is easy to compute a contour plot of its values on M_Σ. For instance, in some structural problems it may be of interest to determine lines of constant stress components. Similarly, the foldlines on M represent contour lines with respect to a measure of the orientation of the projection of the tangent spaces onto the parameter space. This provides for a simple method of approximating the fold-lines on M which can then be used to compute the fold points themselves by means of one of the numerous local iterative processes available for that purpose. Examples of these, and other post-processing procedures will be given elsewhere.

6. References

[1] E.L.Allgower and P.H.Schmidt, An Algorithm for Piecewise-linear Approximation of an Implicitly Defined Manifold; SIAM J.Numer. Anal. 22, 1985, 322-346

[2] T.F.Coleman and D.C.Sorensen, A Note on the Computation of an Orthonormal Basis for the Null Space of a Matrix; Mathem. Progr. 29, 1984, 234-242

[3] P.Deuflhard and G.Heindl, Affine Invariant Convergence Theorems for Newton's Method and Extensions to Related Methods; SIAM J. Num. Anal. 16, 1979, 1-10

[4] J.P.Fink and W.C.Rheinboldt, Solution Manifolds of Parametrized Equations and Their Discretization Error; Numer. Math. 45, 1984, 323-343

[5] A.Jepson and A.Spence, Folds in Solutions of Two-Parameter Systems and Their Calculations, Part I; SIAM J. Num. Anal. 22, 1985, 347-368

[6] M.Kubicek, I.Stuchl, M.Marek, "Isolas" in Solution Diagrams,; J. of Comp. Physics, 48, 1982, 106-116

[7] R.K.Mehra, W.C.Kessel, and J.V.Carroll, Global Stability and Control Analysis of Aircraft at High Angles of Attack; ONR report CR-215-248-1,2,3, June 1977, pp78-79.

[8] W.C.Rheinboldt, Numerical Methods for a Class of Finite Dimensional Bifurcation Problems; SIAM J. Num. Anal. 17, 1980, 221-237

[9] W.C.Rheinboldt, Numerical Analysis of Parametrized Nonlinear Equations, J. Wiley and Sons, New York, NY, 1986

[10] W.C. Rheinboldt, On the Computation of Multi-Dimensional Solution Manifolds of Parametrized Equations, Numer. Math., submitted

[11] W.C.Rheinboldt and J.V.Burkhardt, A Locally Parametrized Continuation Process, ACM Trans.on Math. Softw., 9,1983,236-246

[12] M.J.Todd, The Computation of Fixed Points and Applications; Springer Verlag, New York, NY 1976

[13] J.J.Tyson, The Belousov-Zhabotinskii Reaction, Lecture Notes in Biomathematics, Vol. 10, Springer Verlag, New York, NY 1976

W. C. Rheinboldt
University of Pittsburgh
Department of Mathematics and Statistics
Pittsburgh, PA. 15260

International Series of
Numerical Mathematics, Vol. 79

Numerical computation of origins for Hopf bifurcation in a two-parameter problem

D. Roose
Department of Computer Science, K. U. Leuven, Belgium

1. Introduction

We consider nonlinear evolution equations of the form

$$\frac{du}{dt} = F(u,\lambda,\gamma) \qquad F: \mathbb{R}^n \times \mathbb{R} \times \mathbb{R} \to \mathbb{R}^n \tag{1}$$

where λ and γ denote real parameters. In order to obtain a qualitative description of the solution set of (1) as function of the parameters, one can determine the "bifurcation set" by computing paths of turning points, Hopf bifurcation points and (in some cases) regular bifurcation points. This can be done by a continuation procedure applied to a suitable determining system, starting from a known branching point for the one-parameter problem $du/dt = F(u,\lambda,\gamma^f)$ (γ^f fixed) (see e.g. [9,17]).

A branch of Hopf bifurcation points can bifurcate from a branch of turning points and in this paper we present a method for the computation of the origin of such a branch of Hopf points. Starting from these "origins for Hopf bifurcation" (or "B-points") one can easily compute branches of Hopf points. The knowledge of these origins allows also the investigation of "global Hopf bifurcation".

In the next section we discuss briefly Hopf bifurcation in a one-parameter problem and we give the conditions for an emanating branch of Hopf points in a two-parameter problem. In section 3 we introduce a direct method for the calculation of an origin of such a branch and we indicate how this method can be implemented efficiently. Finally, in section 4 we illustrate our method with results for a model problem.

2. Hopf bifurcation

Sufficient conditions for the occurrence of Hopf bifurcation in a one-parameter problem of the form $du/dt = G(u,\lambda)$ are [4] :

(H1) There exists a smooth branch of steady state solutions $u^{(s)}(\lambda)$ with $u^{(s)}(\lambda^h) = u^h$;

(H2) G is sufficiently smooth in a neighbourhood of (u^h, λ^h);

(H3) The Jacobian matrix $G_u(u^{(s)}(\lambda),\lambda)$ with $\lambda \in (\lambda^h - \epsilon, \lambda^h + \epsilon)$ has a pair of simple complex conjugate eigenvalues $\mu(\lambda)$, $\overline{\mu(\lambda)}$ such that $\mu(\lambda) = \alpha(\lambda) \pm i\omega(\lambda)$ with $\omega(\lambda^h) = \omega^h > 0$, $\alpha(\lambda^h) = 0$ and $\alpha'(\lambda^h) \neq 0$;

(H4) $G_u(u^h, \lambda^h)$ has no eigenvalues of the form $ki\omega^h$, $k \in \{0,2,3,4, \cdots\}$.

The steady state solution (u^h, λ^h) is called a Hopf bifurcation point. At a Hopf point, a branch of time periodic solutions bifurcates from the branch of steady state solutions. Since Hopf bifurcation points are structurally stable in a one-parameter problem, branches of Hopf points can exist in a two-parameter problem of the form (1).

Several methods are proposed in the literature for the numerical determination of a Hopf bifurcation point. A classification and references can be found in [13]. All these iterative methods require a sufficiently good estimate of the Hopf point as a starting point to insure convergence. The performance and the reliability of some methods with respect to the starting point is investigated in [6] and [14].

However, the detection of a Hopf bifurcation point during a continuation process along a branch of steady state solutions is a much more difficult problem than the detection of a turning point or regular bifurcation point. One can compute all the eigenvalues of $G_u(u,\lambda)$ at each continuation point, but this is very expensive compared to the cost of the continuation, especially when the Jacobian matrix has large dimension and is sparse or banded (see [8] and [15] for more details).

Guckenheimer and Holmes [3,pp.364-374] analyse the unfolding of a degenerate vector field with normal form

$$\begin{aligned} \frac{dx}{dt} &= y \\ \frac{dy}{dt} &= x^2 + xy \end{aligned} \qquad (x, y \in \mathbb{R}) \qquad (2)$$

The analysis shows that in a two-parameter problem of the form (1), a branch of Hopf bifurcation points emanates from a singular steady state solution $(u^o, \lambda^o, \gamma^o)$ for which the Jacobian matrix $F_u(u^o, \lambda^o, \gamma^o)$ has a double eigenvalue zero and a one-dimensional nullspace. Also a branch of simple quadratic turning points passes through $(u^o, \lambda^o, \gamma^o)$. Thus at $(u^o, \lambda^o, \gamma^o)$ a branch of Hopf points bifurcates from the branch of quadratic turning points and the branching is one-sided. (The unfolding of this singularity is also studied by other authors; see [1] and [3] for references.)

We will call such a steady state solution $(u^o, \lambda^o, \gamma^o)$ an *"origin for Hopf bifurcation"* or, following Fiedler [1,2], a *"B-point"*.

Thus at an origin for Hopf bifurcation the right and left nullvectors, denoted by ϕ^o and ψ^o, are orthogonal [7], i.e.

$$F_u(u^o,\lambda^o,\gamma^o)\phi^o = 0 \qquad ||\phi^o|| = 1 \tag{3}$$

$$F_u^T(u^o,\lambda^o,\gamma^o)\psi^o = 0 \qquad ||\psi^o|| = 1 \tag{4}$$

$$\psi^{oT}\phi^o = 0 \tag{5}$$

Then the Jacobian matrix $F_u(u^o,\lambda^o,\gamma^o)$ has also a generalized eigenvector ζ^o satisfying

$$F_u(u^o,\lambda^o,\gamma^o)\zeta^o = \phi^o \qquad \phi^{oT}\zeta^o = 0 \tag{6}$$

Note that also other normalization conditions for ζ^o can be used [7].
Assuming that the origin for Hopf bifurcation is a simple quadratic turning point with respect to λ, also the following conditions must be satisfied (see e.g. [9,17])

$$\psi^{oT}F_\lambda(u^o,\lambda^o,\gamma^o) \neq 0 \tag{7}$$

$$\psi^{oT}F_{uu}(u^o,\lambda^o,\gamma^o)\phi^o\phi^o \neq 0 \tag{8}$$

3. The calculation of an "origin of Hopf bifurcation"

Based on the theoretical results of the previous section, the following procedure can be used to compute the origin of a branch of Hopf points in a two-parameter problem :

— determine branches of quadratic turning points by continuation applied to a suitable determining system [9,17]

— during the continuation process, calculate at each turning point $(u^{t_i},\lambda^{t_i},\gamma^{t_i})$

$$c^{t_i} = |\cos(\psi^{t_i},\phi^{t_i})| \tag{9}$$

where ϕ^{t_i} and ψ^{t_i} are the right and left nullvectors of $F_u(u^{t_i},\lambda^{t_i},\gamma^{t_i})$, which can be computed by inverse iteration.

— If this value (9) attains a minimum which is close to zero, probably a nearby turning point (u^o,λ^o,γ^o) is an origin for Hopf bifurcation, which can be computed by solving the determining system given below.

Consider the augmented system

$$C(y) = 0 \qquad C : \mathbb{R}^{3n+2} \to \mathbb{R}^{3n+2} \tag{10a}$$

with $y = (u,\lambda,\gamma,\phi,\zeta)$ and

$$C(u,\lambda,\gamma,\phi,\zeta) = \begin{cases} F(u,\lambda,\gamma) \\ F_u(u,\lambda,\gamma)\phi \\ \phi^T\phi - 1 \\ F_u(u,\lambda,\gamma)\zeta - \phi \\ \zeta^T\phi \end{cases} \tag{10b}$$

The system $C(y) = 0$ is a regular determining system for an origin for Hopf bifurcation (u^o,λ^o,γ^o) since $(u^o,\lambda^o,\gamma^o,\phi^o,\zeta^o)$ is a solution of (10) and since the following theorem holds [15].

Theorem : Let (u^o,λ^o,γ^o) be an origin for Hopf bifurcation lying on a branch of quadratic turning points $(u(\gamma),\lambda(\gamma),\gamma)$ with respect to λ of $F(u,\lambda,\gamma) = 0$ and parametrized by γ. Then the Jacobian matrix $C_y(y^o)$ of (10) evaluated at $y^o = (u^o,\lambda^o,\gamma^o,\phi^o,\zeta^o)$ is nonsingular if

$$\left.\frac{d(\ \psi(\gamma)^T\phi(\gamma)\ )}{d\gamma}\right|_{\gamma=\gamma^o} \neq 0 \tag{11}$$

where $\phi(\gamma)$ and $\psi(\gamma)$ denote the nullvectors of $F_u(u(\gamma),\lambda(\gamma),\gamma)$ and $F_u^T(u(\gamma),\lambda(\gamma),\gamma)$, respectively.

□

Note that condition (11) only means that the inner product $\psi(\gamma)^T\phi(\gamma)$ must be nonzero along the branch of quadratic turning points for $\gamma \in [\gamma^o-\epsilon,\gamma^o+\epsilon] \setminus \gamma^o$.

If the determining system (10) is solved by Newton's method, the computation of the Newton correction requires the solution of a linear system of dimension $3n+2$ but one can take advantage of the structure of the Jacobian matrix to reduce the computational work. In [15] we describe a procedure to compute the Newton correction which only requires :

a) evaluation of the residual and the first and second derivatives of F

b) computation of 6 matrix-vector-products

c) 1 LU-decomposition of the $n\times n$-matrix $A = F_u(u,\lambda,\gamma)$

d) 9 backsubstitutions.

This procedure is very efficient, but can be unstable, since systems have to be solved with the matrix A, which becomes singular at (u^o,λ^o,γ^o). The numerical stability can be improved [15], but our numerical experiments indicate that the procedure is fairly reliable in practice. Indeed, the matrix A will only be very ill-conditioned in the last Newton iteration steps and the relative accuracy of the last (small) Newton corrections is not crucial if the solution must not be known to full machine-precision.

In [15] we also indicate how starting values for ϕ^o and ζ^o can be obtained, if sufficiently good starting values for u^o, λ^o, γ^o are available, e.g. by continuation of a branch of quadratic turning points as outlined above.

4. Application to a model problem : behaviour of a tubular reactor

Axial dispersion of mass and heat in a tubular nonadiabatic reactor where a simple first order chemical reaction takes place can be described by [5]

$$\begin{aligned} \frac{\partial Y}{\partial t} &= \frac{1}{Pe}\frac{\partial^2 Y}{\partial x^2} - \frac{\partial Y}{\partial x} - Da\, Y \exp(\gamma - \frac{\gamma}{T}) \\ \frac{\partial T}{\partial t} &= \frac{1}{Pe}\frac{\partial^2 T}{\partial x^2} - \frac{\partial T}{\partial x} - \beta(T - T^0) + B\, Da \exp(\gamma - \frac{\gamma}{T}) \end{aligned} \tag{12a}$$

with boundary conditions

$$\begin{aligned} x = 0 &: \frac{\partial Y}{\partial x} = Pe(Y-1), \ \frac{\partial T}{\partial x} = Pe(T-1) \\ x = 1 &: \frac{\partial Y}{\partial x} = 0 \qquad , \ \frac{\partial T}{\partial x} = 0 \end{aligned} \tag{12b}$$

Here Y and T represent concentration and temperature, respectively. The bifurcation behaviour of this system with respect to Da (the Damkohler number) has been studied extensively (see e.g. [5]). We will consider β (the heat transfer coefficient) as the second parameter.

Using the customary $O(h^2)$-discretization on an equidistant mesh, this system can be approximated by a system of ordinary differential equations of the form (1), with Da and β playing the role of the parameters previously denoted by λ and γ. In our calculations we used 32 discretization points ($h = 1/31$).

Heinemann and Poore [5] computed bifurcation diagrams with respect to Da for $Pe = 5$, $\gamma = 25$, $B = 0.5$, $T^0 = 1$ and for various values of β. For $\beta = 2$ the diagram has four quadratic turning points.

Starting from these quadratic turning points we calculated branches of turning points. Therefore we used the continuation code PITCON of Rheinboldt and Burkardt [12] to compute solutions of the determining system for quadratic turning points, introduced by Seydel [16] and Moore & Spence [10] (see [15] for more details). In Fig. 1 we show the projection of the computed branches of turning points on the parameter plane, using $12Da - \beta$ and β as coordinate axes. Note that the particular form of this "bifurcation set", containing three cusps, suggests that probably a parabolic umbilic catastrophe is present in this system (compare with Fig. 9.24 in [11]).

We computed origins for Hopf bifurcation on these branches of turning points, using the algorithm given in section 3. At the points $T_1^{(m)}, \ldots, T_4^{(m)}$ (see Table I) the testfunction (9) attained a minimum close to zero. Starting from these points, four origins for Hopf bifurcation $O_1, \ldots, O_4$ were computed by solving the determining system (10) with a damped Newton iteration. The results are given in Table I. Note that O_3 lies very close to one of the cusp points.

Table I : Origins for Hopf bifurcation O_i of Eq. (12) with $Pe = 5$, $\gamma = 25$, $B = 0.5$, $T^0 = 1$ (32 discretization points). $T_i^{(m)}$: used as starting points for the computation of O_i.

	Da	β	$Y(0)$	$T(0)$
O_1	0.0771280	1.2170586	0.8783655	1.0366636
O_2	0.0838718	1.3702861	0.6170820	1.1316478
O_3	0.1629411	2.4712925	0.9323363	1.0163622
O_4	0.0445415	0.8196301	0.9726650	1.0087249
$T_1^{(m)}$	0.0678	1.081	0.8801	1.038
$T_2^{(m)}$	0.0904	1.463	0.6119	1.131
$T_3^{(m)}$	0.1633	2.476	0.9335	1.016
$T_4^{(m)}$	0.0409	0.743	0.9736	1.009

In order to determine the emanating branches of Hopf points we first computed a solution $(u^{h_1}, Da^{h_1}, \beta^{h_1}, \omega_{\text{fixed}}, p^{h_1})$ of the following determining system for Hopf points [13,14]

$$\begin{cases} F(u,Da,\beta) = 0 \\ ([F_u(u,Da,\beta)]^2 + \omega^2)p = 0 \\ p^T p = 1 \\ q^T p = 0 \end{cases} \tag{13}$$

Here q is an fixed vector, chosen orthogonal on the starting vector for p. If the fixed value ω_{fixed} is small (e.g. $\omega_{\text{fixed}} = 0.1$) u^o, Da^o, β^o and ϕ^o will be good starting values for u^{h_1}, Da^{h_1}, β^{h_1} and p^{h_1}, especially since the derivative of ω along the branch of Hopf points becomes infinite at the origin of the branch [14]. Starting from this Hopf point, we used PITCON to compute a branch of solutions of the determining system (13).

Fig. 1 shows that the branches of Hopf points starting at O_1 and O_2 end at O_4 and O_3, respectively. Both branches have a turning point with respect to β, respectively at $\beta = 1.79$ and $\beta = 13.1$. Projected on the parameter plane the branches of Hopf points bifurcate tangentially from the branches of turning points. This is in agreement with the theoretical results of Guckenheimer and Holmes [3]. The projection of these

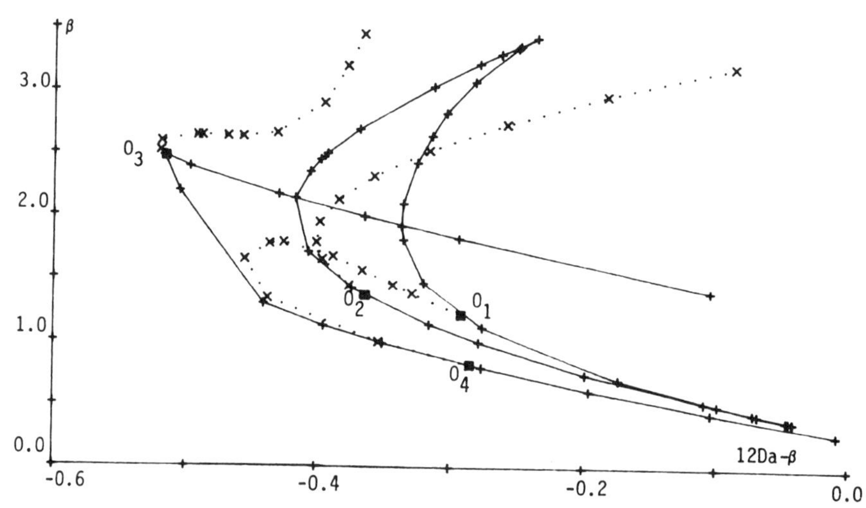

Fig. 1 : Branches of quadratic turning points and Hopf points for Eq. (12) with $Pe = 5$, $\gamma = 25$, $B = 0.5$, $T^0 = 1$. Projection on the $(12Da-\beta,\beta)$-plane.
—— : branches of turning points ; . . . . : branches of Hopf points

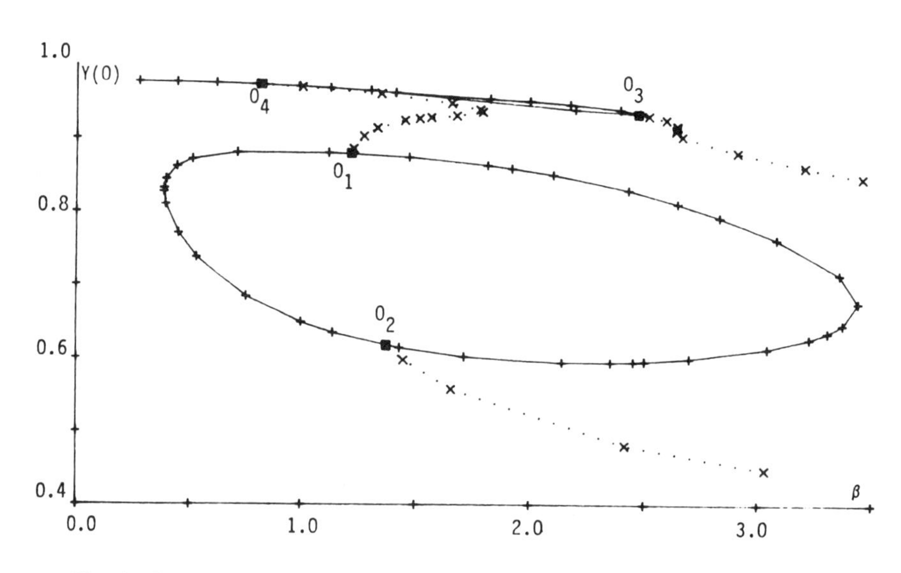

Fig. 2 : Branches of quadratic turning points and Hopf points for Eq. (12) with $Pe = 5$, $\gamma = 25$, $B = 0.5$, $T^0 = 1$. Projection on the $(\beta,Y(0))$-plane.
—— : branches of turning points ; . . . . : branches of Hopf points

branches on the $(\beta, Y(0))$-plane is shown in Fig. 2.

Considering Da as a "distinguished bifurcation parameter", we can determine from Figs. 1 and 2 the number and the location of the Hopf points in the bifurcation diagrams with respect to Da for a fixed value of β. During the construction of the bifurcation diagrams with respect to Da, Heinemann and Poore [5] detected Hopf bifurcation by computing all the eigenvalues of F_u at each continuation point. However they only found Hopf points lying on the branch $O_2 O_3$. So they concluded e.g. that no Hopf bifurcation occurs for $\beta = 1$. Note that in order to find the Hopf points for e.g. $\beta = 1.5$, rather small continuation steps must be taken; otherwise one could "jump" over the two Hopf points in one continuation step.

5. Conclusion

We can conclude that the calculation of origins for Hopf bifurcation, using the procedure outlined above, allows an efficient analysis of Hopf bifurcation for a two-parameter problem.

Firstly, the determination of these origins is very useful for the computation of branches of Hopf points for a two-parameter problem, especially since it is difficult (or expensive) to detect Hopf bifurcation in a one-parameter problem. Note however that in a two-parameter problem also closed branches of Hopf bifurcation points can exist, which do not contain an origin for Hopf bifurcation and thus cannot be determined using our approach.

Secondly, the location of the origins for Hopf bifurcation is important for the analysis of "global Hopf bifurcation" along a one-dimensional closed curve Γ in a two-parameter space, as shown by Fiedler [1]. He introduces the "B-index" of an origin for Hopf bifurcation (or B-point) and proves that global Hopf bifurcation occurs along Γ if the sum of the B-indices of all B-points inside Γ is nonzero. The determination of the B-index requires the calculation of all eigenvalues of $B_u(u^o, \lambda^o, \gamma^o)$. In [2] Fiedler and Kunkel give a numerical procedure to locate B-points and to compute the B-index very efficiently, but only in the special context of equations of the form $D(\gamma)\, du/dt = F(u, \lambda)$ where $D(\gamma)$ represents the diagonal matrix diag $[1, \ldots, \gamma, \ldots, 1]$.

References

[1] B. Fiedler (1986) *Global Hopf bifurcation of two-parameter flows*, Arch. Rat. Mech. Anal. 94, 59-81.

[2] B. Fiedler and P. Kunkel (1986) *A quick multiparameter test for periodic solutions*, These proceedings.

[3] J. Guckenheimer and P. Holmes (1983) *Nonlinear oscillations, dynamical systems, and bifurcations of vector fields*, Applied Mathematical Sciences 42 (Springer, New York).

[4] B. Hassard, N. Kazarinoff and Y-H. Wan (1981) *Theory and applications of Hopf bifurcation* (Cambridge University Press, Cambridge).

[5] R.F. Heinemann and A.B. Poore (1981) *Multiplicity, stability and oscillatory dynamics of the tubular reactor*, Chem. Engng. Sci. 36, 1411-1419.

[6] M. Holodniok and M. Kubicek (1984) *New algorithms for the evaluation of complex bifurcation points in ordinary differential equations. A comparative numerical study*, Appl. Math. Comp. 15, 261-274.

[7] G. Iooss and D.D. Joseph (1980) *Elementary stability and bifurcation theory* (Springer, New York).

[8] A.D. Jepson (1981) *Numerical Hopf bifurcation*, Ph.D. Thesis, Part II, California Institute of Technology, Pasadena.

[9] A.D. Jepson and H.B. Keller (1984) *Steady state and periodic solution paths : their bifurcations and computations*, In : T. Küpper, H.D. Mittelmann and H. Weber (eds.), Numerical methods for bifurcation problems, ISNM 70 (Birkhäuser, Basel), 219-246.

[10] G. Moore and A. Spence (1980) *The calculation of turning points of non-linear equations*, SIAM J. Numer. Anal. 17, 567-576.

[11] T. Poston and I.N. Stewart (1978) *Catastrophe theory and its applications* (Pitman, London).

[12] W.C. Rheinboldt and J.V. Burkardt (1983) *A locally parametrized continuation process*, ACM-TOMS 9, 215-241.

[13] D. Roose and V. Hlavacek (1985) *A direct method for the computation of Hopf bifurcation points*, SIAM J. Appl. Math. 45, 879-894.

[14] D. Roose (1985) *An algorithm for the computation of Hopf bifurcation points in comparison with other methods*, J. Comp. Appl. Math. 12&13, 517-529.

[15] D. Roose (1986) *Numerical determination of an emanating branch of Hopf bifurcation points in a two-parameter problem*, Report TW 82, Dept. Computer Science, K.U. Leuven.

[16] R. Seydel (1979) *Numerical computation of branch points in nonlinear equations*, Numer. Math. 33, 339-352.

[17] A. Spence and A. Jepson (1984) *The numerical calculation of cusps, bifurcation points and isola formation points in two parameter problems*, In : T. Küpper, H.D. Mittelmann and H. Weber (eds.), Numerical methods for bifurcation problems, ISNM 70 (Birkhäuser, Basel), 502-514.

Dirk Roose, Department of Computer Science, K. U. Leuven, Celestijnenlaan 200A, B-3030 Leuven, Belgium.

International Series of
Numerical Mathematics, Vol. 79

BIFURCATIONS OF THE EQUILIBRIUM OF A SPHERICAL DOUBLE PENDULUM AT A MULTIPLE EIGENVALUE

Alois Steindl, Hans Troger
Institut für Mechanik, Technische Universität Wien

1. Introduction

The loss of stability of the downhanging equilibrium position of tubes conveying fluid and allowing for a three-dimensional motion has been studied in [1-3], where both continuous and discrete models for the tube have been used. This problem is particularly interesting from the standpoint of stability theory because due to certain symmetries in the system which we will explain below the critical eigenvalues (i.e. eigenvalues with zero real part; see chapter 3) have double multiplicity. As we want to analyse the nonlinear problem in its post-bifurcational behavior we shall perform a reduction of the n-dimensional problem to a system of bifurcation equations on the center manifold, the dimension of which is equal to the number of critical eigenvalues. Hence at the first glance, due to the doubled dimension, a very complicated situation is found for the derivation of the bifurcation equations. However, due to the symmetry properties, which also the bifurcation equations on the center manifold must fulfill, they are pretty simple, such that these two effects, namely the higher multiplicity of the critical eigenvalue and the symmetry properties of the bifurcation equations compensate each other in some sense.

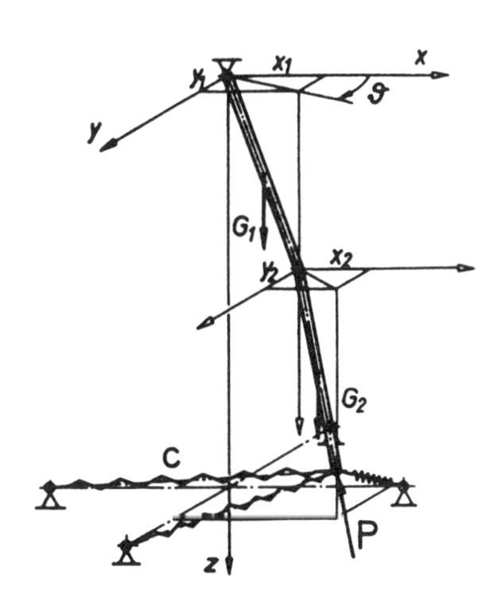

Fig.1: Spherical double pendulum with viscoelastic joints, follower force P and elastic endsupport C

Extending the model used in [3] by an elastic end support (Fig. 1) we study the problem by a discrete model of two rigid rods of length l_1 and l_2 with viscoelastic joints under the action of a follower force P, which is a simpler loading than that given by flowing fluid, but basically has the same qualitative properties.

The end support is important for two purposes. One is to have a model for practical applications. For example for fire fighting purposes and for tubes used for filling up purposes, where the end of the tube must be kept at a certain position, we have tube systems with such elastic end supports. The other is of theoretical interest, because it is important for us that by varying the stiffness of the end support we can control the structure of the critical eigenvalues at loss of stability. Hence we are able to create more degenerate situations, which are quite interesting, firstly, to be analysed theoretically and, secondly, to see their practical implications. The mathematical apparatus to study such cases can be found in [4,5].

2. Equations of motion and symmetry properties

The spherical pendulum of Fig. 1 has four degrees of freedom. It is not possible to use spherical polar coordinates because the angle θ shown in Fig. 1 would jump by the amount of π if the rod moves through the z-axis. Hence we use, as it also has been done in [3], the distances x_1, x_2, y_1, y_2, as shown in Fig. 1, as our variables. In these variables the equations of motion can be given in the form ([3])

$$M\ddot{x} + D\dot{x} + Nx = \begin{Bmatrix} K_1(\xi,\eta) \\ K_2(\xi,\eta) \end{Bmatrix}, \quad M\ddot{y} + D\dot{y} + Ny = \begin{Bmatrix} K_3(\xi,\eta) \\ K_4(\xi,\eta) \end{Bmatrix} \tag{1}$$

where the nonlinear terms up to third degree are collected in the $K_i(\xi,\eta)$. $x = (x_1,x_2)^T$, $y = (y_1,y_2)^T$ and

$$\xi = (x_1,\dot{x}_1,x_2,\dot{x}_2)^T \quad \text{and} \quad \eta = (y_1,\dot{y}_1,y_2,\dot{y}_2)^T \tag{2}$$

are already nondimensional variables. The matrices M,D, N are the mass, damping and stiffness matrix, respectively. Here M and D are symmetric matrices whereas N is not symmetric. This is due to the follower force loading which is basically a nonconservative type of loading. A comment seems to be appropriate concerning the damping forces. We consider two different types of damping mechanisms. One is internal damping created by the viscoelastic joints and proportional to the relative angular velocities of the rods. The other is external damping proportional to the absolut velocities of the rods. When calculating the eigenvalues in chapter 3 we shall see that it is necessary to consider both types of damping. The nonlinear functions K_i of third order have the following properties

$$K_3(\xi,\eta) = K_1(\eta,\xi) \quad \text{and} \quad K_4(\xi,\eta) = K_2(\eta,\xi) \ . \tag{3}$$

That we only retain third order terms in the K_i confines our analysis to motions adjacent to the downhanging equilibrium position. Next we rewrite (1) as a system of equations of first order. Introducing (2) into (1) we obtain

$$\begin{pmatrix} A & 0 \\ 0 & A \end{pmatrix} \begin{pmatrix} \dot{\xi} \\ \dot{\eta} \end{pmatrix} = \begin{pmatrix} B & 0 \\ 0 & B \end{pmatrix} \begin{pmatrix} \xi \\ \eta \end{pmatrix} + (0,K_1(\xi,\eta),0,K_2(\xi,\eta),\ldots,K_4(\xi,\eta))^T \tag{4}$$

where A and B are 4×4 matrices obtained from (1) and the additional equations $\dot{x}_1 = x_2, \ldots$ etc.

As already mentioned in the Introduction, it is of basic importance for the mathematical treatment of this problem that it possesses certain symmetry properties. These properties are pretty obvious, because we see from Fig. 1 that a rotation of the pendulum by an angle θ around the z-axis does not change anything. To have the required symmetry property for the endsupport we either must assume to have many springs or if there are only a few as in Fig. 1 they must be very long (theoretically infinitely long). Furthermore also a reflection at a plane through the z-axis leaves the problem invariant. These two properties, however, are just the actions of the group O(2). Hence the problem which we are treating is invariant under this group action.

In general a system expressed by its equations of motion $\dot{u} = G(u,\lambda)$ is said to be equivariant under the action of a group $\tilde{G}$ if with a repesentation T_g of each element $g \in \tilde{G}$ the relationship

$$T_g G(u,\lambda) = G(T_g u,\lambda) \tag{5}$$

is satisfied. A representation of the group of rotations in our problem is given by the following rotation matrix (8×8)

$$R(\theta) = \begin{pmatrix} \cos\theta & & 0 & & \sin\theta & & 0 & \\ 0 & \ddots & \cos\theta & & 0 & \ddots & \sin\theta & \\ -\sin\theta & & 0 & & \cos\theta & & 0 & \\ 0 & \ddots & -\sin\theta & & 0 & \ddots & \cos\theta & \end{pmatrix} \ . \tag{6}$$

From (5) it follows for the linear part

$$R(\theta)\begin{pmatrix} A & 0 \\ 0 & A \end{pmatrix} = \begin{pmatrix} A & 0 \\ 0 & A \end{pmatrix} R(\theta) \tag{7}$$

and in identical form also a relation for B in (4) holds. With $\zeta = (\xi, \eta)^T$ the equivariance of the nonlinear part in (4) is expressed by

$$R(\theta)K(\zeta) = K(R(\theta)\zeta) \ . \tag{8}$$

That relation (7) holds both for the matrices formed with A and B follows from their block diagonal form. Relation (8) can be checked by calculations, but is also obvious because a rotation of the coordinate system by an angle θ can not change the equations of motion.

Also the reflectional symmetry can be shown easily. A physical interpretation of the reflectional symmetry is given in [3].

3. Linear stability analysis and critical cases

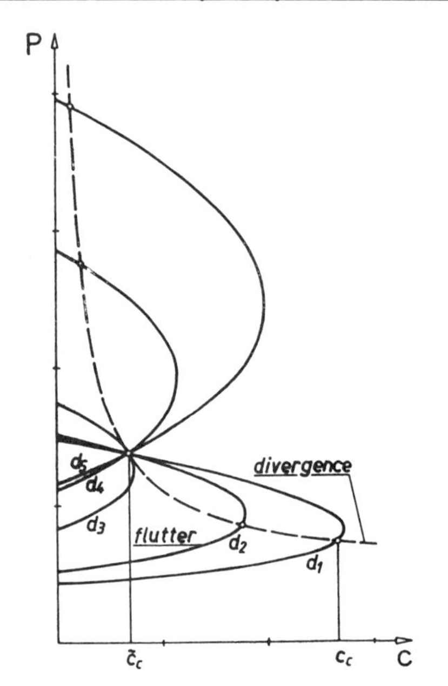

Fig.2: Stability boundaries in the P, C parameter space depending on the damping coefficient d ($d_1 < d_2 < d_3 < \ldots$)

As we study the stability of the downhanging equilibrium position the linearization of (4) is simply the linear part in (4). This linear part, however, decomposes due to the symmetry of the system into two identical parts. Hence we only have to study an eigenvalue problem $(A\mu - B)\xi = 0$ and the multiplicity of each eigenvalue has to be doubled. For the calculation of the eigenvalues we fix all parameters of the problem except the loading which we vary quasistatically, i.e. we always calculate the eigenvalues for fixed values of P. Obviously if we start with P = 0, all eigenvalues will have negative real parts and hence the equilibrium position will be asymptotically stable. Increasing P we reach a critical value P_c for which at the first time one or several eigenvalues with zero real part show up which we call the critical eigenvalues. These parametervalues yield the stability boundary. For our problem it is quite interesting to know how these critical values P_c and the structure of the critical eigenvalues change if, for example, we vary the stiffness C of the end support. The results shown in Fig. 2 are calculated for a fixed ratio of external to internal damping but for different absolute values of damping d. We see that for C big, i.e. for a stiff spring we get a zero root with multiplicity two. For C small, i.e. for a soft spring we get a purely imaginary pair of roots again with multiplicity two. Furthermore there exists a critical value C_c for which a more complicated eigenvalue structure arises. Before we proceed with our analysis we rewrite (4) in the form

$$\dot{\zeta} = H(\lambda)\zeta + F(\zeta,\lambda) \tag{9}$$

where $\zeta = (\xi,\eta) \in R^8$ and $\lambda = (P,C) \in R^2$. We transform (9) into Jordan normal form by means of $\zeta = Lv$ where L is composed by the real and imaginary parts of the eigen- and principal vectors. Thus (9) can be written as $\dot{v} = Jv + L^{-1}F(Lv, \lambda)$ where $J = L^{-1}HL$ is a matrix in Jordan form

$$J = \begin{pmatrix} J_c & 0 \\ 0 & J_s \end{pmatrix} . \tag{10}$$

In (10) J_c includes the eigenvalues with zero real part and J_s has only eigenvalues with negativ real parts. Depending on the stiffness C and the damping d, defined above, we get the following different cases:

(a) One zero root with multiplicity two

$$J_c = \begin{pmatrix} 0 & 0 \\ 0 & 0 \end{pmatrix} \tag{11}$$

This case is obtained for big C.

(b) One pair of purely imaginary roots with multiplicity two

$$J_c = \begin{pmatrix} 0 & \omega_o & 0 & 0 \\ -\omega_o & 0 & 0 & 0 \\ 0 & 0 & 0 & \omega_o \\ 0 & 0 & -\omega_o & 0 \end{pmatrix} \tag{12}$$

This case is obtained for small C.

(c) One zero root and one pair of purely imaginary roots with multiplicity two

$$J_c = \begin{pmatrix} 0 & \omega_o & 0 & & & \\ -\omega_o & 0 & 0 & & 0 & \\ 0 & 0 & 0 & & & \\ & & & 0 & \omega_o & 0 \\ & 0 & & -\omega_o & 0 & 0 \\ & & & 0 & 0 & 0 \end{pmatrix} \tag{13}$$

This case occurs for the special value $C = C_c$ (Fig. 2).

Case (c) occurs if the value of external damping is small compared to internal damping, i.e. if d is small. If we increase external damping (d big) we get case ($C = \tilde{C}_c$ in Fig. 2)

(d) Two zero roots with multiplicity two

$$J_c = \begin{pmatrix} 0 & 1 & 0 & 0 \\ 0 & 0 & 0 & 0 \\ 0 & 0 & 0 & 1 \\ 0 & 0 & 0 & 0 \end{pmatrix} \tag{14}$$

Whereas the cases (a) and (b) occur generically with respect to C, the case (c) is a special case concerning the stiffness C. However, if we vary for this special case the external damping then (c) and (d) are generic cases and we get a special more degenerate case

(e) Three zero roots with multiplicity two

$$J_c = \begin{pmatrix} 0 & 1 & 0 & & & \\ 0 & 0 & 1 & & 0 & \\ 0 & 0 & 0 & & & \\ & & & 0 & 1 & 0 \\ & 0 & & 0 & 0 & 1 \\ & & & 0 & 0 & 0 \end{pmatrix} \tag{15}$$

This system is found if the flutter and the divergence curves in Fig. 2 have a point of tangency.

4. Bifurcation equations

As the cases (a) and (b) are already treated in [3] and [6] we focus on case (c). The resulting motion will be determined by the coupling of flutter and divergence. From Center Manifold Theory ([7]) we know that the bifurcation system on the Center Manifold will be six-dimensional. The most convenient way of expressing the symmetry properties is to introduce three complex variables w_1, w_2, w_3 by

$$w_1 = v_1 + iv_2 , \quad w_2 = v_3 + iv_4 , \quad w_3 = v_5 + iv_6 \tag{16}$$

which we transform once more to ([8])

$$z_1 = w_1 + iw_2 , \quad z_2 = \bar{w}_1 + i\bar{w}_2 , \quad z_3 = w_3 . \tag{17}$$

The transformation from (16) to (17) simplifies the analysis further because with (17) the linear matrix J_c is diagonal and the symmetry properties are further simplified. A rotation of (w_1,w_2,w_3) by an angle of θ corresponds to a multiplication of z_1 and z_3 with $e^{i\theta}$ and of z_2 with $e^{-i\theta}$. I.e. we have

$$(w_1,w_2,w_3) \to (e^{i\theta}w_1,e^{i\theta}w_2,e^{i\theta}w_3) \Longleftrightarrow (z_1,z_2,z_3) \to (e^{i\theta}z_1,e^{-i\theta}z_2,e^{i\theta}z_3). \tag{18}$$

The reflectional symmetry yields

$$(w_1,w_2,w_3) \to (\bar{w}_1,\bar{w}_2,\bar{w}_3) \Longleftrightarrow (z_1,z_2,z_3) \to (z_2,z_1,\bar{z}_3) . \tag{19}$$

We denote the cubic terms in the bifurcation equations by

$$z_1^{m_1}\bar{z}_1^{m_2}z_2^{m_3}\bar{z}_2^{m_4}z_3^{m_5}\bar{z}_3^{m_6} \tag{20}$$

where $\quad m_1 + m_2 + m_3 + m_4 + m_5 + m_6 = 3 .$ (21)

Substituting (18) into (20) we obtain terms of the form

$$e^{i(m_1-m_2-m_3+m_4+m_5-m_6)\theta}\tilde{z}_1^{m_1}\tilde{\bar{z}}_1^{m_2}\tilde{z}_2^{m_3}\tilde{\bar{z}}_2^{m_4}\tilde{z}_3^{m_5}\tilde{\bar{z}}_3^{m_6} . \tag{22}$$

From invariance of the bifurcation equations under a rotation it follows from (22) that

$$m_1 - m_2 - m_3 + m_4 + m_5 - m_6 = \pm 1 \tag{23}$$

must hold, where +1 applies for the equations having $\dot{z}_1$ and $\dot{z}_3$ on the left hand side and -1 for $\dot{z}_2$.

From (21) and (22) we obtain, for example, the following terms in the first ($\dot{z}_1$) equation: $z_1^2\bar{z}_1$, $z_1^2 z_2$, $z_1^2\bar{z}_3$, $|z_1|^2\bar{z}_2$, $z_1|z_2|^2$, $z_1\bar{z}_2\bar{z}_3$, $|z_1|^2 z_3$, $z_1 z_2 z_3$, $z_1|z_3|^2$, $\bar{z}_1\bar{z}_2^2$, $|z_2|^2 z_2$, $\bar{z}_2^2\bar{z}_3$, $\bar{z}_1\bar{z}_2 z_3$, $|z_2|^2 z_3$, $\bar{z}_2 z_3^2$, $\bar{z}_1 z_3^2$, $z_2 z_3^2$, $|z_3|^2 z_3$. Applying now the Normal Form Theorem ([4]) it is well known that we can omit all terms by a nonlinear transformation of variables, except those for which a resonance condition holds. The resonance conditions are respectively

$$1 = m_1 - m_2 + m_3 - m_4 \; ; \quad 0 = m_1 - m_2 + m_3 - m_4 \; . \tag{24}$$

Hence the bifurcation equations take the following form

$$\begin{aligned}
\dot{z}_1 &= (\lambda+i\omega)z_1 + A_1|z_1|^2 z_1 + A_2 z_1|z_2|^2 + A_3 z_1|z_3|^2 + A_4 z_2 z_3^2 \\
\dot{z}_2 &= (\lambda+i\omega)z_2 + A_2|z_1|^2 z_2 + A_1|z_2|^2 z_2 + A_3|z_3|^2 z_2 + A_4 z_1 \bar{z}_3^2 \\
\dot{z}_3 &= \mu z_3 + A_5|z_1|^2 z_3 + \bar{A}_5|z_2|^2 z_3 + A_6|z_3|^2 z_3 + A_7 z_1\bar{z}_2\bar{z}_3
\end{aligned} \tag{25}$$

where we have written down already the unfolded bifurcation equations. The unfolding parameters are λ and μ. Further the coefficients are in general complex, i.e. $A_j = c_j + id_j$ and because of the reflectional symmetry we have $d_6 = d_7 = 0$. Introducting polar coordinates $z_j = r_j e^{i\phi_j}$ we obtain from (25) with $\psi = \phi_1 - \phi_2 - 2\phi_3$

$$\begin{aligned}
\dot{r}_1 &= (\lambda + c_1 r_1^2 + c_2 r_2^2 + c_3 r_3^2)r_1 + r_2 r_3^2(c_4\cos\psi + d_4\sin\psi) \\
\dot{r}_2 &= (\lambda + c_2 r_1^2 + c_1 r_2^2 + c_3 r_3^2)r_2 + r_1 r_3^2(c_4\cos\psi - d_4\sin\psi) \\
\dot{r}_3 &= (\mu + c_5(r_1^2 + r_2^2) + c_6 r_3^2 + c_7 r_1 r_2\cos\psi)r_3 \\
\dot{\psi} &= (d_1-d_2-2d_5)(r_1^2 - r_2^2) + r_3^2 d_4\cos\psi\left(\frac{r_2}{r_1} - \frac{r_1}{r_2}\right) - \\
&\quad - r_3^2 c_4\sin\psi\left(\frac{r_1}{r_2} + \frac{r_2}{r_1}\right) - 2c_7 r_1 r_2\sin\psi \; .
\end{aligned} \tag{26}$$

We have the following steady state solutions of (26):

(1) $r_1 = r_2 = r_3 = 0$: trivial solution, TS

(2) $r_1 \neq 0$, $r_2 = r_3 = 0$: rotating pendulum (rotating wave, RW)

(3) $r_1 = r_2$, $r_3 = 0$: oscillating pendulum in a plane (standing wave, SW)

(4) $r_1 = r_2 = 0$, $r_3 \neq 0$: statically buckled pendulum, SS

(5) $r_1 \neq 0$, $r_2 = o(|r_3|)$, $r_3 \neq 0$: statically buckled and rotating pendulum, SSRW (27)

(6) $r_1 = r_2 \neq 0$, $r_3 \neq 0$: oscillation of the pendulum about a statically buckled position. There exist two solutions SSW0, SSWπ as explained below.

As an example we take now case (6). Setting $r_1 = r_2$ we get from (16)

$$\dot{\psi} = 0 = (-2c_4 r_3^2 - 2c_7 r_1^2)\sin\psi \tag{28}$$

which has two solutions $\psi = 0$ (the oscillation is in the plane given by the buckled pendulum, denoted by SSW0) and $\psi = \pi$ (the pendulum oscillates orthogonal to this plane, denoted by SSWπ). The amplitude equations are

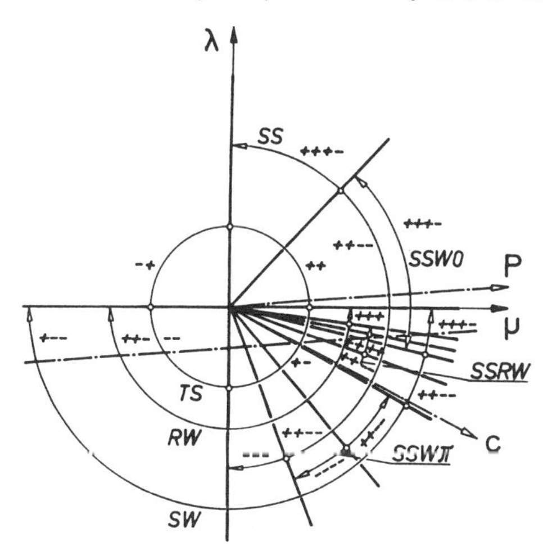

Fig.3: Bifurcation diagram showing the domains of existence of the different solutions (see (27)) and their stability indicated by - or + signs designating stable or unstable eigenvalues

$$\lambda + (c_1 + c_2)r_1^2 + (c_3 + c_4\cos\psi)r_3^2 = 0$$
$$\mu + (2c_5 + c_7\cos\psi)r_1^2 + c_6 r_3^2 = 0 \,. \tag{29}$$

Setting r_1^2 and $r_2^2 \geq 0$ we obtain the values of λ and μ where the solutions SSWπ and SSW0 exist. These solutions are shown in Figs. 3 and 4 among all other solutions which we have found. To check the stability we have to calculate the real parts of the eigenvalues of the Jacobian at the solutions. In Fig. 3 the bifurcation diagram, i.e. the partition of the parameter space $\mu - \lambda$, is given. The domain of existence of the different solutions according to (27) is indicated by the sections of a

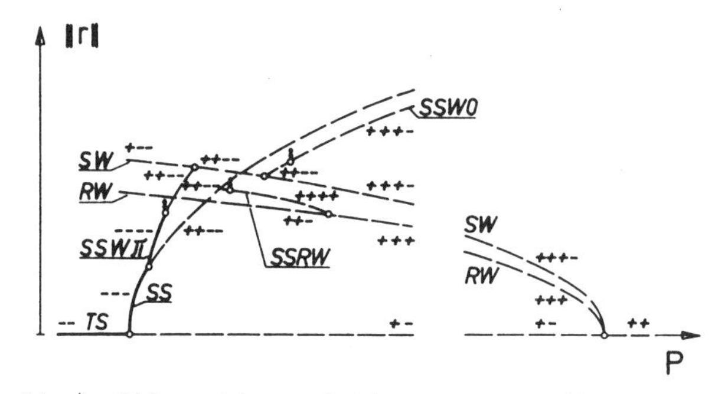

Fig.4: Bifurcation solutions corresponding to the broken line parallel to the P-axis in Fig.3

circle and the stability is given by + and - signs. A minus sign corresponds to a stable eigenvalue and a plus sign to an unstable one. Furthermore also the relationship to the physical parameters P and C is indicated. If we start in the quadrant $\mu < 0$, $\lambda < 0$ then the trivial solution (TS) is stable. We move now along the broken line parallel to the P axis and in the direction of increasing values of P. Then the first bifurcation occurs at $\mu = 0$, ($\lambda < 0$), where the pendulum buckles statically (SS). All eigenvalues are stable. In Fig. 4 the diagram of the bifurcated solution is drawn corresponding to the variation of P according to the broken line in Fig. 3. Increasing P further we reach the next bifurcation where the solution SS becomes unstable and bifurcates into the stable solution SSWπ. I.e. we have a standing oscillation about a buckled state. The next bifurcation is a degenerate Hopf bifurcation because we only retained terms of third order in our analysis. To get a complete picture of this bifurcation we would have to include higher order terms.

5. Acknowledgement

This work has been supported by the Austrian "Fonds zur Förderung der wissenschaftlichen Forschung" under project P 5519.

6. References

[1] Bajaj A.K., P.R. Sethna, T.S. Lundgren, Hopf Bifurcation Phenomena in Tubes Carrying a Fluid, SIAM J. Appl. Math. 39 (1980) 213 - 230.

[2] Bajaj A.K., Bifurcating Periodic Solutions in Rotationally Symmetric Systems, SIAM J. Appl. Math. 42 (1982) 1078 - 1098.

[3] Bajaj A.K., P.R. Sethna, Bifurcations in Three-Dimensional Motions of Articulated Tubes, Part 1: Linear Systems and Symmetry, J. Appl. Mech. 49 (1982) 606 - 611, Part 2: Nonlinear Analysis, J. Appl. Mech. 49 (1982) 612 - 618.

[4] Golubitsky M., I. Stewart, Hopf Bifurcation in the Presence of Symmetry, Archive Rat. Mech. Anal. 87 (1984) 107 - 165.

[5] Dangelmayr G., E. Knobloch, The Takens-Bogdanov Bifurcation with O(2)-Symmetry, Phil. Trans. Roy. Soc. London, to appear.

[6] Lindtner E., A. Steindl, H. Troger, Stabilitätsverlust der gestreckten Lage eines räumlichen Doppelpendels mit elastischer Endlagerung unter einer Folgelast, ZAMM 67 (1987), in print.

[7] Carr J., Application of Centre Manifold Theory, Appl. Math. Sciences 35, Springer-Verlag, New York, Heidelberg, Berlin, 1981.

[8] van Gils S.A., Hopf Bifurcation and Symmetry: Travelling and Standing Waves on the Circle, preprint.

Institut für Mechanik
TU-Wien
Karlsplatz 13
A-1040 Wien
Austria

International Series of
Numerical Mathematics, Vol. 79

Multiple solutions of the semiconductor equations.

Herbert Steinrück and Richard Weiss

Institut für Angewandte und Numerische Mathematik,
Technische Universität Wien

1) Introduction

Semiconductor devices have become the basis of modern electronics and the mathematical modelling of such devices is an essential tool for their design. The carrier flow in semiconductors is modelled by a system of partial differential equations derived from Maxwell's equations and the convection diffusion mechanism gouverning the charge transport.

An analysis of these equations yields insight into the structure of the carrier flow within the device, its dependence on the various parameters characterizing a particular device, and also leads to a mathematical derivation of the current voltage characteristic. While in devices like diodes and transistors there is a unique stationary carrier flow for any applied voltage, there are devices, called thyristors, for which multiple steady states exist, and this fact is crucial for their performance.

In this paper we will consider the question of multiple solutions of the semiconductor equations employing a one-dimensional model. In particular, we will examine the bifurcation occuring when certain parameters of the device are varied. While this was still an open problem at the time of the conference, it has meanwhile been resolved by H.Steinrück [4]. The aim of the present paper is to give a brief introduction into the techniques required to derive the 'bifurcation equation' from the underlying boundary value problem and to present a summary of the results of [4].

The paper is organized as follows: In section 2 we will present the one-dimensional model and give a comparison of diodes and thyristors. The singular perturbation techniques required for the analysis of these equations are then discussed in section 3, and section 4 is devoted to the bifurcation phenomenon.

2) The one-dimensional model

An essential step in the analysis of the semiconductor equations is the scaling of independent and dependent variables, which, in the stationary case, leads to a singularly perturbed elliptic system.

A discussion of the derivation of the equations from physical principles and of the scaling procedure can be found in Markowich [2]. The scaled one-dimensional problem, which we will use as the basis of our analysis, is of the following form:

(2.1) $\lambda^2\psi'' = n - p - C(x)$ (Poisson's equation)

(2.1) $J_n' = R$ (electron continuity equation)

(2.3) $J_p' = -R$ (hole continuity equation)

(2.4) $\mu_n(n'-n\psi') = J_n$ (electron current relation)

(2.5) $-\mu_p(p'+p\psi') = J_p$ (hole current relation)

The equations are valid on the scaled interval [0,1]. The dependent variables and their meanings are:

ψ : (scaled) electrostatic potential

$n>0$: (scaled) electron concentration

$p>0$: (scaled) hole concentration

J_n : (scaled) electron current density

J_p : (scaled) hole current density

The normed characteristic Debye length λ, whose square multiplies ψ'' in Poisson's equation (2.1) is a small parameter, typically of the magnitude 10^{-7}-10^{-3}. $C(x)$ is the (scaled) doping profile, a given function, most essential for the performance of a particular device. It is scaled so that its maximum value equals one. The relevance of $C(x)$ will be discussed further at the end of this section. The quantity R in (2.2),(2.3) is the scaled recombination generation rate. The most common recombination phenomenon, Shockley-Read-Hall recombination, is modelled by

(2.6) $$R = \frac{np-\delta^4}{T(n+p+2\delta^2)}$$

where δ^2 is a small quantity, typically between 10^{-7} and 10^{-3} and T is the (scaled) lifetime of a charge carrier. μ_n,μ_p in (2.4),(2.5) are the (scaled) electron and hole mobilities, they are of magnitude 1 and will subsequently assumed to be equal to 1.

The boundary conditions for (2.1)-(2.5) are

(2.7) $\psi(0) = \psi_{BI}(0)\ , \quad \psi(1) = \psi_{BI}(1) - U$

where

$$\psi_{BI}(x) = \ln(\frac{C(x)+\sqrt{C^2(x)+4\delta^4}}{2} \qquad x = 0,1$$

is the so-called scaled built-in-potential and U is the scaled voltage applied to the device,

(2.8) $n(x) = \frac{1}{2}(\ C(x)+\sqrt{C^2(x)+4\delta^4} \qquad x = 0,1$

(2.9) $p(x) = \frac{1}{2}(-C(x)+\sqrt{C^2(x)+4\delta^4} \qquad x = 0,1$

The quantity δ^2 is the same as in (2.6).

For our purposes the doping concentration is modelled as a piecewise constant function with jumps at the so-called pn-junctions. A fairly typical doping concentration for a diode, i.e. a device with two differently doped regions, is shown in fig.2.1.

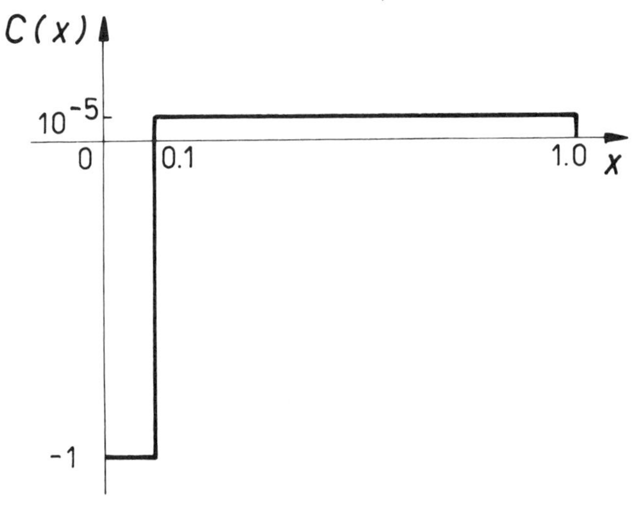

Fig. 2.1

A thyristor is a device with four differently doped regions, and a typical doping concentration is depicted in fig.2.2.

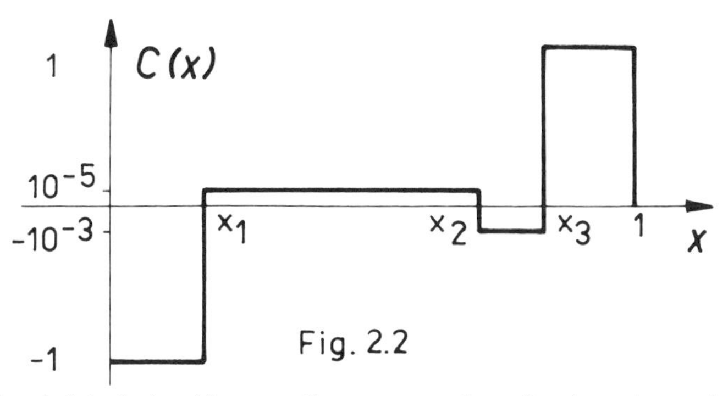

Fig. 2.2

A first insight into the performance of a device is gained through its current voltage characteristic, where the total current J is plotted versus the applied voltage U. A typical current voltage characteristic for a diode is given in fig.2.3.

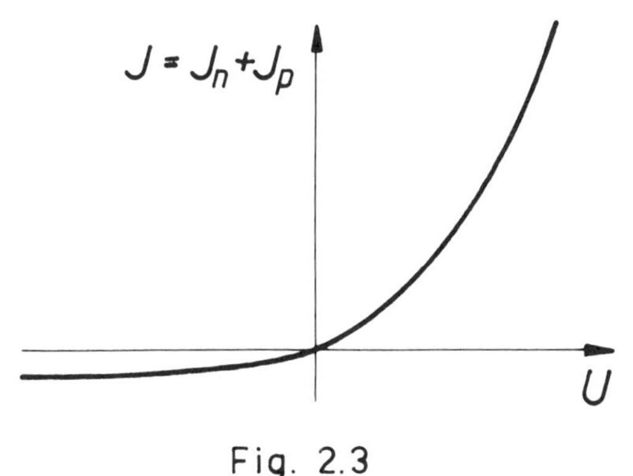

Fig. 2.3

We observe that there is a unique value of J (corresponding to a unique solution of the boundary value problem) for each value of U. The state $U = 0$, $J = 0$ is called 'thermal equilibrium'; the diode is in 'forward bias' when $U > 0$ and in 'reverse bias' when $U < 0$.

A typical current voltage characteristic of a thyristor is given in fig.2.4, and we observe that it differs markedly from that of a diode.

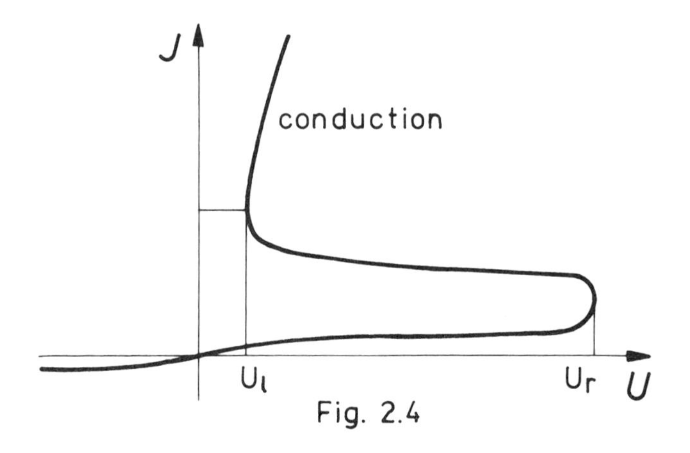

Fig. 2.4

There are three different values of J (three different solutions of the boundary value problem) for $U_l \leqq U < U_r$. The lower branch is called the blocking branch, the current on this branch is very small, of magnitude $\delta^4 \sim 10^{-14}$, the upper branch, where a large current is flowing, is called the conduction branch. The middle branch is unstable. The thyristor serves as an electronic switch: When U is gradually increased from zero, the system is in the blocking state. In this state, a small perturbation of the system (induced by a small gate current, which cannot be taken into account in a one-dimensionale model) causes the system to jump to the conduction state. When the gate current is turned off the system remains in this state. To achieve a reduction of J in this state, U must be reduced appropriately.

3) The singular perturbation analysis

In the sequel we assume that we are dealing with a thyristor, i.e.

$$C(x) = C_i \;, \quad x_{i-1} < x < x_{i+1} \;, \quad i = 1,2,3$$

where $x_o = 0$, $x_4 = 1$, $C_1 < 0$, $C_2 > 0$, $C_3 < 0$, $C_4 > 0$. Since in (2.1) $\lambda^2 \ll 1$, the boundary value problem (2.1)-(2.9) is singularly perturbed. The approximate solution of this problem via singular perturbation theory proceeds in two steps. First the 'reduced solution' is constructed. This is the solution of the boundary value problem when λ^2 in (2.1) is set equal to zero.

Hence denoting the reduced solution by $\bar{\psi},\bar{n},\bar{p},\bar{J}_n,\bar{J}_p$ we obtain:

(3.2) $\bar{n} - \bar{p} - C_i = 0$

(3.3) $(\bar{n}'-\bar{n}\bar{\psi})' = R \quad , \quad x_{i-1} < x < x_i \quad , \quad i = 1,2,3,4$

(3.4) $(\bar{p}' + \bar{p}\bar{\psi})' = R$

Since $C(x)$ is discontinuous at the pn-junction x_i, $i = 1,2,3$, the reduced solution is discontinuous there as well; its continuous parts are connected through the following interface conditions, which are obtained from singular perturbation analysis, Markowich [2].

(3.5) $[\ln \bar{n} - \bar{\psi}]_{x_i} = 0 \quad , \quad [\ln \bar{p} + \bar{\psi}]_{x_i} = 0 \quad ,$

(3.6) $[\bar{n}' - \bar{n}\bar{\psi}']_{x_i} = 0 \quad , \quad [\bar{p}' + \bar{p}\bar{\psi}']_{x_i} = 0 \quad , \quad i = 1,2,3$

Here $[y]_x := y(x+) - y(x-)$. The reduced solution also satisfies the boundary conditions (2.7)-(2.9).

In the second step of the singular perturbation procedure the reduced solution has to be corrected by layer terms at each junction. These layer corrections are uniquely defined when the reduced solution is given. So, if the boundary value problem should have more than one solution, these solutions can be generated only by the reduced problem.

To bring the reduced problem in a more convenient form we introduce

$$\sigma = \bar{n} + \bar{p} \qquad J = \bar{J}_n + \bar{J}_p$$

and obtain from (3.2),(3.3),(3.4)

$$\sigma' - C_i\bar{\psi}' = \bar{J}_n - \bar{J}_p \ ,$$

$$\sigma\bar{\psi}' = -J \ .$$

This yields

(3.7) $\sigma'' - \dfrac{C_i J}{\sigma^2}\sigma' = 2R(\sigma) \quad , \quad x_{i-1} < x < x_i \quad , \quad i = 1,2,3,4$

with

$$R(\sigma) = \frac{1}{4T}\,\frac{\sigma^2 - C^2 - 4\delta^4}{\sigma + 2\delta^2} \ ,$$

and

(3.8) $\bar{\psi}' = -\dfrac{J}{\sigma}$.

The interface conditions are

$$(3.9) \qquad [\sigma^2 - C^2]_{x_i} = 0$$

$$(3.10) \qquad [\sigma' + \frac{CJ}{\sigma}]_{x_i} = 0$$

$$(3.11) \qquad [\bar{\psi}]_{x_i} = -[\ln(\sigma - C)]_{x_i} \quad , \quad i = 1,2,3$$

and the boundary conditions are

$$(3.12) \qquad \sigma(x) = \sqrt{C^2(x) + 4\delta^4} \quad , \quad x = 0,1$$

$$(3.13) \qquad \bar{\psi}(0) = \psi_{BI}(0) \quad , \quad \bar{\psi}(1) = \psi_{BI}(1) - U \; .$$

When the total current J ist prescribed σ (the sum of the carrier concentrations $\bar{n}$ and $\bar{p}$) can be determined from (3.7),(3.9), (3.10),(3.12). Then the potential $\bar{\psi}$ can be computed from (3.8), (3.11),(3.13). Therefore we will perform our analysis in terms of the parameter J and not in terms of U.

4) The bifurcation analysis

In this section we will qualitatively discuss the current voltage characteristic of a thyristor. Let us again consider a typical current voltage curve of a thyristor:

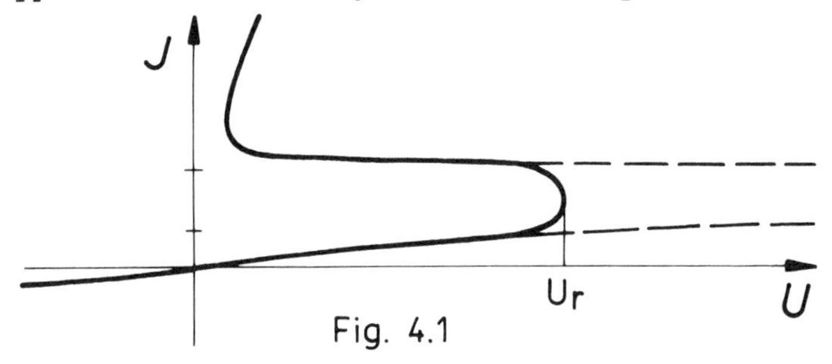

Fig. 4.1

In a realistic thyristor, the lower and the middle branch are connected at the break-through voltage U_r. The connection of these solution branches is caused by the avalanche effect, see Gerlach [1] and Sze [5]. In our simplified model we have neglected this effect, therefore we cannot expect the middle and the lower branch to be connected. The analysis will show that for our model we have to modify the current voltage characteristic by continuing the middle and the lower branch to infinity (the dashed lines in fig.4.1). Rubinstein [3] was the first to obtain such a current voltage curve.

If we consider the modified curve and follow the lower or the middle branch as $U \to \infty$, the current density will tend to a finite value, it 'saturates'. If we follow the current voltage curve as $U \to -\infty$, the current density will saturate as well. Therefore we expect that our simplified thyristor model will have three saturation currents, two positive ones, and a negative one.

Now we consider again the reduced problem (3.6)-(3.12). The system splits into two parts:

i) A boundary value problem for σ when the total current density J is prescribed.

ii) A problem for the potential ψ, which can be determined from σ by integration.

It can be shown (see Steinrück [4]) that the problem for σ, linearized at $J = 0$, is uniquely solvable and therefore, in a vicinity of $J = 0$, $\sigma(x,J)$ depends smoothly on J and the device parameters. So we can expand,

$$\sigma(x,J) = \sigma_0(x) + \sigma_1(x)J + \frac{1}{2}\sigma_2(x)\frac{J^2}{2} + O(J^3) \tag{4.1}$$

where

$$\sigma_0(x) = \sqrt{C^2(x)+4\delta^4} \tag{4.2}$$

is the solution of the equilibrium problem $(J=0)$.

When examining the question, what equations are responsible for the occurance of the saturation currents, one has to consider the interface conditions for $\bar{\psi}$:

$$\bar{\psi}(x_i+) - \bar{\psi}(x_i-) = \ln(\sigma(x_i-)-C(x_i-)) - \ln(\sigma(x_i+)-C(x_i+)). \tag{4.3}$$

It is clear that $U(J)$ tends to infinity if $\sigma(x_i-,J)-C(x_i-)$ or $\sigma(x_i+,J)-C(x_i+)$ tend to zero. Therefore the saturation currents can be determined as the solution of

$$\sigma(x_i-,J) - C(x_i-) = 0 \quad \text{or} \quad \sigma(x_i+,J) - C(x_i+) = 0 \qquad i = 1,2,3. \tag{4.4}$$

We only have to consider those equations of (4.4) where the doping profile is positive, because $\sigma(x)$, a sum of densities, is always positive. Since $C(x)$ changes sign at x_i, only one of the equations (4.4) for a given i can cause a saturation current. Let us consider for example x_2: Then we know from (3.1) that $C(x_2-) = C_2 > 0$, and we can determine a saturation current from

(4.5) $\sigma(x_2-,j) - C_2(x_2-) = 0$.

Using (4.1),(4.2) and

(4.6) $\sigma_o(x_2-) - C_2 = \frac{2\delta^4}{C_2} + 0(\delta^6)$,

we obtain the following approximation to a saturation current:

(4.7) $J = -\frac{2\delta^4}{C_2\sigma_1(x_2-)} + 0(\delta^6)$.

Analogously we get saturation currents caused by the other junctions,

(4.8) $J = -\frac{2\delta^4}{C_2\sigma_1(x_1+)} + 0(\delta^6)$, $J = -\frac{2\delta^4}{C_4\sigma_1(x_3+)} + 0(\delta^6)$.

In the course of the analysis of the linearized problem for σ at $J = 0$ it is shown that $\sigma_1(x_1+)$ and $\sigma_1(x_3+)$ are always positive, if the doping profile $C(x)$ satisfies (3.1). The corresponding saturation currents are negative. Therefore the middle junction at x_2 has to be responsible for a positive saturation current.

The interesting case occurs when $\sigma_1(x_2-)$ is zero. Then, in determining the saturation current, we have to use the expansion (4.1) up to the second order term:

(4.9) $\frac{\sigma_2(x_2-)}{2}J^2 + \sigma_1(x_2-)J + \frac{2\delta^4}{C_2} = 0$.

Equation (4.6) determines the solvability of (4.5) when $\sigma_2(x_2-) \neq 0$ and $\sigma_1(x_2-)$, δ^4 are small. The solutions of (4.4) are

$$J = -\frac{\sigma_1}{\sigma_2} \pm \sqrt{\frac{\sigma_1^2}{\sigma_2^2} - \frac{4\delta^4}{C_2\sigma_2}}$$

where $\sigma_1 = \sigma_1(x_2-)$, $\sigma_2 = \sigma_2(x_2-)$. If σ_2 is positive and σ_1 is negative, both possible saturation currents are positive. This situation corresponds to the modified current voltage curve in fig.4.1. Hence, for small values of σ_1 and δ^4, the solution behaviour of positive saturation currents is characterized by a fold catastrophe.

The following pictures show how the current voltage characteristic behaves when the device parameters are varied in a vicinity of the critical values for which the discriminant of the quadratic equation (4.9) is zero.

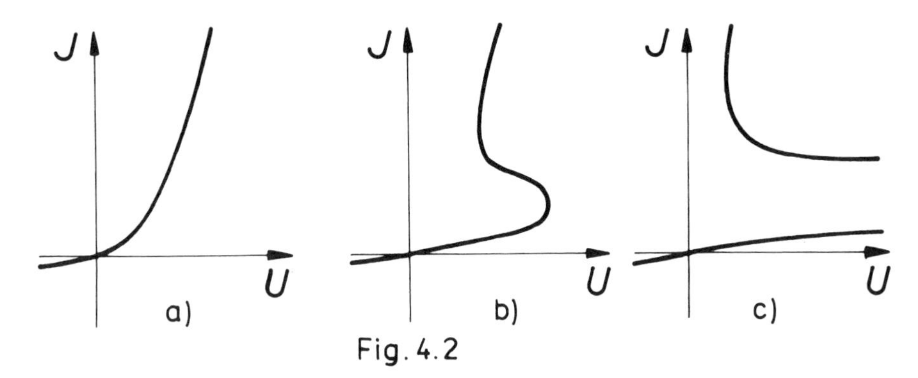

Fig. 4.2

The qualitative form of the curves can be obtained by computing U(J) approximately using the expansion (4.1) of $\sigma(x,J)$ at J = 0, Steinrück [4].

In fig. 4.2c we have a characteristic with two positive saturation currents. If the discriminant of (4.9) passes through zero, the positive saturation currents vanish and the middle and lower solution branch are connected. However, this is possible only for a 'small' range of device parameters. If we change the device parameters further, the lower and middle branch vanish and we get a characteristic similar to a characteristic of a diode. For a detailed discussion see Steinrück [4].

References

[1] W.Gerlach, Thyristoren, Springer, Berlin-Heidelberg-New York, 1981.

[2] P.A.Markowich, The Stationary Semiconductor Device Equations, Springer, Wien-New York, 1986.

[3] I.Rubinstein, Multiple Steady States in One-Dimensional Electrodiffusion with Local Electroneutrality, to appear in SIAM J. of Appl.Math..

[4] H.Steinrück, Multiple solutions of the steady state semiconductor device equations (manuscript).

[5] S.M.Sze, Physics of Semiconductor Devices, Willey, New York, 1981.

International Series of
Numerical Mathematics, Vol. 79

DELAY AS BIFURCATION PARAMETER

G. Stépán

Budapest University of Technology, Budapest, Hungary

Introduction

It is well-known that the equilibrium positions of dynamical systems are usually destabilized by increasing delay. However, opposite effects can also be experienced in special cases. In practice such situations may occur in some models of population dynamics, in the case of machine tool vibrations, in controls of manipulators or in man-machine systems. The equilibrium of such a system may get back its stability if the delay is great enough in spite of the fact that it loses its stability at a small value of the time lag. Thus, it is interesting to investigate these problems from the point of view of Hopf bifurcations because the trivial solutions may have a lot of bifurcation points with respect to the delay.

These dynamical systems are described by retarded functional differential equations (RFDE). However, even the stability investigation of the trivial solutions causes serious difficulties in these infinite dimensional problems.

This paper presents a stability criterion for linear RFDEs and uses it for examples where a great number of bifurcations occurs. The analysis of a predator-prey model will show the existence of sub- and supercritical Hopf bifurcations as well as that of some degenerate ones with codimension 1 when the bifurcation parameter is the delay in the system.

1. Stability Investigations

If we want to deal with a bifurcation analysis we always have to find the critical values of the bifurcation parameter. This means the stability investigation of the zero solution of the linear RFDE:

$$\dot{x}(t) = \int_{-r}^{0} [d\eta(\vartheta)]\, x(t+\vartheta) \qquad (1)$$

($x:R \mapsto R^N$ is a function, the N dimensional matrix η is a function of bounded variation, $\vartheta \in [-r,0]$, r is the length of retardation and t stands for time). Its characteristic function is

$$D(\lambda) = \det\,(\lambda I - \int_{-r}^{0} e^{\lambda\vartheta} d\eta(\vartheta)) \qquad (2)$$

(I_{NxN} is the unit matrix). There are several methods in the literature (see Hale(1977) or McDonald(1978)) for the investigation of (2) as to whether the real parts of all the zeros of (2) are negative. However, these methods cannot be used generally. Therefore, a new method is applied here. Theorems 1 and 2 contain this necessary and sufficient criterion.

Let the functions M and S be defined as follows:

$$M(y) = \mathrm{Re}D(iy), \qquad S(y) = \mathrm{Im}D(iy) \qquad (3)$$

($i=\sqrt{-1}$, $y \in R_+$). Let $\mu_1 \geq \mu_2 \geq \ldots \geq \mu_m \geq 0$ denote the non-negative zeros of M. It is easy to see that the number m of these zeros is finite if the dimension N of the RFDE (1) is even (N=2n, n is integer). Similarly, $\sigma_1 \geq \sigma_2 \geq \ldots \geq \sigma_s = 0$ denote the non-negative zeros of S where s is finite if N=2n+1.

Theorem 1. The zero solution of the N=2n dimensional RFDE (1) is asymptotically stable if and only if

$$S(\mu_k) \neq 0 \ , \quad k=1,\ldots,m \quad \text{and} \tag{4}$$

$$\sum_{k=1}^{m}(-1)^k \operatorname{sign} S(\mu_k) = (-1)^n n \ . \tag{5}$$

Theorem 2. The trivial solution of the N=2n+1 dimensional RFDE (1) is asymptotically stable if and only if

$$M(\sigma_k) \neq 0 \ , \ k=1,\ldots,s-1 \quad \text{and} \quad M(0)>0 \quad \text{and} \tag{6}$$

$$\sum_{k=1}^{s-1}(-1)^k \operatorname{sign} M(\sigma_k) + \frac{1}{2}((-1)^s+(-1)^n) + (-1)^n n = 0 \ . \tag{7}$$

To prove these theorems each of the steps of the proof of a similar theorem in Stépán(1979) has to be repeated. Let us see the applications now!

Example 1. The first order scalar equation

$$\dot{x}(t) = -bx(t-r_1) - bx(t-r_2) \tag{8}$$

with two discrete delays r_1 and r_2 has a quite simple and "conventional" stability chart (see Fig.1). The stability regions are determined as follows.

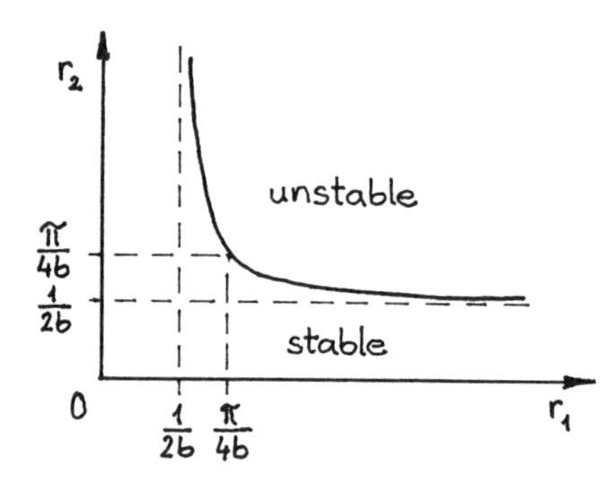

Fig. 1. Stability chart of a first order equation

Theorem 3. The zero solution of (8) is asymptotically stable iff

$$b>0 \quad \text{and} \quad \frac{\pi}{r_1+r_2} - 2b\cos\left(\frac{r_1-r_2}{r_1+r_2}\frac{\pi}{2}\right) > 0 \ . \tag{9}$$

Proof. According to (2) and (3) we need the functions

$$D(\lambda) = \lambda + be^{-r_1\lambda} + be^{-r_2\lambda},$$

$$M(y) = 2b\cos((r_1+r_2)y/2)\cos((r_1-r_2)y/2),$$

$$S(y) = y - 2b\sin((r_1+r_2)y/2)\cos((r_1-r_2)y/2).$$

In theorem 2 we have n=0 and (6) gives

$$M(0) = 2b > 0. \tag{10}$$

If the number s of the zeros of S is even then

$$S'(0) < 0 \tag{11}$$

and condition (7) has the form

$$\sum_{k=1}^{s-1} (-1)^k \operatorname{sign} M(\sigma_k) = -1,$$

which is equivalent to

$$S(\pi/(r_1+r_2)) > 0, \tag{12}$$

where $\pi/(r_1+r_2)$ is the less positive zero of M. If s is odd then

$$\sum_{k=1}^{s-1} (-1)^k \operatorname{sign} M(\sigma_k) = 0$$

(see (7)) which is fulfilled only for s=1, i.e. when

$$S'(0) > 0. \tag{13}$$

(11) and (13) can be omitted while (10) and (12) are equivalent to (9) in theorem 3.

Example 2. The second order scalar equation

$$\ddot{x}(t) + 6x(t) - x(t-r_1) - x(t-r_2) = 0 \tag{14}$$

has a stability chart of a very complicated structure (see Fig.2, the stable regions are shaded). This result is proved only in two special cases here.

Theorem 4. The zero solution of (14) is asymptotically stable iff

for $r_2=0$: $0 \le r_1 < \pi/\sqrt{6}$ or $\pi < r_1 < 3\pi/\sqrt{6}$ or $2\pi < r_1 < 7\pi/\sqrt{6}$; (15)

for $r_2=r_1$: $0 \le r_1 < \pi/\sqrt{8}$ or $\pi < r_1 < 3\pi/\sqrt{8}$. (16)

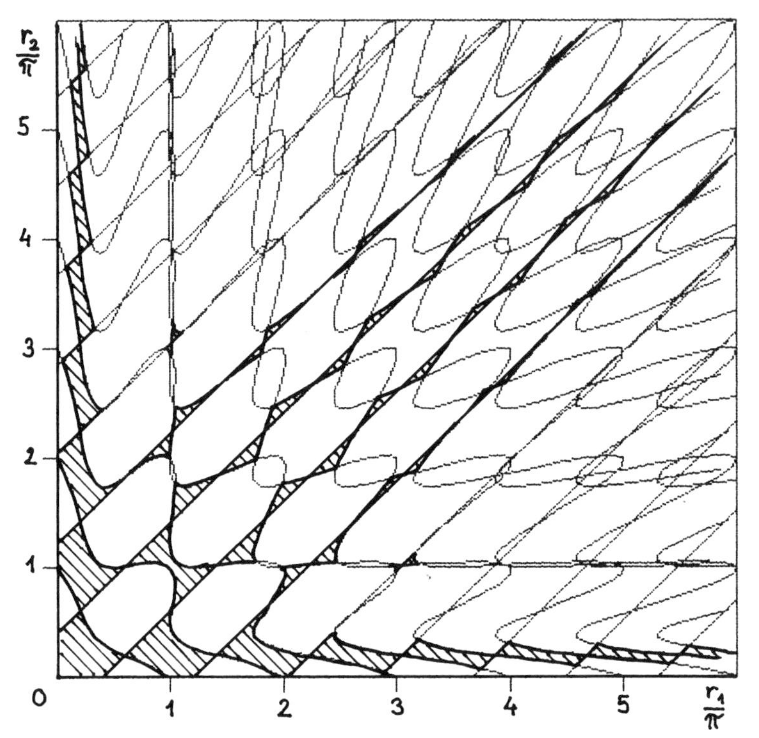

Fig. 2. Stability chart of a second order equation

Proof. Theorem 1 is used with n=1 and according to (2) and (3)

$$M(y) = -y^2 + 6 - \cos(r_1 y) - \cos(r_2 y) , \qquad (17)$$

$$S(y) = \sin(r_1 y) + \sin(r_2 y) . \qquad (18)$$

For the investigation of condition (5) the sign variation of S is needed. Its determination is easy if $r_2=0$ or $r_2=r_1$ because S=0 for $y_j=j\pi/r_1$, $j=0,1,\ldots$. Thus, (5) is fulfilled iff

$$M(y_j) > 0 \quad \text{for} \quad j=0,1,\ldots,2h \quad \text{and}$$

$$M(y_j) < 0 \quad \text{for} \quad j=2h+1,2h+2,\ldots$$

are true for a non-negative integer h. These inequalities give (15) or (16).

(14) is similar to a model of a manipulator controlled by a human

operator where r_1 is the delay in the information transmission and r_2 is the delay of the operator's reflex (see Stépán (1986a)). In the same paper a stability chart for a third order scalar equation is also presented.

Example 3. The predator-prey model of Lotka-Volterra-McDonald (see McDonald(1978)) is described by the nonlinear RFDE

$$\begin{aligned} \dot{x}_1(t) &= -x_1(t)/2 - x_2(t) - x_1^2(t)/2 - x_1(t)x_2(t) \\ \dot{x}_2(t) &= -\frac{1}{2}\int_{-r}^{0} x_1(t+\vartheta)w(\vartheta)d\vartheta - x_2(t)\int_{-r}^{0} x_1(t+\vartheta)w(\vartheta)d\vartheta \end{aligned} \tag{19}$$

where all the parameters of the model has been fixed except the length r of the delay and its weight function $w(\vartheta)$.

Theorem 5. The zero solution of (19) with

$$w(\vartheta) = (\pi/(2r))\sin(\pi\vartheta/r) , \quad \vartheta\in[-r,0]$$

is asymptotically stable iff

$$r < r_{cr} = 2\mu_1(\pi^2-\mu_1^2)/(\pi^2\sin\mu_1) \tag{20}$$

where μ_1 is the only positive zero of the equation

$$-\mu_1^2 + r^2\pi^2(1+\cos\mu_1)/(4(\pi^2-\mu_1^2)) = 0 . \tag{21}$$

Proof. Theorem 1 can be used with n=1 and (see (2))

$$D(\lambda)=\begin{cases}\lambda^2+\lambda r/2+r^2\pi^2(e^{-\lambda}+1)/(4(\lambda^2+\pi^2)) & \text{if } \lambda\neq\pm i\pi \\ -\pi^2\pm i\pi/2\mp ir^2\pi/8 & \text{if } \lambda=\pm i\pi \end{cases} .$$

There exists only m=1 positive zero of M(y)=ReD(iy) and it is μ_1 according to (21), $\mu_1\in(0,\pi)$. Condition (5) gives

$$S(\mu_1) = \mathrm{Im}D(i\mu_1) = 1 \quad \text{i.e.} \quad S(\mu_1) > 0$$

which is equivalent to (20).

In a similar way the stability chart of Fig.3 has been determined when the weight function in (19) has the form

$$w(\vartheta) = -p\delta(\vartheta) - (1-p)\delta(\vartheta+r) \tag{22}$$

(δ is the Dirac function, $p\in[0,1]$ is the weight of the "present").

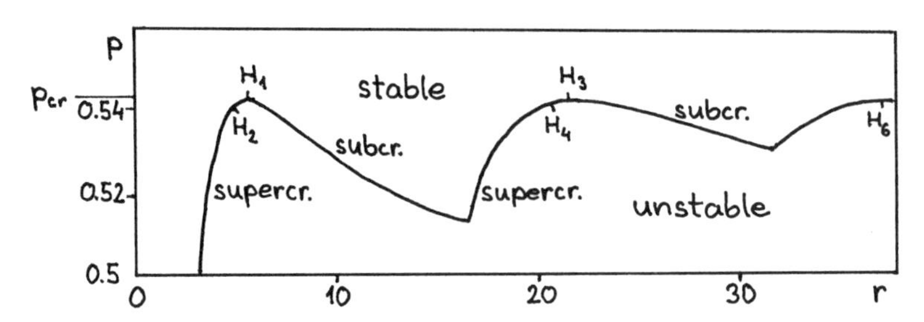

Fig. 3. Stability chart for a predator-prey model

2. Hopf bifurcations

It is shown by Fig.3 that there are a lot of bifurcation points with respect to the delay r in (19) with weight function (22) if $0.52<p<p_{cr}=0.543...$. After the transformation of (19) into an operator differential equation (see Hale(1977)), the existences of Hopf bifurcations can be proved by means of the method of Hassard et al.(1981). For example the bifurcation diagram of Fig. 4 shows stable and unstable periodic solutions when p=0.535 (see Stépán(1986b)).

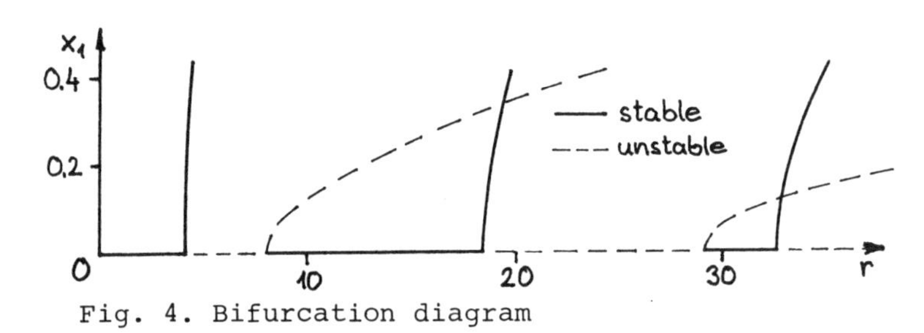

Fig. 4. Bifurcation diagram

However, using the classification of Golubitsky and Langford (1981), we can find both types of the degenerate Hopf bifurcations of codimension 1 at the points $H_1, H_3, ...$ and $H_2, H_4, ...$ (see them in Fig.3). It is easy to get the bifurcation diagram at H_1 (it is shown in Fig.5) but it is only our guess that the Hopf bifurcation at H_2 (where the super- and subcriticality are changed) is supercritical and the system is globally stable.

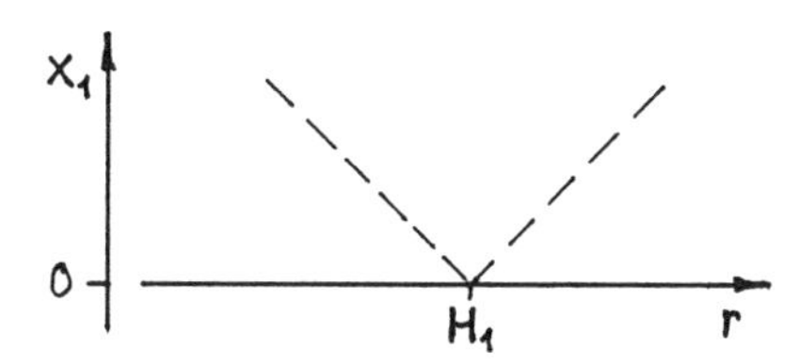

Fig. 5. Degenerate Hopf bifurcation for $p=p_{cr}$

References

Golubitsky,M.,Langford,W.F. (1981) Classification and unfoldings of degenerate Hopf bifurcations, J. of Differential Equations, 41, 375-415.

Hale,J.K. (1977) Theory of Functional Differential Equations (Springer,Berlin).

Hassard,B.D.,Kazarinoff,N.D.,Wan,Y.H. (1981) Theory and Applications of Hopf Bifurcations, "London Math. Soc. Lecture Note Series", 41, (Cambridge University Press, Cambridge).

McDonald,N. (1978) Time Lags in Biological Models, "Lecture Notes in Biomathematics", 27, (Springer,Berlin).

Stépán,G. (1979) On the stability of linear differential equations with time lag, in Colloquia Math. Soc. J. Bolyai, 30, 971-984 (North Holland, Amsterdam).

Stépán,G. (1986a) The role of delay in robot dynamics, in Proc. of "Ro.man.sy.'86", 150-157

Stépán,G. (1986b) Great delay in a predator-prey model, Nonlinear Analysis TMA, to appear.

Dr.G. Stépán, Department of Engineering Mechanics, Budapest University of Technology, H-1521 Budapest, Hungary.

International Series of
Numerical Mathematics, Vol. 79

BIFURCATION FROM THE CONTINUOUS SPECTRUM IN $L^p(\mathbb{R})$

C.A. STUART
Département de Mathématiques
Ecole Polytechnique Fédérale de Lausanne
LAUSANNE (Switzerland)

1. INTRODUCTION

The study of small solutions of the nonlinear eigenvalue problem:

$$\Delta u(x) + \lambda u(x) + q(x)\,|u(x)|^{\sigma} u(x) = o \quad \text{for } x \in \mathbb{R}^N$$

$$\lim_{|x|\to\infty} u(x) = o$$

where $\sigma > o$ and $q\colon \mathbb{R}^N \to \mathbb{R}$ are given, leads to questions involving bifurcation from the continuous spectrum since its linearisation about $u \equiv o$ has the interval $[o,\infty)$ as spectrum. In most of the previous results [1,2,3] it is assumed that

either (i) $\lim_{|x|\to\infty} q(x) = o$

or (ii) $q(x) = Q(|x|)\ \ \forall x \in \mathbb{R}^N$ and $N \geqq 2$.

Here we describe more recent work dealing with cases where (i) and (ii) do not hold. It is assumed throughout that $N = 1$ and that

(H) $q \in C(\mathbb{R})$ is such that $\lim_{x\to\pm\infty} q(x) = q(\pm\infty)$ exist and $o < A \leqq q(x) \leqq B \quad \forall x \in \mathbb{R}$.

We can assume without loss of generality that $q(-\infty) \leqq q(+\infty)$.

The earliest work on this problem is the treatment by Küpper and Reimer [4] of the case where q is constant. Subsequently the case where q is even and decreasing on (o,∞) was treated by Stuart [5] using variational methods and by

Toland [6] using degree theory. The results presented here do not require q to satisfy these restrictive conditions.

To shorten the discussion we deal only with positive solutions, the set of all positive classical solutions being denoted by

$$\mathcal{S} = \{(\lambda,u) \in \mathbb{R} \times C^2(\mathbb{R}): \lim_{|x|\to\infty} u(x) = o \text{ and } u(x) > o \text{ and } u''(x) + \lambda u(x) + q(x)u(x)^{\sigma+1} = o \quad \forall x \in \mathbb{R}\}.$$

The variational approach leads naturally to weak solutions so we set

$$\mathcal{S} = \{(\lambda,u) \in \mathbb{R} \times H^1(\mathbb{R}): u(x) > o \quad \forall x \in \mathbb{R} \text{ and } \langle u',v'\rangle = \langle \lambda u + qu^{\sigma+1},v\rangle \quad \forall v \in H^1(\mathbb{R})\}$$

where $\langle .,. \rangle$ denotes the usual inner product on $L^2(\mathbb{R})$.

<u>Lemma 1</u>: Let q satisfy (H). Then

(i) $\mathcal{S} = S$ and for $(\lambda,u) \in S$

(ii) $\lambda < o$ and $u \in H^2(\mathbb{R}) \cap C^2(\mathbb{R}) \cap L^p(\mathbb{R})$ for $1 \leqq p \leqq \infty$

(iii) $\lim_{|x|\to\infty} e^{t|x|} u(x) = o \quad \forall t < \sqrt{-\lambda}$, $\lim_{x\to\pm\infty} \frac{u'(x)}{u(x)} = \mp\sqrt{-\lambda}$.

<u>Proof</u>: The main point is to show that if $(\lambda,u) \in \mathcal{S}$ then u and $u' \in L^2(\mathbb{R})$. For this, set $p(x) = \lambda + q(x)u(x)^\sigma$. Since $\lambda < o$, $\exists z > o$ such that $p(x) \leqq \frac{\lambda}{2} < o \quad \forall x > z$.

Thus $u''(x) > o$ for $x > z$ and so $\lim_{x\to\infty} u'(x)$ exists. Since $\lim_{x\to\infty} u(x) = o$ we must have $\lim_{x\to\infty} u'(x) = o$ and $u'(x) < o \quad \forall x > z$. Thus for $x > z$,

$$-\frac{1}{2}\{u'(x)^2\}' = p(x)u(x)u'(x) = \frac{1}{2}p(x)\{u(x)^2\}' \geqq \frac{1}{4}\lambda\{u(x)^2\}'$$

and so, $u'(x)^2 \geqq -\frac{1}{4}\lambda u(x)^2$. Hence for $x > z$, $u'(x) \geqq \frac{1}{2}\sqrt{-\lambda}\, u(x)$ and this implies that $\{e^{1/2\sqrt{-\lambda}x} u(x)\}' \leqq o$ for $x > o$.

Thus $\lim_{x\to\infty} e^{1/2\sqrt{-\lambda}x} u(x)$ exists and $u \in L^p(o,\infty)$ for $1 \leqq p \leqq \infty$.

A similar argument applies on $(-\infty,o)$ and the remainder of the proof is completed as in Theorem 2.1 of [7].

Definition: For $1 \leq p \leq \infty$ there is L^p-bifurcation at $\mu \in \mathbb{R} \iff \exists \{(\lambda_n, u_n)\} \subset S$ such that $\lambda_n \to \mu$ and $|u_n|_p \to o$ where $|\cdot|_p$ denotes the usual norm on $L^p(\mathbb{R})$.

As we shall see $\mu = o$ is the only point at which L^p-bifurcation can occur. Whether or not it occurs depends upon σ and q. We begin by collecting some 'a priori' bounds for solutions.

Lemma 2: Let q satisfy (H). Then

(i) $C k^{2/\sigma} \leq |u|_\infty \leq D k^{2/\sigma}$

(ii) $E k^{2/\sigma - 1/p} \leq |u|_p \qquad \forall\, 1 \leq p \leq \infty$

(iii) $k^2 |u|_p^p \leq B |u|_{p+\sigma}^{p+\sigma} \qquad \forall\, 1 \leq p < \infty$

whenever $(-k^2, u) \in S$ with $k > o$ where B,C,D and E are positive constants depending only on q and σ.

Proof: See [7].

Theorem 1: Let q satisfy (H).

(a) At $\mu \neq o$ there is no L^p-bifurcation for $1 \leq p \leq \infty$.

(b) At $\mu = o$ there is no L^p-bifurcation for $1 \leq p \leq \frac{\sigma}{2}$.

Proof: This follows from Lemmas 1 and 2.

For $\mu = o$ and $\frac{\sigma}{2} < p \leq \infty$ we must impose additional assumptions on q. The next result gives conditions implying that there is no L^p-bifurcation at $\mu = o$. Then in Theorem 3 we give a variety of conditions which are sufficient to ensure that L^p-bifurcation at $\mu = o$ does occur if $\frac{\sigma}{2} < p \leq \infty$.

Theorem 2: Let q satisfy (H) and also

(H1) $q \in C^1(\mathbb{R})$, $q(-\infty) < q(+\infty)$ and $\exists\, \alpha > o$ such that $q'(x) \geq o \quad \forall |x| \geq \alpha$.

Then at $\mu = o$ there is no L^p-bifurcation for $1 \leq p \leq \infty$.

Proof: For $(\lambda,u) \in S$, $\int_{-\infty}^{\infty} q' u^{\sigma+2}\,dx = 0$ and so $\int_{-\alpha}^{\alpha} q' u^{\sigma+2}\,dx \leq 0$.

Thus
$$0 \geq \int_{-\alpha}^{\alpha} q'(x)\{(\frac{u(x)}{u(\alpha)})^{\sigma+2} - 1 + 1\}dx$$
$$= \int_{-\alpha}^{\alpha} q'(x)\{(\frac{u(x)}{u(\alpha)})^{\sigma+2} - 1\}dx + q(\alpha) - q(-\alpha).$$

However α can be choosen so that $q(\alpha) - q(-\alpha) > 0$.

If L^p-bifurcation occurs at $\mu = 0$ for $\frac{\sigma}{2} < p \leq \infty$ we would have a sequence $(\lambda_n, u_n) \subset S$ such that

$$\int_{-\alpha}^{\alpha} q'(x)\{(\frac{u_n(x)}{u_n(\alpha)})^{\alpha+2} - 1\}dx \to 0 \quad \text{as} \quad n \to \infty.$$

This leads to a contradiction. The details are given in [7].

Theorem 3: Let q satisfy (H) and one of the following additional conditions:

(H2) $q(+\infty) = q(-\infty)$ and $\exists\, t \in \mathbb{R}$ such that $q(t-y) + q(t+y) \gneqq 2\,q(+\infty)$ $\forall y \in \mathbb{R}$.

(H3) $q(-\infty) < q(+\infty)$ and $\forall h > 0$ $\lim_{x\to\infty} e^{hx}\{q(x) - q(+\infty)\} = +\infty$.

(H4) $\exists\, t \in \mathbb{R}$ such that q is even about $x = t$ (i.e. $q(t-y) = q(t+y)$ $\forall y \in \mathbb{R}$).

Then at $\mu = 0$ there is L^p-bifurcation if and only if $\frac{\sigma}{2} < p \leq \infty$.

Remarks 1: In all cases it is enough to prove that $\exists\, C, \delta > 0$ such that

$$\text{(E)} \quad \begin{cases} \text{for each } k \in (0,\delta)\ \exists\,(-k^2, u_k) \in S \\ \text{with } |u_k|_2 \leq C\,k^{2/\sigma - 1/2}. \end{cases}$$

Indeed if this is establised then by Lemma 2(i) and interpolation we obtain

$$|u_k|_p \leq F\,k^{2/\sigma - 1/p} \quad \text{for} \quad 2 \leq p \leq \infty.$$

A similar estimate for $1 \leq p \leq 2$ can be deduced from Lemma 2(iii) and the result then follows immediately.

2. The proof of (E) using (H2) or (H3) depends upon a variational approach as in [7] whereas for (H4) the variational method may fail. The proof of (E) using (H4) follows from the implicit function theorem after a suitable rescaling of the variables as in [10], but this method may fail for (H2) or (H3).

2. THE VARIATIONAL APPROACH

Let q satisfy (H) and for $k > o$ set

$$H = H^1(\mathbb{R}) \text{ with } \langle\langle u,v\rangle\rangle = \int_{\mathbb{R}}\{u'v'+k^2 uv\}dx \text{ and } \|u\| = \langle\langle u,u\rangle\rangle^{1/2}.$$

Let
$$q^*(x) = \begin{cases} q(-\infty) & \text{for } x < o \\ q(+\infty) & \text{for } x \geq o \end{cases}$$

$$\phi(u) = \int_{\mathbb{R}} q|u|^{\sigma+2}\, dx \quad \text{for } u \in H$$

and ϕ_* and ϕ_∞ are defined similarly with q replaced by q^* and $q(x) \equiv q(+\infty)$ respectively.

For $u \in H\backslash\{o\}$, $J_k(u) = \dfrac{\|u\|^2}{\phi(u)^{2/\sigma+2}}$, $m_k = \inf\{J_k(u): u \in H\backslash\{o\}\}$ and J_k^∞ and m_k^∞ are defined similarly with ϕ replaced by ϕ_∞.

Lemma 2.1: Let q satisfy (H).

(i) $\phi \in C^1(H,\mathbb{R})$ and for $u,v \in H$,

$$\phi'(u)v = \langle\langle \nabla\phi(u),v\rangle\rangle = (\sigma+2)\int_{\mathbb{R}} q|u|^\sigma uv\, dx.$$

(ii) $o < (\frac{1}{B})^{\frac{2}{\sigma+2}}\, k^{\frac{\sigma+4}{\sigma+2}} \leq m_k \leq 2(\frac{\sigma+2}{2A})^{\frac{2}{\sigma+2}}\, k^{\frac{\sigma+4}{\sigma+2}}$.

(iii) If $J_k(w_k) = m_k$ for $w_k \in H\backslash\{o\}$ then $(-k^2,u_k) \in S$, $m_k = \phi(u_k)^{\frac{\sigma}{\sigma+2}}$ and

$|u_k|_2 \leq Ck^{\frac{2}{\sigma}-\frac{1}{2}}$ where $u_k = t_k|w_k|$, $t_k = \left\{\dfrac{\|w_k\|^2}{\phi(w_k)}\right\}^{1/\sigma}$ and $C = \sqrt{2}\left\{\dfrac{(\sigma+2)}{A}\right\}^{1/\sigma}$.

Proof:

(i) Standard; see [5], for example.

(ii) For $u \in H$, $|u|_\infty^2 \leq |u'|_2\,|u|_2$, $|u|_{\sigma+2}^{\sigma+2} \leq |u|_\infty^\sigma\,|u|_2^2$ and $|u|_2 \leq k^{-1}\|u\|$.

Thus $\phi(u) \leq B|u|_{\sigma+2}^{\sigma+2} \leq B\|u\|^{\sigma+2}\, k^{-(\frac{\sigma+4}{2})}$ and so $J_k(u) \geq (\frac{1}{B})^{\frac{2}{\sigma+2}} k^{\frac{\sigma+4}{\sigma+2}}$ for $u \neq o$.

Setting $v_k(x) = e^{-k|x|}$ we have $\|v_k\|^2 = 2k^2|v_k|_2^2 = 4k^2 \int_0^\infty e^{-2kx}\,dx = 2k$ and $\phi(v_k) \geq 2A \int_0^\infty e^{-(\sigma+2)kx}\,dx = \frac{2A}{(\sigma+2)k}$.

Since $m_k \leq J_k(v_k)$ the result follows.

iii) Since $J_k(w) = J_k(|w|)$ we can assume that $w_k \geq 0$. Then since $J_k'(w_k)v = 0$ $\forall v \in H$ we have

$$\langle\langle w_k, v\rangle\rangle - \frac{1}{(\sigma+2)} \frac{\|w_k\|^2}{\phi(w_k)} \phi'(w_k)v = 0.$$

Setting $u_k = t_k w_k$ this implies that u_k is a solution of $u'' - k^2u + qu^{\sigma+1} = 0$ and $u_k \geq 0$. Since $u_k \not\equiv 0$ we conclude that $u_k(x) > 0$ $\forall x \in \mathbb{R}$ and so $(-k^2, u_k) \in S$. Then $\|u_k\|^2 = \phi(u_k)$ and $m_k = J_k(w_k) = J_k(u_k) = \phi(u_k)^{\frac{\sigma}{\sigma+2}} = \|u_k\|^{\frac{2\sigma}{\sigma+2}}$.

Since $|u_k|_2 \leq k^{-1}\|u_k\| = k^{-1}\, m_k^{\frac{\sigma+2}{2\sigma}}$, the result follows from (ii).

Remarks:

1. Using Lemma 2.1 and the remark following Theorem 3, it is enough to prove that m_k is attained for $0 < k < \delta$.

2. For $q(x) \equiv q(+\infty)$ $\forall x \in \mathbb{R}$, the problem has an explicit solution

$$\tilde{u}_k(x) = k^{2/\sigma} \left\{ \frac{(\sigma+2)}{2q(+\infty)} \right\}^{1/\sigma} \operatorname{ch}^{-2/\sigma}\left(\frac{\sigma}{2} kx\right)$$

and as in [7] we find that

$$m_k^\infty = J_k^\infty(\tilde{u}_k) = 2k^{\frac{\sigma+4}{\sigma+2}} \left\{ \frac{B(\frac{2}{\sigma}, \frac{2}{\sigma})}{\sigma(\sigma+4)q(+\infty)^{2/\sigma}} \right\}^{\sigma/(\sigma+2)}$$

where B is the Beta function.

3. However in many circumstances m_k is not attained, as in shown by the following lemma.

Lemma 2.2: Let q satisfy (H).

(i) If $q(x) \leq q(+\infty)$ $\forall x \in \mathbb{R}$ and $q \not\equiv q(+\infty)$ then $J_k(u) > m_k \geq m_k^\infty$ $\forall u \in H\backslash\{0\}$.

(ii) If $q \in C^1(\mathbb{R})$ with $q'(x) \geq 0$ $\forall x \in \mathbb{R}$ and $q \not\equiv q(+\infty)$ then $S = \phi$.

Proof:

(i) Suppose that $J_k(u) = m_k$ for some $u \in H\backslash\{o\}$. By Lemma 2.1 we can suppose that

$$u(x) > o \quad \forall x \in \mathbb{R} \quad \text{and we set}$$
$$u_t(x) = u(x-t) \quad \forall x,t \in \mathbb{R}.$$

Then $\|u_t\| = \|u\| \quad \forall t \in \mathbb{R}$ and

$$\lim_{t\to\infty} \phi(u_t) = \lim_{t\to\infty} \int_{\mathbb{R}} q(x+t)u(x)^{\sigma+2}\, dx = \int_{\mathbb{R}} q(+\infty)u(x)^{\sigma+2}\, dx > \phi(u).$$

Hence $\lim_{t\to\infty} J_k(u_t) < J_k(u) = m_k$. This contradiction implies that m_k is not attained

(ii) This is a trivial consequence of the identity

$$\int_{\mathbb{R}} q'(x)u(x)^{\sigma+2}\, dx = o \quad \text{which holds} \quad \forall(\lambda,u) \in S.$$

Lemma 2.3: Let q satisfy (H). If $m_k < m_k^\infty$ then $\exists\, w_k \in H\backslash\{o\}$ such that $J_k(w_k) = m_k$.

Proof: Let $C = \{u \in H: \phi(u) = 1\}$. Then $\exists\ \{u_n\} \subset C$ such that $J_k(u_n) = \|u_n\|^2 \to m_k$ and $u_n \rightharpoonup u_\infty$ weakly in H.

By Ekeland's ε-variational principle [8] we can assume further that $\|(J|_C)'(u_n)\| \to o$. This means that

$$\frac{\| u_n - \langle\langle u_n, \nabla\phi(u_n)\rangle\rangle \nabla\phi(u_n) \|}{\|\nabla\phi(u_n)\|^2} \to o.$$

Since $\langle\langle u_n, \nabla\phi(u_n)\rangle\rangle = (\sigma+2)\phi(u_n) = \sigma+2$ we conclude that

$$\lim \|\nabla\phi(u_n)\|^2 = (\sigma+2)^2/m_k$$

and
$$\|u_\infty\|^2 = \lim \langle\langle u_n, u_\infty\rangle\rangle = \lim \langle\langle \frac{(\sigma+2)\nabla\phi(u_n)}{\|\nabla\phi(u_n)\|^2}, u_\infty\rangle\rangle$$

$$= \frac{m_k}{(\sigma+2)} \lim \phi'(u_n)u_\infty = m_k \lim \int_{\mathbb{R}} q|u_n|^\sigma u_n u_\infty dx$$

$$= m_k \int_{\mathbb{R}} q|u_\infty|^\sigma u_\infty u_\infty\, dx = m_k\phi(u_\infty).$$

If $u_\infty \neq o$, $m_k \leqq J_k(u_\infty) = \dfrac{\|u_\infty\|^2}{\phi(u_\infty)^{2/\sigma+2}} = m_k \phi(u_\infty)^{\sigma/\sigma+2}$ and so $\phi(u_\infty) \geqq 1$. But then $J_k(u_\infty) \leqq \|u_\infty\|^2 \leqq \lim \|u_n\|^2 = m_k$ and so $J_k(u_\infty) = m_k$.

If $u_\infty = o$ then $\lim[\phi_*(u_n)-1] = \lim[\phi_*(u_n)-\phi(u_n)] = \lim \int_{\mathbb{R}}(q^*-q)|u_n(x)|^{\sigma+2}dx = o$ as in Lemma 3.2 of [5] and

$$m_k^\infty \leqq \frac{\|u_n\|^2}{\phi_\infty(u_n)^{2/\sigma+2}} \leqq \frac{\|u_n\|^2}{\phi_*(u_n)^{2/\sigma+2}} \qquad \forall n \in \mathbb{N}.$$

Thus $m_k^\infty \leqq \lim \|u_n\|^2 = m_k$.

Since $m_k^\infty > m_k$ we conclude that $u_\infty \neq o$ and the proof is complete.

<u>Remark</u>: Since m_k^∞ is known it is sufficient to find a test function u so that $J_k(u) < m_k^\infty$. Conditions (H2) and (H3) imply that a suitable translation of $\tilde{u}_k$ is successful.

For example, for $u(x) = \tilde{u}_k(x-c)$ we have

$$m_k^\infty = \frac{\|u\|^2}{\phi_\infty(u)^{2/\sigma+2}} \quad \text{and} \quad J_k(u) = \frac{\|u\|^2}{\phi(u)^{2/\sigma+2}}.$$

But
$$\begin{aligned}\phi(u) - \phi_\infty(u) &= \int_{\mathbb{R}}\{q(x) - q(+\infty)\}\tilde{u}_k(x-c)^{\sigma+2}\,dx \\ &= \int_o^\infty \{q(-y+c) + q(y+c) - 2q(+\infty)\}\tilde{u}_k(y)^{\sigma+2}\,dx .\end{aligned}$$

Hence (H2) implies that $\phi(u) > \phi_\infty(u)$ and $m_k \leqq J_k(u) < m_k^\infty$.

As in [7] a similar but longer calculation leads to the same conclusion when (H3) holds.

<u>Remark</u>:

1. By Lemma 2.2(i) we see that condition (H4) does not imply that m_k is attained and so another approach is necessary to treat this case.
2. The variational approach can be adapted to treat cases where (H) does not hold but $\lim_{|x|\to\infty} q(x) = o$.

3. THE RESCALING APPROACH

Let q satisfy (H) and (H4). Then $q(-\infty) = q(+\infty)$ and for $k > o$, $(-k^2, u_k) \in S \iff u_k(x) = k^{2/\sigma} v_k(kx)$ where v_k satisfies

$(P)_k$ $\quad v \in H^2(\mathbb{R})$ and $v'' = v\{1-q_k|v|^\sigma\}$

with $v_k(x) > o$ for $x \in \mathbb{R}$ where $q_k(x) = q(\frac{x}{k})$.

Setting $q_o(x) \equiv q(+\infty)$ $\forall x \in \mathbb{R}$ we see that $\lim_{k\to o} q_k(x) = q_o(x)$ $\forall x \neq o$, which suggests treating $(P)_k$ as a perturbation of

$(P)_o$ $\quad v \in H^2(\mathbb{R})$ and $v'' = v\{1-q_o|v|^\sigma\}$.

As in section 2 we find that, up to translation, $(P)_o$ has a unique positive solution

$$v_o(x) = \{\frac{\sigma+2}{2q(+\infty)}\}^{1/\sigma} \operatorname{ch}^{-2/\sigma}(\frac{\sigma x}{2}).$$

For small $k > o$, the implicit function theorem yields solutions of $(P)_k$ near v_o. The degeneracy caused by the translation invariance of $(P)_o$ is removed by restricting the discussion to even functions. Let

$$Y = L^2(o,\infty) \cong \{u \in L^2(\mathbb{R}) : u(x) = u(-x) \text{ a.e.}\}$$

$$X = \{u \in H^2(o,\infty) : u'(o) = o\} \cong \{u \in H^2(\mathbb{R}) : u(x) = u(-x) \text{ a.c.}\}$$

For $\quad (k,u) \in \mathbb{R} \times X$ we set, $Lu = u'' - u$

$$F(k,u) = q_{|k|}|u|^\sigma u, \quad G(k,u) = Lu + F(k,u).$$

Then $v_o \in X$ and $G(o,v_o) = o$. The smoothness of G required by the implicit function theorem [9] is given by the following result.

<u>Lemma 3.1</u>: Let q satisfy (H) and (H4).

(i) $F: \mathbb{R} \times X \to Y$ is continuous, $D_uF(k,u)w = (\sigma+1)q_{|k|}|u|^\sigma w$ $\forall u,w \in X$ and $D_uF: \mathbb{R} \times X \to B(X,Y)$ is continuous.

(ii) $L \in B(X,Y)$ is an isomorphism.

(iii) $G(k,v_k) = o$ for $(k,v_k) \in \mathbb{R} \times X \iff v_k$ satisfies $(P)_k$ and $v_k(x) > o \ \forall x \in \mathbb{R}$.

Proof: (i) For $w,z \in L^2(o,\infty) \cap L^\infty(o,\infty)$ we have

$$\int_o^\infty |z|^{2\sigma} w^2 dx \leq \begin{cases} |z|_{2\sigma}^{2\sigma} \, |w|_\infty^2 & \text{if } \sigma \geq 1 \\ |z|_2^{2\sigma} \, |w|_{2/1-\sigma}^2 & \text{if } o < \sigma < 1. \end{cases}$$

Also $X \subset L^p(o,\infty)$ $\forall p \in [2,\infty]$ and $\exists$ C_p such that $|u|_p \leq C_p \|u\|_X$ for $2 \leq p \leq \infty$. For (k,u) and $(h,v) \in \mathbb{R} \times X$,

$$|F(k,u) - F(k,v)|_2^2 = \int_o^\infty q_{|k|}^2 \{|u|^\sigma u - |v|^\sigma v\}^2 dx$$

$$\leq B^2 \int_o^\infty \{\int_o^1 (\sigma+1) |tu + (1-t)v|^\sigma (u-v) dt\}^2 dx$$

$$\leq B^2 (\sigma+1)^2 \int_o^1 \int_o^\infty |tu + (1-t)v|^{2\sigma} |u-v|^2 dx \, dt$$

$$\leq CB^2 (\sigma+1)^2 \|u-v\|_X^2 \int_o^1 \|v + t(u-v)\|_X^{2\sigma} dt \quad \text{and}$$

$$\lim_{k\to h} |F(k,v) - F(h,v)|_2^2 = \lim_{k\to h} \int_o^\infty (q_{|k|} - q_{|h|})^2 |v|^{2(\sigma+1)} dx = o$$

by dominated convergence. This proves the continuity of F at (h,v) and a standard calculation shows that for $k \in \mathbb{R}$, $F(k,\cdot)\colon X \to Y$ is Fréchet differentiable with $D_u F(k,u)w = (\sigma+1) q_{|k|} |u|^\sigma w$.

For the continuity of $D_u F$ at (h,v), let (k,u) and $(h,v) \in \mathbb{R} \times X$. For $w \in X$ we have

$$|[D_u F(k,u) - D_u F(k,v)]w|_2^2 \leq B^2 (\sigma+1)^2 \int_o^\infty \{|u|^\sigma - |v|^\sigma\}^2 w^2 dx$$

$$\leq B^2 (\sigma+1)^2 \, ||u|^\sigma - |v|^\sigma|_\infty^2 \, \|w\|_X^2$$

and

$$|[D_u F(k,v) - D_u F(h,v)]w|_2^2 = \int_o^\infty (\sigma+1)^2 \, (q_{|k|} - q_{|h|})^2 \, |v|^{2\sigma} w^2 \, dx$$

$$\leq (\sigma+1)^2 \begin{cases} |(q_{|k|} - q_{|h|})^{1/\sigma} v|_{2\sigma}^{2\sigma} \, |w|_2^2 & \text{if } \sigma \geq 1 \\ |(q_{|k|} - q_{|h|})^{1/\sigma} v|_2^{2\sigma} \, |w|_{2/1-\sigma}^2 & \text{if } o < \sigma < 1. \end{cases}$$

Since $\lim_{k\to h} |(q_{|k|} - q_{|h|})^{1/\sigma} v|_p = o$ for $2 \leq p < \infty$ by dominated convergence, this proves the continuity of $D_u F$ at (h,v).

Lemma 3.2: Let q satisfy (H) and (H4). Then

(i) $\exists\ \delta > o$ and $\psi \in C([o,\delta),X)$ such that

$$\psi(o) = v_o \text{ and } G(k,\psi(k)) = o \qquad \forall k \in [o,\delta).$$

Setting $v_k = \psi(k)$ for $k \in [o,\delta)$ we have that

(ii) v_k satisfies $(P)_k$, $v_k(x) = v_k(-x) > o$ $\forall x \in \mathbb{R}$. Furthermore $v_k \in C^2(\mathbb{R})$,

$$\lim_{k\to o} v_k''(o) = \left\{\frac{\sigma+2}{2q(+\infty)}\right\}^{1/\sigma} \left\{1 - \frac{(\sigma+2)q(o)}{2q(+\infty)}\right\}$$

and $$\lim_{k\to o} |v_k - v_o|_p = o \quad \text{for} \quad 2 \leq p \leq \infty.$$

We even have $\lim_{k\to o} |v_k' - v_o'|_\infty = o$, but $\lim_{k\to o} v_k''(o) = v_o''(o)$ if and only if $q(o) = q(+\infty)$.

Proof: As in [10] it can be shown that $D_uG(k,u)$ is an isomorphism <=> $D_uG(k,u)$ in injective.

But $D_uG(o,v_o)w = o$ for $w \in X$ <=> $w'' - w + (\sigma+1)|v_o|^\sigma w = o$ with $w'(o) = o$ and $\lim_{x\to\infty} w(x) = o$.

Since $z = v_o'$ satisfies the same equation and theboundary conditions $z(o) = o$ and $\lim_{x\to\infty} z(x) = o$ we deduce that $w \equiv o$ as in [10].

Part (i) follows immediately from the implicit function theorem.

Setting $v_k(x) = \psi(k)(|x|)$ we have that v_k satisfies $(P)_k$ and $|v_k - v_o|_p \to o$ for $2 \leq p \leq \infty$.

For the positively of v_k on $\mathbb{R}$ it is enough to prove that $\exists\ z > o$ such that $v_k(x) > o$ for $x \geq z$ and $o \leq k \leq \delta$.

Now $\exists\ z > o$ such that $o < v_o(x)^\sigma < \frac{1}{4B}$ $\forall x \geq z$ and so $|v_k(x)| \leq |v_o(x)| + |v_o - v_k|_\infty \leq (\frac{1}{2B})^\sigma$ $\forall x \geq z$ and $o \leq k < \delta$, provided δ is small enough.

Thus $1 - q_k(x)|v_k(x)|^\sigma \geq \frac{1}{2}$ for $x \geq z$ and $o \leq k < \delta$ and consequently $(P)_k$

implies that $-v_k'(x)v_k(x) - \int_x^\infty v_k'(y)^2 dy = \int_x^\infty \{1-q_k|v_k|^\sigma\}v_k^2\, dy$

for $x \geqq z$ and $o \leqq k < \delta$. This proves that $v_k(x) > o$ for $x \geqq z$ and $o \leqq k < \delta$.

Finally we note that

$$\begin{aligned}\lim_{k\to o} v_k''(o) &= \lim_{k\to o} v_k(o)\{1-q_k(o)v_k(o)^\sigma\} = v_o(o)\{1-q(o)v_o(o)^\sigma\}\\ &= \left\{\frac{\sigma+2}{2q(+\infty)}\right\}^{1/\sigma}\left\{1-\frac{(\sigma+2)q(o)}{2q(+\infty)}\right\}\end{aligned}$$

and
$$v_o''(o) = v_o(o)\{1-q(+\infty)v_o(o)^\sigma\} = \frac{\sigma}{2}\left\{\frac{\sigma+2}{2q(+\infty)}\right\}^{1/\sigma}.$$

Remarks:

1. Setting $u_k(x) = k^{2/\sigma} v_k(kx)$ where v_k is given by Lemma 3.2(ii) we have that $|u_k|_2 = k^{\frac{2}{\sigma}-\frac{1}{2}}|v_k|_2$ and $|v_k|_2 \to |v_o|_2$ as $k \to o$. Recalling the remark following Theorem 3 we see that this establishes the result under the hypothesis (H4).

2. Suppose that (H4) holds and that $q(o) < (\frac{2}{\sigma+2})\, q(+\infty)$. Then for $o < k < k_o$, v_k (and hence u_k) has a local minimum at $x = o$ and consequently u_k has at least two local maxima on $\mathbb{R}$. Ruppen [11] has shown that for

$$q(x) = \begin{cases} r_1 & \text{for } |x| < 1 \text{ with } r_1 < (\frac{2}{\sigma+2})\, r_2 \\ r_2 & \text{for } |x| \geqq 1 \end{cases}$$

there is a positive even solution with exactly two maxima on $\mathbb{R}$.

On the other hand, if q satisfies (H4) and is decreasing on (o,∞), it can be shown as in [10] that $u_k < o$ on (o,∞) and so u_k has a unique maximum at $x = o$.

3. The rescaling approach has been studied in much greater detail by Magnus [11].

4. In cases where $q(+\infty) \neq q(-\infty)$ the rescaling approach fails since the function q_o would become

$$q_o(x) = \begin{cases} q(+\infty) & \text{for } x > o \\ q(-\infty) & \text{for } x < o \end{cases}$$

and then $(P)_0$ has no solution. Even in cases where $q(+\infty) = q(-\infty)$ but q is not even the approach may fail because the degeneracy of $D_uF(o,v_o)$ on $H^2(\mathbb{R})$ cannot be eliminated.

REFERENCES

[1] Stuart, C.A.:
Bifurcation for variational problems when the linearisation has no eigenvalues, J. Functional Anal., 38 (1980), 169-187.

[2] Stuart, C.A.:
Bifurcation for Dirichlet problems without eigenvalues, Proc. London Math. Soc., 45 (1982), 169-192.

[3] Stuart, C.A.:
Bifurcation from the essential spectrum, Proc. of Equadiff 82, Springer Lecture Notes in Math., No 1017, 1983.

[4] Küpper, T. and Reimer, D.:
Necessary and sufficient conditions for bifurcation from the continuous spectrum, Nonlinear Anal. TMA, 3 (1979), 555-561.

[5] Stuart, C.A.:
Bifurcation for Neumann problems without eigenvalues, J. Differential Equat., 36 (1980), 391-407.

[6] Toland, J.F.:
Global bifurcation for Neumann problems without eigenvalues, J. Differential Equat., 44 (1982), 82-110.

[7] Stuart, C.A.:
Bifurcation in $L^p(\mathbb{R})$ for a semilinear equation, J. Differential Equat., 64 (1986), 294-316.

[8] Aubin, J.P. and Ekeland, I.:
Applied Nonlinear Analysis, Wiley-Interscience, New York, 1984.

[9] Akilov, G. and Kantorovitch, L.:
Analyse Fonctionnelle, Vol.II, Editions Mir, Moscow, 1981.

[10] Stuart, C.A.:
A global branch of solutions to a semilinear equation on an unbounded interval, Proc. Roy. Soc. Edinburgh, 101A (1985), 273-282.

[11] Ruppen, H.-J.:
Le Problème de Dirichlet Non linéaire sur $\mathbb{R}^N$, Thèse No 593 (1985), Dépt. de Math., EPFL, Lausanne.

[12] Magnus, R.J.:
On the asymptotic properties of solutions of a differential equation in the case of bifurcation without eigenvalues, preprint 1985.

C. A. Stuart
Ecole Polytechnique
Département de Mathématiques
CH-1015 Lausanne

International Series of
Numerical Mathematics, Vol. 79

BIFURCATION PROBLEMS IN RAILWAY VEHICLE DYNAMICS

Hans True
Laboratory of Applied Mathematical Physics
Technical University of Denmark, DK-2800 Lyngby, Denmark

Introduction

Railway Vehicle Dynamics is a challenging and interesting application of "finite but many-dimensional" nonlinear dynamics. In this article we shall describe only one aspect, namely the investigation of lateral oscillations of railway vehicles moving along a straight, horizontal and "ideal" track. In order to describe the problem adequately the mathematical model consists of from four to order of sixty coupled first-order ordinary differential equations. The number depends of course on the design of the vehicle and the desired accuracy, since there are circumstances where approximations can be made which split the system into some decoupled systems.

The general situation is the following: Up to a certain speed the irregular motions of the vehicle will exclusively be a response to irregularities in the track geometry. Above a "characteristic" speed, which for the same vehicle depends on external disturbances, e.g. track geometry, changes in the coefficient of friction, the vehicle will oscillate in a lateral direction. A linear stability analysis will most often yield a characteristic speed above the experimentally determined characteristic speeds. It is therefore logical to look for a subcritical Hopf bifurcation in the problem.

This was indeed found to be the case for the models investigated in this article, but the investigation has been carried on to very high speeds and with wheels with or without flanges in order to examine the dynamics of the wheel-rail interaction.

We are dealing in this article with a fourteenth order system, and the investigation is carried out numerically using a program PATH, developed by Christian Kaas-Petersen. (See references at the end of the article.) Several interesting solutions are found and discussed.

Kinematics

The fundamental guidance system of railways consists of a flanged wheelset rolling on two rails. A clearance between the outer edge of the flange and the inner face of the rail is provided to prevent squeezing of the wheels between the rails. A plane through the wheelset containing its centerline will intersect the wheel tread in a curve called the wheel profile. The angle δ (also termed "the conicity" when measured in radians) between the tangent to the wheel profile and the centerline is made positive in most points on the wheel tread. This makes the wheelset kinematically self-centering. Negative angles, which may develop as a result of wear, make the motion of the wheelset unstable and should be avoided.

As long as the wheelset rolls along a straight track, the flanges ought not contact the rails. Disturbances in the track geometry will push the wheelset off its centered position, but owing to the stabilizing effect of the wheel profile the wheelset will then oscillate horizontally around the centerline of the track. Flange contact will occur only if the disturbances are sufficiently large. When it occurs, the restoring force suddenly grows very fast.

Problems will obviously arise, when a vehicle with a fixed wheelbase negotiates a curve. The longer the wheelbase the smaller the curvature must be to prevent flange contact and squeezing. This presents special dynamical problems that will not be treated in this paper. On the other hand, it is also obvious that the longer the fixed wheelbase is the smaller will the yaw be of a car rolling on a straight track. This leads to a significant improvement in riding quality.

In an effort to solve the conflict between good curving behavior and riding quality the railway companies introduced rail vehicles with two or more wheelsets in an undercarriage. The undercarriage is termed the "truck" in North American terminology and "bogie" in European usage. In bogie vehicles, the body is carried on the bogie, and its weight is transmitted through the frame of the bogie to the wheelsets. A typical passenger bogie car is shown in Figure 1.

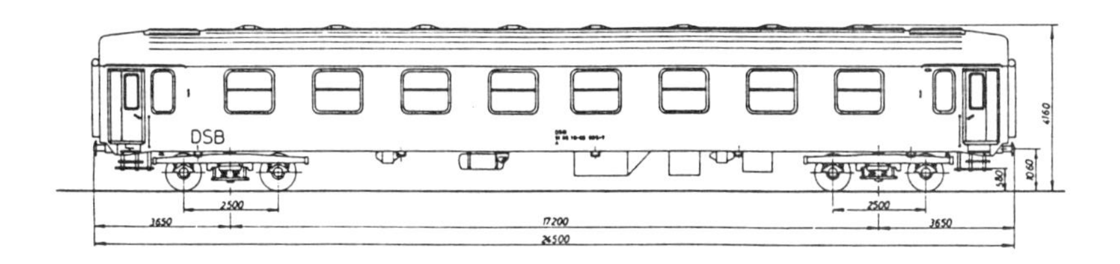

Figure 1. Four-axle Bogie passenger car.

In this article we shall examine the dynamics of bogies and bogie vehicles with no external torques such as torques from motors or applied brakes.

Since we are mainly interested in the lateral and yaw oscillations, we must find the variables that determine the wheelset lateral position relative to the rails. They are:

r_L - instantaneous rolling radius of the left wheel

r_R - instantaneous rolling radius of the right wheel

z_L - instantaneous height of contact point on the left rail

z_R - instantaneous height of contact point on the right rail

δ_L - angle between the contact plane on the left wheel and the axle centerline

δ_R - angle between the contact plane on the right wheel and the axle centerline

ϕ_w - roll angle of the wheelset with respect to the plane of the rails.

These constrained variables and corresponding coordinate systems are illustrated in Figure 2.

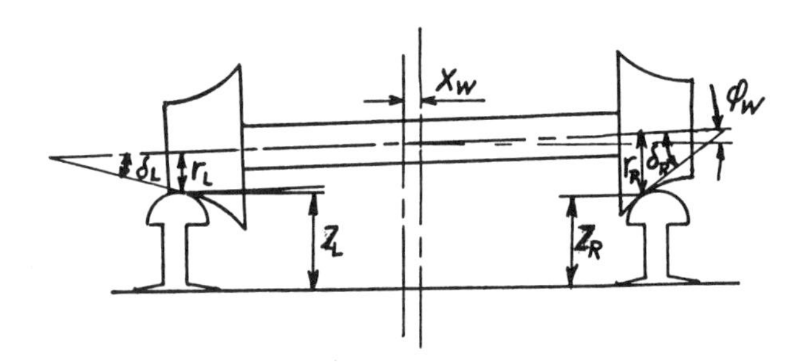

Figure 2. Wheel-rail parameters, rear-view.

The difference in rolling radii will turn the wheelset and force it to roll towards its centered position. The important influence of this so-called "effective conicity" has been recognized for quite some time. It is the cause of a kinematic instability, which was treated by KLINGEL [7] in a celebrated paper. In general, the "effective conicity" has a destabilizing effect, meaning that the critical speed for onset of hunting - the name given to the horizontal oscillations - is lower with larger values of conicity.

The contact angle constraints enter the model equations in the description of the magnitude and direction of the contact forces between the wheel and the rail.

The wheel profile therefore strongly influences the stability of the vehicle. With wheel wear, the speed, at which hunting oscillations occur, decreases. Attempts have been made to find wheel tread profiles that change only little as they wear. They must have a profile that resembles the profile of the rail in order to distribute the wear more evenly. It means that the new wheel profiles look like worn ones or - in other words - in order to prevent an aggravation of the stability we make the stability worse to begin with!

For given profiles of wheel tread and railhead the contact point will move in dependance of the lateral position of the wheelset. The position of a contact point can be calculated as a function of the lateral position of the wheelset as soon as the profiles are known. The lateral position, however, is one of the unknowns in the dynamical problem, so an implicit computational scheme must be applied to solve the equations for the lateral position of the wheelset and the position of the contact points simultaneously.

Variations in the position of the rail have additional influence on the wheelset dynamics. These variations are usually of such a magnitude that they can be treated as disturbance inputs to the model equations.

Dynamics

When the contact forces are modelled we shall adopt two points of view. In the derivation of the relations between the contact forces and the motion between the surfaces of the wheel and the rail it is necessary to take the following aspects into account:

1. The contact between rail and wheel occurs over a finite surface that may not lie in a plane;
2. The relative motion of different points in this surface may be different;
3. The stress distributions across the surface are not uniform.

On the other hand, when the equations of motion for the wheelset are formulated, it is useful to consider the contact as occurring at a single point, that the wheel and the rail have a relative velocity at this point, and that as long as the two bodies remain in contact this relative velocity must lie in a plane that is tangent to the bodies at the contact point.

When we adopt the contact point assumption, we shall define the wheel-rail contact forces in the way it is shown in Figure 3.

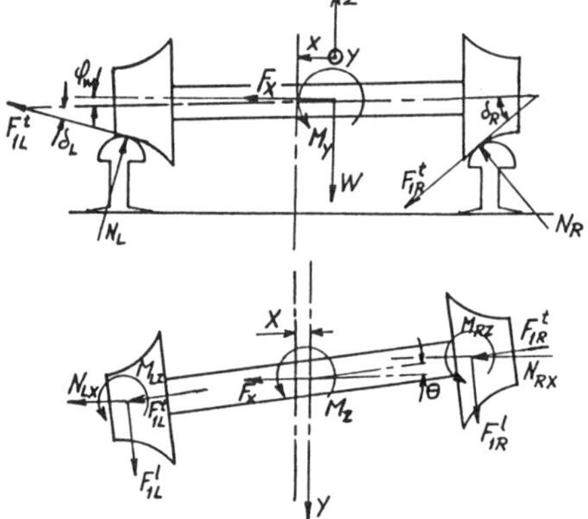

Figure 3. Wheel-rail contact forces.

We shall assume that the contact point lies in a vertical plane that contains the centerline of the wheelset, and that the contact plane must lie perpendicular to that vertical plane. The tangential forces in the contact plane may be projected in a direction parallel to the direction of wheelset travel and in a direction orthogonal to it in the contact plane. The tangential force parallel to the direction of wheelset travel is called the longitudinal creep force, and the other component is called the lateral creep force. The creep forces are very important in rail vehicle dynamics. They are created by the difference in strain rates of the two bodies in contact. The relative strain rate normalized by the forward velocity of the wheel is termed the creepage. In general, the creepage consists of a part stemming from pure translational sliding - the translational creepage - and a part due to the yaw motion of the wheel - called the spin creepage.

The Creep Forces

The objective is to formulate a functional relationship between the creepages and the creep forces and moments. When the wheel and the rail come into contact under a normal load, the normal forces are distributed over a small area of contact between the wheel and the rail. As shown in Figure 4, these normal stresses must be zero at the boundaries of the contact zone and reach a maximum at the center of the contact area. When the wheel rolls and simultaneously experiences a lateral or longitudinal force, tangential or shear stresses are generated in the contact plane between the bodies. Under such rolling conditions, the two bodies remain locked together over a portion of the contact region as they would, if rolling took place without slip. All the relative slip between the two bodies occurs in the remainder of the contact zone. This situation is depicted in Figure 4. As a result of this division of the contact zone into a locked and a slipped region, the shear stress distribution across the contact zone has a form similar to that shown in Figure 4. Imposition of a spin moment about an axis normal to the contact surface further complicates the shear stress distribution and alters the shape of the locked and slipped regions.

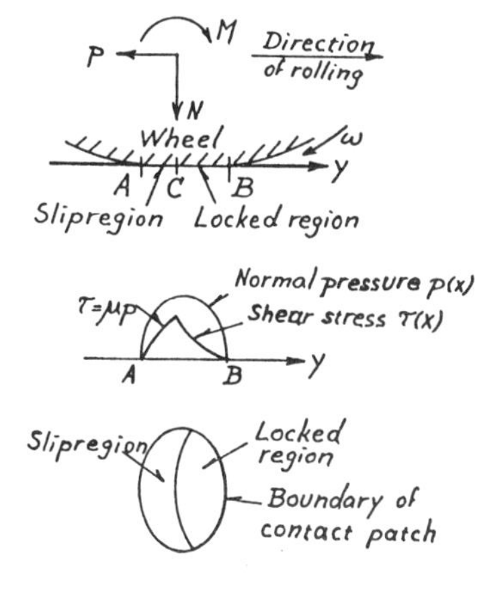

Figure 4. Wheel-rail contact mechanics.

The general form of the relationship between creepages and creep forces is shown in Figure 5. The function is nonlinear owing to the limitation of the forces imposed by the Coulomb friction law. The asymptotic limit corresponding to the Coulomb friction law is attained in the limit of pure sliding.

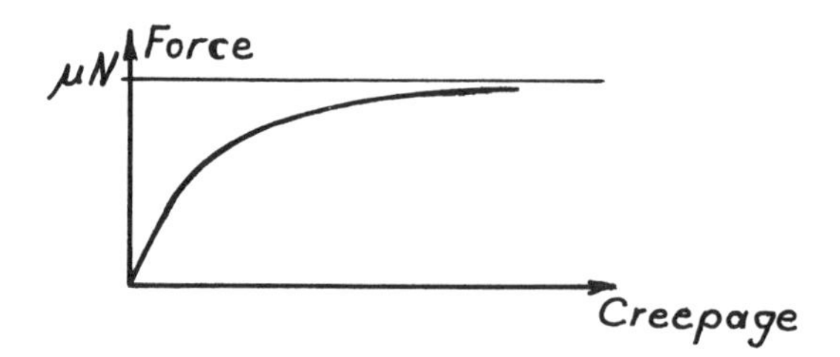

Figure 5. General form of the creepage-creep force relationship.

Several solutions to this problem have been presented over the years, but it is now generally recognized that the best creep theories have been formulated by KALKER [4], [5], [6]. Kalker has formulated a linear relationship, which has been widely used because of its simplicity. It is, however, unable to describe the bifurcation phenomena in the problem. Kalker's nonlinear theory yields results which compare very well with measurements made on test runs on railways. It is, however, very complicated to use, and its numerical implementation is expensive to run and demands a large computer. JOHNSON [9] has developed a nonlinear creep force theory that accounts for the limitation imposed by the coefficient of friction for the case of rolling without relative spin between the two bodies. This theory yields satisfactory results as long as the angle between the centerline of the wheelset and the contact surface is not "too large". It means that curving behavior of the wheelset and oscillations that lead to flange contact are poorly described by Johnson's theory. The theory is much simpler to implement than Kalker's nonlinear theory, and it yields very satisfactory results as long as the wheelset moves on a straight track and its oscillations are mainly lateral not leading to flange contact. Johnson's theory is also able to yield a satisfactory description of the bifurcation phenomena because of its qualitatively correct relationship between the lateral creepage and the creep force, but spin creep is neglected. For these reasons, Johnson's theory is used to obtain the results presented in this paper.

Kalker's and Johnson's theories have in common that the shape and dimensions of the contact area as well as the distribution of the normal stress are given by the Hertz solution for two elastic bodies in contact. The contact surface in this solution is plane and elliptical in shape. For details, the interested reader is referred to the references given in this article.

In order to calculate the lateral displacements we shall of course need to know the lateral creep forces. It must be noted, however, that the longitudinal creepage has a reciprocal effect on the lateral creep forces. Furthermore, the effect of spin creepage is to produce a lateral force. This contribution to the total lateral force reduces the longitudinal creep force - sometimes quite substantially, so altogether there exists a rather complicated relationship between all the creepages and the components of the creep forces.

The Normal Forces

The normal forces between the wheel and the rail enter the equations of motion for the wheelset in two distinct ways. First - as discussed in the last section - the creep forces depend nonlinearly on the normal forces. Secondly, the lateral resultant of the two normal forces acting on the left and right wheels respectively will be different from zero, when the left and right contact angles differ. This difference of the lateral components of the normal forces is often referred to as the gravitational stiffness. It is approximately equal to the axle load multiplied by half the difference in contact angles plus the wheelset roll angle for small contact angles. At larger contact angles trigonometric functions of the angles enter the expressions for these terms thereby making the lateral resultant force a nonlin ear function of the contact angles.

In most cases the gravitational stiffness force acts in a direction to move the wheelset towards its centered position. If the gravitational stiffness is large owing to a large change in contact angle difference with lateral displacement, significant restoring forces will result. Thus gravitational stiffness can be termed a stabilizing effect.

When the wheel tread is conical with the same conicity on both wheels the angle of contact will not change with lateral displacement of the wheelset until the wheelset is moved laterally far enough for the contact point to move to the flange. Consequently, such wheelsets exhibit no gravitational stiffness unless flange contact occurs.

The Equations of Motion

Two models are examined. The first one consists of a bogie car on two two-axle simple bogies with flangeless wheels. The second one is a model of a complex bogie with flanged wheels. The terms "simple" and "complex" refer to the models COOPERRIDER introduced in [1]. We have, however, also included roll of the complex bogie frame and of the car body in our calculations, so our models both have seven degrees of freedom. They are: Lateral and yaw motion of each wheelset and lateral, yaw, and roll motion of the car body and the bogie frame respectively. For this design of car and bogie respectively, the equations of motion in these degrees of freedom uncouple from the other equations of motion when we assume that the vertical displacements are sufficiently small. The mathematical model therefore consists of a system of 14 first-order differential equations. The models and the nomenclature used are shown in Figures 6 and 7.

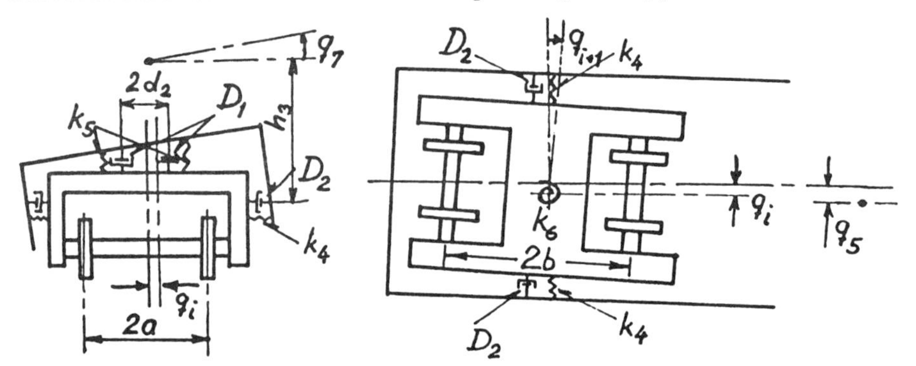

Figure 6. Model af the car resting on simple bogies. Front bogie i = 1, rear bogie i = 3.

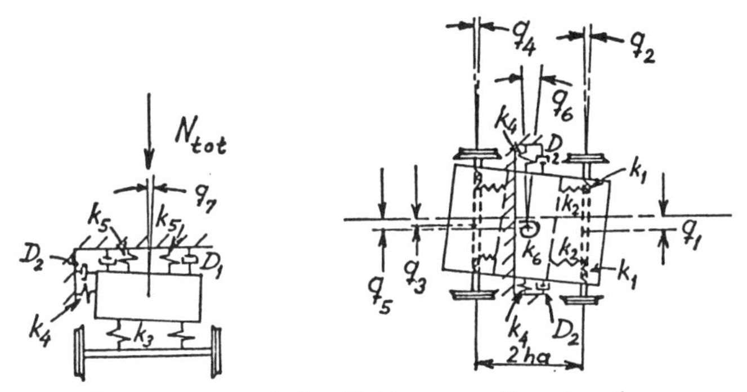

Figure 7. Model of the complex bogie.

The derivation of the equations can be found in COOPERRIDER [1] and KAAS-PETERSEN and TRUE [3], where also the used parameter values are given. We shall only remark here that all parts of the vehicle are treated as rigid and their elasticity is included in the suspension elements. Friction is included only in the creep forces, and flange forces are modelled as very stiff linear springs with a dead band. The wheels are assumed to be conical and the rail heads an arc of a circle. We use JOHNSON's theory [9] to evaluate the creep forces. For details the interested reader is referred to COOPERRIDER [1] or TRUE and KAAS-PETERSEN [8].

The Method of Solution

A computer is used to solve the systems of 14 ordinary differential equations of first order. The program PATH is used in the computations. It is described in KAAS-PETERSEN [2]. Its most important feature is that it uses a mixture of integration in time and Newton iteration to find the time-dependent solutions, whereby computational work is kept to a minimum. The program determines stable and unstable periodic solutions with the same ease and accuracy. In case of a chaotic behaviour an approximation to the maximal Liapunov exponent is calculated. The program PATH can also trace biperiodic solutions, but when the solutions are unstable, problems with the path-following technique may arise. Periodic solutions are treated as equilibrium solutions in a transversal plane to the trajectory in phase space under a Poincaré map. In order to examine the linear stability the linearized operator is computed numerically, and its eigenvalues are determined. The bifurcation points are determined according to the well-known conditions - either the real part of the eigenvalue passes through zero, or the modulus of a Floquet multiplier passes through one. Only simple bifurcation points occur.

Some Results

In order to illustrate various kinds of bifurcation, some results are presented as bifurcation diagrams, where the maximal amplitude is plotted versus the speed. The speed of the vehicle is the bifurcation parameter. The zero solution is always a solution for all speeds, but it is only stable up to the first bifurcation point V_L. On all other branches full line indicates a stable solution and dotted line an unstable solution.

Figures 8 -10 show results of calculations of a car supported by two identical bogies with wheels without flanges. The steady motion becomes unstable at the bifurcation point V_L = 63.86 m/s. A subcritical Hopf bifurcation takes place, and this periodic solution becomes stable in a saddle-node bifurcation at V_C = 38.28 m/s. The new finite-amplitude stable oscillation loses stability in a subcritical bifurcation of a quasi-periodic motion (Neimark bifurcation) at V_B = 70.67 m/s. The unstable quasi-periodic branch meets with a stable periodic oscillation with large amplitude at V_S = 52.95 m/s, which remains stable as far as we have traced it in parameter space. We need further investigations to demonstrate what happens at V_S, but there is an indication that a bifurcation takes place there.

Figures 11 and 12 show results of calculations of a complex bogie rolling on flanged wheels. The most interesting feature is the occurrence of chaotic behaviour at high velocities, when flange contact occurs (i.e. lateral amplitudes larger than 0.009 m). It should be noticed that so far we have only observed chaotic motion of vehicles, when flange contact occurs. We have calculated an approximation to the largest Liapunov exponent, which in our case seems to converge towards a value definitely above zero. There are, however, other interesting bifurcation phenomena which are now objects of further study in order to obtain a clearer picture of the possible dynamical behaviour of this bogie.

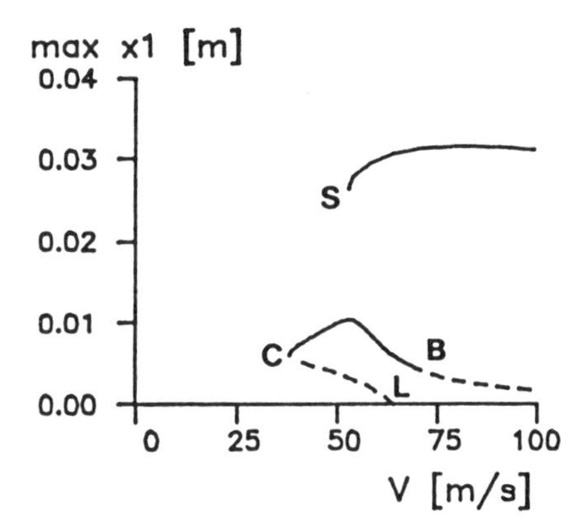

Figure 8. Maximal lateral displacement of the center of the front bogie versus speed. (Car on two simple bogies. No flanges on the wheel.)

Figure 9. Unstable biperiodic solution at V = 69.5 m/s. Lateral displacement of the center of the front bogie versus time.

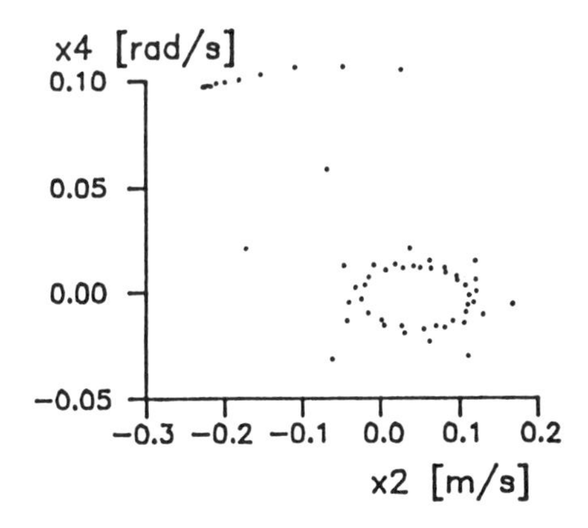

Figure 10. Stroboscopic plot of the unstable biperiodic solution at V = 69.5 m/s. The projection onto the x_2x_4-plane of 60 strobes in the return plane $x_6 = 0$. The points wander away from a closed curve towards a point representing a large amplitude periodic solution.

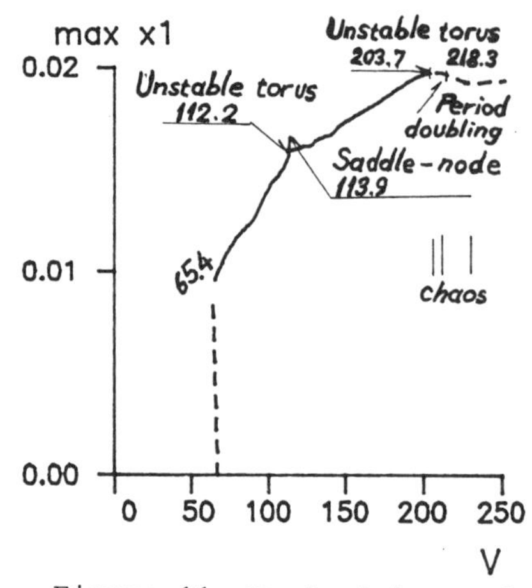

Figure 11. Maximal lateral displacement of the front axle versus speed. (Complex bogie with flanged wheels.)

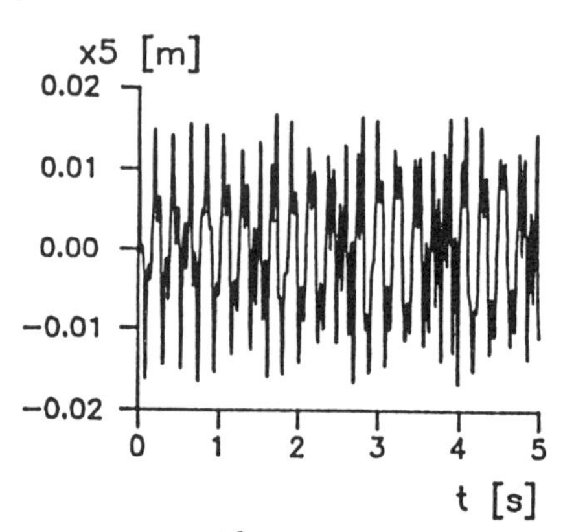

Figure 12. Lateral displacement of the rear axle versus time for V = 225 m/s. Chaotic behaviour. (Complex bogie with flanged wheels.)

References

[1] N.K. Cooperrider: The hunting behavior of conventional railway trucks, ASME J. Eng. Industry, Vol. 94, 1972, pp. 752-762.

[2] Chr. Kaas-Petersen: PATH - User's Guide. Lab. Appl. Math. Phys., The Technical University of Denmark, Lyngby, Denmark, May 1985.

Further information may be found in:

Chr. Kaas-Petersen: Computation of quasi-periodic solutions of forced dissipative systems, J. Comput. Phys., Vol. 58, No. 3, 1985, pp. 395-408.

Chr. Kaas-Petersen: Computation of quasi-periodic solutions of forced dissipative systems II, J. Comput. Phys., (in print).

Chr. Kaas-Petersen: Chaos in a railway bogie, Acta Mechanica, Vol. 61, 1986, pp. 89-107.

[3] Chr. Kaas-Petersen and H. True, Periodic, bi-periodic and chaotic dynamical behaviour of railway vehicles (to appear in Proc. of 9th IAVSD Symposium on Dynamics of Vehicles on Roads and Tracks, Linköping, Sweden, June 24-28, 1985).

[4] J.J. Kalker: Rolling with slip and spin in the presence of dry friction. Doctoral thesis, Delft, 1969.

[5] J.J. Kalker: Simplified theory of rolling contact, Delft Progress Report, Vol. 1, 1973, pp. 1-10.

[6] J.J. Kalker: The computation of three-dimensional rolling contact with dry friction, Int. Journal for Num. Math. in Eng., Vol. 14, 1979, pp. 1293-1307.

[7] Klingel: Über den Lauf der Eisenbahnwagen auf gerader Bahn. Organ für die Fortschritte des Eisenbahnwesens in technischer Beziehung. Neue Folge XX, Band (1883), Nr. 4, pp. 113-123 and Tafel XXI.

[8] H. True and Chr. Kaas-Petersen: A bifurcation analysis of nonlinear oscillations in railway vehicles, Proc. 8th IAVSD Symposium, Vehicle System Dynamics, 1984, pp.655-665.

[9] P.J. Vermeulen and K.L. Johnson: Contact of nonspherical elastic bodies transmitting tangential forces, J. Appl. Mech., Vol. 31, 1964, pp. 338-340.

Acknowledgements

The numerical computations were done on an IBM 3033 computer at UNIC, Lyngby. Most of the calculations have been made by Christian Kaas-Petersen. The Danish State Railways gave permission to the reproduction of Figure 2 in this article. The second, third, and fourth chapters in this article are based on unpublished lecture notes by N.K. Cooperrider and E. Harry Law.

H. True
The Technical University of Denmark
Laboratory of Applied Mathematical Physics
DK-2800 Lyngby

International Series of
Numerical Mathematics, Vol. 79

BIFURCATION AND POST-BIFURCATION BEHAVIOUR FOR ELASTIC-PLASTIC SOLIDS

Viggo Tvergaard

Department of Solid Mechanics
The Technical University of Denmark, Lyngby, Denmark

Abstract

Elastic-plastic material behaviour represents an important class of nonlinear material response. The effect of such nonlinearities on bifurcation and post-bifurcation analyses is briefly described. The application of the theory is illustrated by analyses of structural buckling as well as examples from the field of materials mechanics, where tensile instabilities are often precursors of fracture.

1. Introduction

The theory of elastic stability is a classical subject in solid mechanics. Bifurcation is also important in the presence of nonlinear material behaviour, such as the elastic-plastic behaviour of structural metals. The understanding of these more complex bifurcation problems has developed during the first half of this century in connection with studies of plastic buckling. For columns SHANLEY (1947) explained the significance of the so-called tangent modulus load as the critical bifurcation point, at which the straight configuration loses its uniqueness, but not its stability.

A more general framework was provided by HILL's (1958, 1961) general theory of uniqueness and bifurcation in elastic-plastic solids, which applies to bifurcation under large strain conditions as well as to the more classical area of structural buckling. This theory is based on introducing an elastic comparison solid, with the property that bifurcation in the actual elastic-plastic solid is excluded as long as uniqueness of the incremental solution is predicted for the comparison solid. The theory applies to a broad class of elastic-plastic material models, but not to all elastic-plastic models of current interest.

The behaviour after bifurcation is often important in order to determine the mode of structural collapse or the mode of material failure. In the elastic range the different types of post-bifurcation behaviour and their effect on imperfection-sensitivity are explained by the theory of KOITER (1945). For the plastic range HILL's (1958) bifurcation theory has been extended into the initial post-bifurcation range by an asymptotic theory due to HUTCHINSON (1973, 1974). This theory is considerably complicated by the necessity to account for elastic unloading regions that start pointwise at bifurcation and subsequently spread into the material. In most cases loss of stability occurs subsequent to bifurcation, at some point on the post-bifurcation solution. So far, all information regarding advanced post-bifurcation behaviour or imperfection-sensitivity in the plastic range relies entirely on numerical analyses.

The present paper will give a brief introduction to the theory of bifurcation and post-bifurcation behaviour in the plastic range. The application of the theory is illustrated by analyses of the plastic buckling of columns and thin shell structures. Other examples refer to the mechanics of materials, where tensile instabilities such as localized necking or shear localization are often precursors of fracture.

2. Material model and equilibrium conditions

The theory to be discussed here will be presented in the framework of a Lagrangian formulation of the field equations. A material point in the solid is identified by the coordinates x^i in the reference configuration, and the displacement components on the reference base vectors are denoted by u^i. The Lagrangian strain tensor, $\eta_{ij} = \frac{1}{2}(G_{ij} - g_{ij})$, is related to the displacements by the expression

$$\eta_{ij} = \frac{1}{2}(u_{i,j} + u_{j,i} + u^k_{,i} u_{k,j}) \tag{1}$$

where G_{ij} and g_{ij} are the metric tensors in the current configuration and in the reference configuration, respectively, with determinants G and g, and $(\ )_{,i}$ denotes covariant differentiation in the reference configuration. Indices range from 1 to 3, and the summation convention is adopted for repeated indices. The contravariant components of the Kirchhoff stress tensor on the embedded deformed coordinates are denoted by τ^{ij}, and are related to the components of the Cauchy stress tensor σ^{ij} (the true stress tensor) by the expression $\tau^{ij} = \sqrt{G/g}\,\sigma^{ij}$.

For an elastic-plastic material the stresses τ^{ij} cannot be expressed as unique functions of the strains $\eta_{k\ell}$. The solutions are path dependent, and the constitutive law can only be expressed as an incremental relationship of the form

$$\dot{\tau}^{ij} = L^{ijk\ell}\dot{\eta}_{k\ell} \tag{2}$$

where $(\dot{\ })$ denotes a small increment. The components of the tensor of instantaneous moduli $L^{ijk\ell}$ are generally strongly nonlinear functions of the stresses τ^{ij} and the stress increments $\dot{\tau}^{ij}$.

For a broad class of elastic-plastic materials the instantaneous moduli $L^{ijk\ell}$ in (2) are of the form

$$L^{ijk\ell} = \mathcal{L}^{ijk\ell} - \mu M_G^{ij} M_F^{k\ell} \tag{3}$$

This type of expression for the moduli follows from the assumption that the plastic part of the strain rate is given by

$$\dot{\eta}^{P}_{ij} = \frac{1}{H} m^{G}_{ij} m^{F}_{k\ell} \overset{\nabla}{\sigma}{}^{k\ell} \tag{4}$$

where $\overset{\nabla}{\sigma}{}^{k\ell}$ is the Jaumann (corotational) rate of the Cauchy stress tensor, and continued plastic loading requires $1/H\, m^{F}_{k\ell} \overset{\nabla}{\sigma}{}^{k\ell} \geq 0$. Then, with the elastic stress-strain relationship given by $\overset{\nabla}{\sigma}{}^{ij} = R^{ijk\ell} \dot{\eta}^{E}_{k\ell}$, and the assumption $\dot{\eta}_{k\ell} = \dot{\eta}^{E}_{k\ell} + \dot{\eta}^{P}_{k\ell}$, the constitutive relationship (2), based on the instantaneous moduli (3) appears, as specified by

$$M^{ij}_{G} = R^{ijk\ell} m^{G}_{k\ell} \quad , \quad M^{k\ell}_{F} = m^{F}_{rs} R^{rsk\ell} \tag{5}$$

$$\mu = \begin{cases} 0 & , \text{ for elastic unloading} \\ \sqrt{G/g}\,[H + m^{F}_{k\ell} M^{k\ell}_{G}]^{-1} & , \text{ for plastic loading} \end{cases} \tag{6}$$

$$L^{ijk\ell} = \sqrt{\frac{G}{g}} \left[R^{ijk\ell} - \frac{1}{2}(\sigma^{ik}G^{j\ell} + \sigma^{jk}G^{i\ell} + \sigma^{i\ell}G^{jk} + \sigma^{j\ell}G^{ik}) + \sigma^{ij}G^{k\ell} \right] \tag{7}$$

Equilibrium states satisfy the principle of virtual work. Due to the incremental nature of the elastic-plastic constitutive relations (2), we use the incremental form of the principle of virtual work

$$\int_{V} \left\{ \dot{\tau}^{ij} \delta\eta_{ij} + \tau^{ij} \dot{u}^{k}_{,i} \delta u_{k,j} \right\} dV = \int_{S} \dot{T}^{i} \delta u_{i} dS \tag{8}$$

Here, V and S are the volume and surface, respectively, of the body in the reference configuration, and T^{i} are the specified nominal surface tractions.

The Euler equations of (8) express the same requirement directly in terms of the equilibrium equations and the corresponding boundary conditions

$$(\dot{\tau}^{ij} + \tau^{kj} \dot{u}^{i}_{,k} + \dot{\tau}^{kj} u^{i}_{,k})_{,j} = 0 \tag{9}$$

$$\dot{u}_i = 0 \quad \text{on} \quad S_U \quad , \quad \dot{T}^i = (\dot{\tau}^{ij} + \tau^{kj}\dot{u}^i_{,k} + \dot{\tau}^{kj}u^i_{,k})n_j = 0 \quad \text{on} \quad S_T \qquad (10)$$

where displacements and tractions are specified on the surface parts S_U and S_T, respectively, and n_j is the surface normal. When (2) is substituted into (9) and (10), using the incremental form of (1), three partial differential equations for the displacement increment fields $\dot{u}_i$ emerge, together with the corresponding boundary conditions.

The instantaneous moduli $L^{ijk\ell}$ appear in most of the coefficients of the partial differential equations (9) for $\dot{u}_i$. The plastic part of these moduli (3) are nonlinear functions of the local stress state, but according to (6) there are also two possible values of the parameter μ in each material point, and the value to be chosen depends on the incremental solution to be found. This is a major complication in solving elastic-plastic problems, and particularly in bifurcation studies.

3. Bifurcation

For materials with plastic incompressibility the last (unsymmetric) term in (7) can be neglected, so that the elastic moduli have the symmetry $\mathcal{L}^{ijk\ell} = \mathcal{L}^{k\ell ij}$. Furthermore, many materials, including most metals, satisfy the condition $M_G^{ij} = M_F^{ij}$ (the plastic strain increment (4) is normal to the yield surface). HILL's (1958, 1961) general theory of uniqueness and bifurcation in elastic-plastic solids refers to materials that satisfy these two conditions.

The equations governing bifurcation are formulated by assuming, at any point of the loading history, that there are at least two distinct solutions $\dot{u}_i^a$ and $\dot{u}_i^b$ corresponding to a given increment of the prescribed quantity. The difference between such two solutions is denoted by $(\tilde{\ }) = (\dot{\ })^a - (\dot{\ })^b$. Then, using (8) the following equation is obtained

$$\int_V \left\{ \tilde{\tau}^{ij} \delta\eta_{ij} + \tau^{ij} \tilde{u}^k_{,i} \delta u_{k,j} \right\} dV - \int_S \tilde{T}^i \delta u_i dS = 0 \tag{11}$$

which must be satisfied by non-zero bifurcation solutions $(\tilde{\ })$. Here, τ^{ij} are the current stresses.

HILL (1958, 1961) makes use of the expression

$$I = \int_V \left\{ \tilde{\tau}^{ij} \tilde{\eta}_{ij} + \tau^{ij} \tilde{u}^k_{,i} \tilde{u}_{k,j} \right\} dV - \int_S \tilde{T}^i \tilde{u}_i dS \tag{12}$$

to prove uniqueness. A *comparison solid*, with fixed instantaneous moduli $L_C^{ijk\ell}$ in (2), is defined by choosing these moduli equal to the plastic branch of the elastic-plastic moduli (specified by (3) and (6)) for every material point currently on the yield surface, and the elastic branch elsewhere. For the comparison solid the following quadratic functional is considered

$$F = \int_V \left\{ L_C^{ijk\ell} \tilde{\eta}_{ij} \tilde{\eta}_{k\ell} + \tau^{ij} \tilde{u}^k_{,i} \tilde{u}_{k,j} \right\} dV - \int_S \tilde{T}^i \tilde{u}_i dS \tag{13}$$

It can be proved, for non-negative μ , that the relation

$$\tilde{\tau}^{ij} \tilde{\eta}_{ij} \geq L_C^{ijk\ell} \tilde{\eta}_{ij} \tilde{\eta}_{k\ell} \tag{14}$$

is satisfied in every material point, and thus $F \leq I$. A non-trivial solution of (11) gives $I = 0$, and therefore the requirement $F > 0$ is a sufficient condition for uniqueness. Equality in (14), leading to $I = 0$ for $F = 0$, is in fact satisfied in many cases. For such cases bifurcation in the elastic-plastic solid can be analysed by solving the much easier bifurcation problem resulting from the variational equation $\delta F = 0$.

For a number of materials normality of the plastic flow rule is not a good approximation. Then $M_G^{ij} \neq M_F^{ij}$, and in this case (14) is not satisfied for the usual comparison solid. RANIECKI and BRUHNS (1981) have proposed an *alternative comparison solid*, which provides a lower bound to the first critical bifurcation point. An application of this method to the ductile fracture of metals indicates that the best such lower bound can give a rather inaccurate estimate of the actual bifurcation point (see TVERGAARD, 1982).

The assumption that a pointed vertex develops on the yield surface is of some importance in metal plasticity. Physical models of polycrystalline metal plasticity imply the occurrence of such a vertex, whereas the experimental evidence of vertex formation is conflicting. It turns out that the question whether or not a vertex forms has a significant influence on bifurcation predictions (for more detailed discussion see HUTCHINSON, 1974; NEEDLEMAN and TVERGAARD, 1982). When a vertex forms on the yield surface, the instantaneous moduli are not given by an expression of the form (3). However, a comparison solid can be found (the total loading moduli), which satisfies the requirement (14) of Hill's theory (see SEWELL, 1972). If the fundamental solution deviates from total loading in some regions, this comparison solid can result in a rather poor lower bound. An *alternative comparison solid* that gives an upper bound has been proposed (TVERGAARD, 1983a), and it is expected that this upper bound gives a rather good approximation of the actual bifurcation point for the elastic-plastic solid.

A full understanding of bifurcation behaviour requires also insight in the post-bifurcation solution and the sensitivity to small imperfections, as is provided in the elastic range by KOITER's (1945) elastic stability theory. HILL's (1958, 1961) bifurcation theory has been extended into the initial post-bifurcation range by HUTCHINSON (1973, 1974), who has obtained an asymptotically exact expression for the prescribed load (or deformation) parameter λ in terms of the bifurcation mode amplitude ξ of the form

$$\lambda = \lambda_c + \lambda_1 \xi + \lambda_2 \xi^{1+\beta} + \dots \quad , \quad \text{for} \quad \xi \geq 0 \qquad (15)$$

Here, λ_c is the value of λ at the first critical bifurcation point, λ_1 is positive, λ_2 is negative, and $0 < \beta < 1$. The asymptotic expansions of growing elastic unloading zones, etc., that lead to the expression (15) are strongly dependent on details of the constitutive law.

For materials that do not satisfy $M_G^{ij} = M_F^{ij}$ in (3), or for materials that develop a vertex on the yield surface, the post-bifurcation expansion (15) does not apply. For such materials all current understanding of the post-bifurcation behaviour relies on numerical solutions.

Bifurcation from a fundamental equilibrium path to a secondary equilibrium path is associated with cases, where the geometry, the loading conditions and the material homogeneity are perfect according to certain requirements. Deviations from perfect conditions often lead to a considerable reduction of the maximum load-carrying capacity, or the critical strain for failure. The current knowledge of such imperfection-sensitivity, for any elastic-plastic material, is based on numerical solutions.

4. Applications

Plastic buckling of columns or plate- and shell structures is one area, in which bifurcation plays a significant role, and another area is the mechanics of materials. Bifurcation analyses for elastic-plastic solids will be illustrated here by a few examples taken from these two areas.

4.1 Structures

The plastic buckling of columns has been studied in detail since the end of the last century. There was a great deal of controversy regarding the location of the first critical bifurcation point (see SEWELL, 1972), until SHANLEY (1947) explained that bifurcation occurs at the tangent modulus load. Subsequently, the interest in columns has moved towards the post-bifurcation behaviour (HUTCHINSON, 1974). It has been found that the non-linear material behaviour coupled with an asymmetric column cross-section can lead to a strongly nonsymmetric post-bifurcation behaviour and imperfection-sensitivity (TVERGAARD and NEEDLEMAN, 1975), even though the elastic behaviour is symmetric. However, most of the interest has been directed towards thin shell structures, for which buckling is a very important failure mode, also in the elastic range (HUTCHINSON and KOITER, 1970; TVERGAARD, 1977a).

As an example, Fig. 1 shows results obtained by TVERGAARD (1977b) for an axially stiffened circular cylindrical shell under axial compression. The mode of instability considered here is local buckling between the stiffeners, and the figure shows the load parameter λ vs. the buckling mode deflection w_1. The bifurcation point at λ_c has been determined based on Hill's theory, and the post-bifurcation behaviour is computed by a numerical solution of the nonlinear field equations. It is noted that the initially increasing load on the post-bifurcation path and the load maximum slightly above λ_c agree with the result $\lambda_1 > 0$ and $\lambda_2 < 0$ in (15). The imperfection-sensitivity is also illustrated in the figure in terms of the reduced load carrying capacity for a shell with an initial geometric imperfection in the shape of the bifurcation mode (amplitude $\bar{\xi}$ normalized by the shell thickness).

A common feature of many structures subject to compressive loading is that the critical buckling mode is

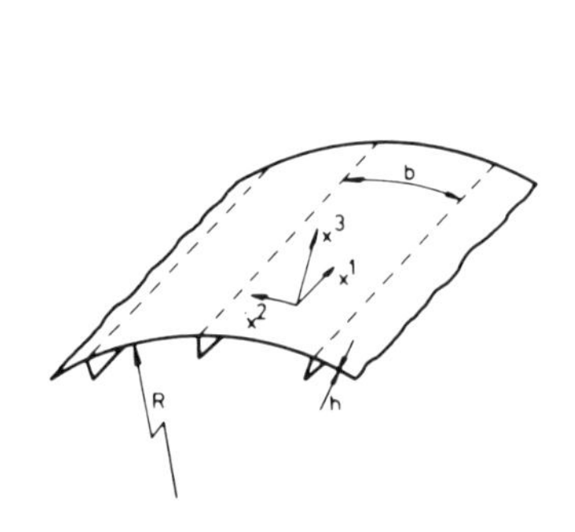

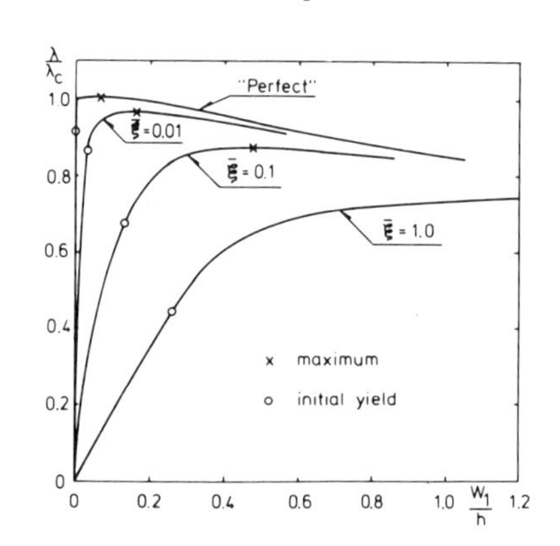

Fig. 1. Local buckling between stiffeners in elastic-plastic cylindrical shell under axial compression. Load λ vs. bifurcation mode deflection w_1 (TVERGAARD, 1977b).

periodic in the direction of compression. However, the final mode of collapse in such structures is often a localized one, involving only the growth of one or a few buckles. TVERGAARD and NEEDLEMAN (1980) have shown that this is a common feature of a wide variety of structures with the property that the load reaches a maximum. The onset of localization occurs just after the load maximum, as a bifurcation from a periodic mode into a non-periodic mode. This is illustrated in Fig. 2 for the relatively simple example of an elastic column on a softening elastic-plastic foundation. The initial elastic foundation stiffness is K_1, and the tangent stiffness for deflections larger than w_0 is K_2. The initial imperfections referred to in Fig. 2 are sinusoidal, with m half-waves along the column, and the arrows mark the point of bifurcation into the localized mode. In particular for the perfect column the first bifurcation leads to the periodic mode of deflection, and localization occurs at a secondary bifurcation point.

Among the structures susceptible to localization are elastic-plastic plate strips or elastic-plastic multi-bay columns under axial compression, and tubes subject to pure bending (TVERGAARD and NEEDLEMAN, 1980; NEEDLEMAN and TVERGAARD, 1982),

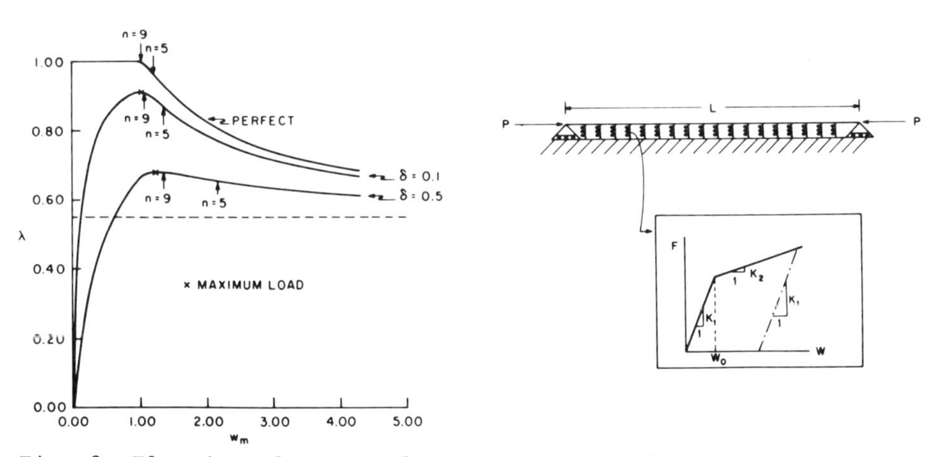

Fig. 2. Elastic column on elastic-plastic softening foundation, where $\lambda = P/P_c$ and δ is the amplitude of a periodic imperfection, normalized by w_0. Arrows mark points of bifurcation into a localized mode (TVERGAARD and NEEDLEMAN, 1980).

or railway tracks subject to thermal buckling (TVERGAARD and NEEDLEMAN, 1981). For axially compressed circular cylindrical shells it has been found that localization of an axisymmetric buckling mode explains the transition from a diamond mode of collapse for relatively thin shells to an axisymmetric mode for thicker shells (TVERGAARD, 1983b).

4.2 Materials mechanics

Tensile instabilities such as localized necking or shear localization are important failure mechanisms, e.g. in various metal forming processes. Thus, during stretching of a thin metal sheet bifurcation from the uniform thickness fundamental mode may occur, leading to a mode of deformation where all additional deformation is concentrated in a narrow band of material (the neck), while elastic unloading has occurred outside. Then, a little additional overall straining gives large strains inside the narrow band, which eventually lead to fracture (e.g. see STÖREN and RICE, 1975; TVERGAARD, 1980).

A full numerical analysis of necking in a sheet subject to plane strain conditions has been carried out by TVERGAARD, NEEDLEMAN and LO (1981). The material modelled develops a vertex on the yield surface, and the fundamental solution satisfies total loading, so that bifurcation into the necking mode is

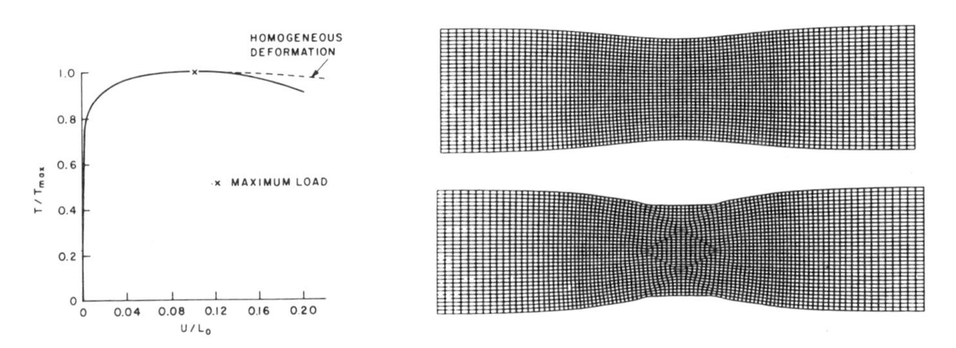

Fig. 3. Tensile load T vs. elongation for a plane strain tensile test. Necking occurs slightly after load maximum, and two subsequent deformed meshes are shown (TVERGAARD, NEEDLEMAN and LO, 1981).

directly covered by Hill's theory. The deformations after the onset of necking are illustrated by the deformed meshes shown in Fig. 3. At the first stage the deformations vary smoothly inside the neck, whereas at the second stage an additional localization into narrow shear bands has occurred, induced by the large strains in the neck. This last stage of deformation indicates the final failure mode, by shear fracture across the neck region, which is often observed in practice.

The bifurcation discussed in section 3 refers to a nonuniqueness that occurs in the elliptic range, whereas the occurrence of shear bands is associated with loss of ellipticity of the governing differential equations. Still, the onset of a shear band in a homogeneously deformed solid is also a bifurcation. For the material considered in Fig. 3 HUTCHINSON and TVERGAARD (1981) have made a detailed study of shear band bifurcation, including the post-bifurcation behaviour and the imperfection-sensitivity. Fig. 4 shows an example of the results obtained in terms of the angle of shearing ω inside the band vs. the logarithmic strain ε in the homogeneously strained material outside. It is clear from the figure that shear localization can lead to intense straining of the material in the band without further straining outside, which will result in fracture.

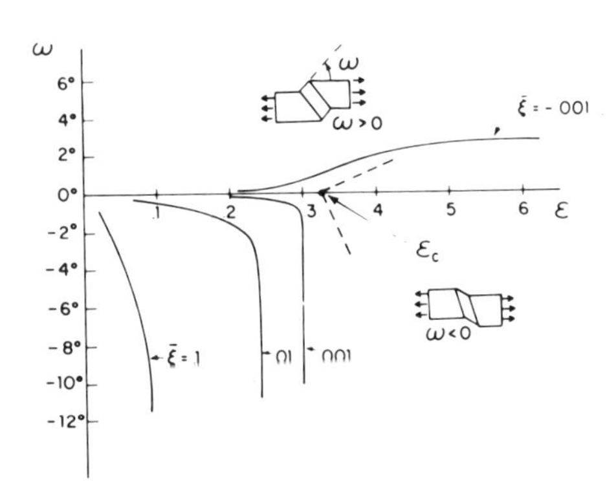

Fig. 4. Growth of a shear band in a solid that develops a corner on the yield surface. Dashed lines show initial post-bifurcation slopes for a homogeneous solid, and the imperfection amplitude $\bar{\xi}$ represents a material inhomogeneity (HUTCHINSON and TVERGAARD, 1981).

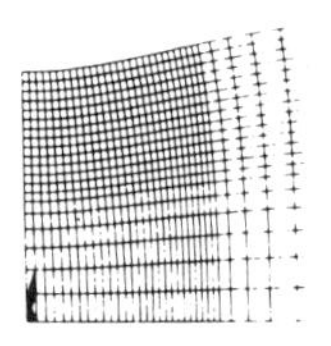
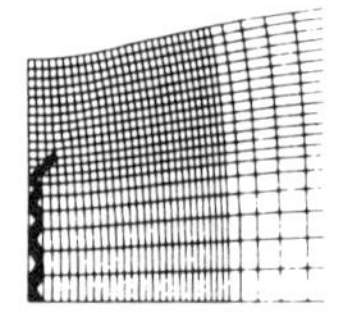
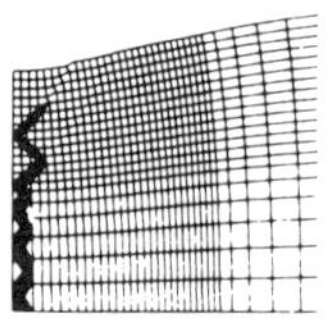

Fig. 5. Deformed meshes in the neck of a round tensile test specimen, showing the growth of a macroscopic crack (painted black) during post-bifurcation (TVERGAARD and NEEDLEMAN, 1984).

An elastic-plastic material model that incorporates failure by the nucleation and growth of small voids has also been used to analyse tensile instabilities. For this model of a porous ductile material the instantaneous moduli are of the form (3), sometimes with $M_G^{ij} \neq M_F^{ij}$, so that normality of the plastic flow rule is not necessarily satisfied, and Hill's bifurcation theory is not directly applicable (see TVERGAARD, 1982). Fig. 5 shows results of a finite element analysis for a round bar under uniaxial tension (TVERGAARD and NEEDLEMAN, 1984). After some straining bifurcation into a necking mode occurs, and the deformed meshes in Fig. 5 illustrate three stages of the deformation in the neck region. The areas painted black in the figures are those in which final failure by coalescence of voids has been predicted by the constitutive relations. Thus, bifurcation leads to localized necking, and during the neck development (the post-bifurcation solution) a crack starts in the centre of the neck and grows into a cup-cone fracture, as is often observed in experiments.

Other cases, in which tensile instabilities represent an important part of the failure mechanism, include rotating turbine disks, tubes under internal pressure and spherical tanks. Some such cases have been further discussed by TVERGAARD (1980) and NEEDLEMAN and TVERGAARD (1984).

References

Hill, R. (1958) A general theory of uniqueness and stability in elastic-plastic solids. J. Mech. Phys. Solids 6, 236-249.

Hill, R. (1961) Bifurcation and uniqueness in nonlinear mechanics of continua. In: Problems of Continuum Mechanics, Society of Industrial and Applied Mathematics, Philadelphia, 155-164.

Hutchinson, J.W. (1973) Postbifurcation behavior in the plastic range. J. Mech. Phys. Solids 21, 163-190.

Hutchinson, J.W. (1974) Plastic buckling. Adv. Appl. Mech. (Ed. C.S. Yih), Vol. 14, Academic Press, New York, 67-144.

Hutchinson, J.W. and W.T. Koiter (1970) Postbuckling theory. Appl. Mech. Rev. 23, 1353-1366.

Hutchinson, J.W. and V. Tvergaard (1981) Shear band formation in plane strain. Int. J. Solids Struct. 17, 451-470.

Koiter, W.T. (1945) Over de stabiliteit van het elastisch evenwicht. Thesis, Delft, H.J. Paris, Amsterdam. English translations (a) NASA TT-F10, 833, (1967), (b) AFFDL-TR-70-25 (1970).

Needleman, A. and V. Tvergaard (1982) Aspects of plastic post-buckling behaviour. In: Mechanics of Solids, The Rodney Hill 60th Anniversary Volume (Eds. H.G. Hopkins and M.J. Sewell), Pergamon Press, Oxford, 453-498.

Needleman, A. and V. Tvergaard (1984) Finete element analysis of localization in plasticity. Finite Elements - Special Problems in Solid Mechanics, Vol. 5 (Eds. J.T. Oden and G.F. Carey), Prentice-Hall, Inc., 94-157.

Raniecki, B. and O.T. Bruhns (1981) Bounds to bifurcation stresses in solids with non-associated plastic flow law at finite strain. J. Mech. Phys. Solids 29, 153-172.

Sewell, M.J. (1972) A survey of plastic buckling. In: Stability (Ed. H. Leipholz), University of Waterloo Press, 85-197.

Shanley, F.R. (1947) Inelastic column theory. J. Aeronaut. Sci. 14, 261-267.

Stören, S. and J.R. Rice (1975) Localized necking in thin sheets. J. Mech. Phys. Solids 23, 239-264.

Tvergaard, V. (1977a) Buckling behaviour of plate and shell structures. In: Theoretical and Applied Mechanics, Proc. 14th IUTAM Congress (Ed. W.T. Koiter), North-Holland, Amsterdam, 233-247.

Tvergaard, V. (1977b) Buckling of elastic-plastic cylindrical panel under axial compression. Int. J. Solids Struct. 13, 957-970.

Tvergaard, V. (1980) Bifurcation and imperfection-sensitivity at necking instabilities. ZAMM 60, T26-T34.

Tvergaard, V. (1982) Influence of void nucleation on ductile shear fracture at a free surface. J. Mech. Phys. Solids 30, 399-425.

Tvergaard, V. (1983a) Plastic buckling of axially compressed circular cylindrical shells. Int. J. Thin-Walled Struct. 1, 139-163.

Tvergaard, V. (1983b) On the transition from a diamond mode to an axisymmetric mode of collapse in cylindrical shells. Int. J. Solids Struct. 19, 845-856.

Tvergaard, V. and A. Needleman (1975) On the buckling of elastic-plastic columns with asymmetric cross-sections. Int. J. Mech. Sci. 17, 419-424.

Tvergaard, V. and A. Needleman (1980) On the localization of buckling patterns. J. Appl. Mech. 47, 613-619.

Tvergaard, V. and A. Needleman (1981) On localized thermal track buckling. Int. J. Mech. Sci. 23, 577-587.

Tvergaard, V. and A. Needleman (1984) Analysis of the cup-cone fracture in a round tensile bar. Acta Metallurgica 32, 157-169.

Tvergaard, V., A. Needleman and K.K. Lo (1981) Flow localization in the plane strain tensile test. J. Mech. Phys. Solids 29, 115-142.

V. Tvergaard
The Technical University of Denmark
Department of Solid Mechanics
DK-2800 Lyngby

International Series of
Numerical Mathematics, Vol. 79

SYMMETRY-BREAKING AT POSITIVE SOLUTIONS OF ELLIPTIC EQUATIONS

A. Vanderbauwhede

Institute for Theoretical Mechanics, State University, Ghent, Belgium

1. Introduction

We consider a Dirichlet boundary value problem

$$\begin{aligned} &\Delta w(x) + f(w(x),\lambda) = 0 , && x \in \Omega , \\ &w(x) = 0 , && x \in \partial\Omega , \end{aligned} \qquad (1)$$

where $\Omega = \{x \in \mathbb{R}^n \mid |x|<1\}$ is the unit ball in $\mathbb{R}^n$ $(n \geq 2)$, and $f : \mathbb{R} \times \mathbb{R} \to \mathbb{R}$ is smooth. If (w_0,λ_0) is a solution of (1) such that $w_0(x) > 0$ for all $x \in \Omega$, then a theorem of GIDAS, NI and NIRENBERG (1979) implies that w_0 is spherically symmetric. In this paper we give some results on the structure of the solution set of (1) near (w_0,λ_0) in the case where w_0 has a neighborhood (in an appropriate space) containing nonpositive functions. In particular we give sufficient conditions for symmetry-breaking at (w_0,λ_0); by definition this means that each neighborhood of (w_0,λ_0) contains solutions of (1) which are not spherically symmetric (and hence also nonpositive). Our results complement and improve the work of SMOLLER and WASSERMAN (1984, 1986), who considered the case $f(w,\lambda) = \lambda\tilde{f}(w)$. Other related work was published by CERAMI (1986), POSPIECH (1986) and BUDD (1986). Full details and proofs are given elsewhere (VANDERBAUWHEDE, 1986).

2. Preliminaries

The problem (1) has clearly a rotational symmetry; in order to formulate this symmetry precisely we rewrite (1) in a more abstract form. Classical solutions of (1) will belong to the space $C_0^{2,\alpha}(\bar{\Omega})$ for all $\alpha \in (0,1)$; so we fix some $\alpha \in (0,1)$, put $W := C_0^{2,\alpha}(\bar{\Omega})$ and $Z := C^{0,\alpha}(\bar{\Omega})$, and define $F : W \times \mathbb{R} \to Z$ by

$$F(w,\lambda)(x) := \Delta w(x) + f(w(x),\lambda) \quad , \quad \forall x \in \Omega . \tag{2}$$

We want to describe the solution set (in $W \times \mathbb{R}$) of the equation

$$F(w,\lambda) = 0 . \tag{3}$$

Now the group $O(n) := \{\gamma \in \mathcal{L}(\mathbb{R}^n) \mid \gamma^T\gamma = I\}$ acts on the space $C^0(\bar{\Omega})$ (and on its subspaces W and Z) as follows :

$$\gamma.w := w \circ \gamma^{-1} \quad , \quad \forall \gamma \in O(n) \; , \; \forall w \in C^0(\bar{\Omega}) . \tag{4}$$

The symmetry of (1) is then expressed by the $O(n)$-equivariance of F :

$$F(\gamma.w,\lambda) = \gamma.F(w,\lambda) \quad , \quad \forall \gamma \in O(n) . \tag{5}$$

Consequently solutions of (3) come in orbits of the form $\{(\gamma.w,\lambda) \mid \gamma \in O(n)\}$, where (w,λ) is a particular solution. We can describe the symmetry of a mapping $w \in C^0(\bar{\Omega})$ (and in particular of a solution of (3)) by the isotropy subgroup of w :

$$\Sigma_w := \{\gamma \in O(n) \mid \gamma.w = w\} . \tag{6}$$

For each $\nu \in S^{n-1} := \{x \in \mathbb{R}^n \mid |x| = 1\}$ we denote its isotropy subgroup $\{\gamma \in O(n) \mid \gamma\nu = \nu\}$ by Σ^ν; it is a subgroup of $O(n)$ isomorphic to $O(n-1)$. If $w \in C^0(\bar{\Omega})$ is such that $\Sigma_w = \Sigma^\nu$ for some $\nu \in S^{n-1}$, then we say that w is <u>axisymmetric</u>.

Spherically symmetric solutions of (3) are solutions $(w,\lambda) \in W \times \mathbb{R}$ with $\Sigma_w = O(n)$; if this is the case then we can write $w(x) = \bar{w}(|x|)$, where $\bar{w}(r)$ is a solution of

$$w'' + \frac{n-1}{r} w' + f(w,\lambda) = 0$$
$$w'(0) = 0 \quad , \quad w(1) = 0 \ . \tag{7}$$

It was shown by SMOLLER and WASSERMAN (1984) that for each $(p,\lambda) \in \mathbb{R}^2$ the inital value problem

$$w'' + \frac{n-1}{r} w' + f(w,\lambda) = 0$$
$$w'(0) = 0 \quad , \quad w(0) = p \tag{8}$$

has a unique solution $\tilde{w}(r;p,\lambda)$, depending smoothly on all variables.
So spherically symmetric solutions of (3) can be found by solving the equation

$$\tilde{w}(1;p,\lambda) = 0 \ . \tag{9}$$

3. Conditions for symmetry-breaking

Suppose that $(w_0,\lambda_0) \in W \times \mathbb{R}$ is a solution of (3), with $w_0(x) > 0$, $\forall x \in \Omega$; then w_0 is spherically symmetric, and hence $w_0(x) = \tilde{w}(|x|;p_0,\lambda_0)$, with $p_0 := w_0(0)$; also $\tilde{w}_r(1;p_0,\lambda_0) \leq 0$. If $\tilde{w}_r(1;p_0,\lambda_0) < 0$ then w_0 has a neighborhood $\tilde{\omega}$ in W such that $w(x) > 0$ for all $x \in \Omega$ and for all $w \in \tilde{\omega}$; hence we have $\Sigma_w = O(n)$ for all solutions (w,λ) of (3) in $\tilde{\omega} \times \mathbb{R}$. This gives the following.

Lemma 1. A necessary condition for symmetry-breaking at (w_0,λ_0) is that $\tilde{w}_r(1;p_0,\lambda_0) = 0$.
If this necessary condition is satisfied then one can prove the following results.

Theorem 1. If $\tilde{w}_r(1;p_0,\lambda_0) = 0$ then there exists a neighborhood ω of (w_0,λ_0) in $W \times \mathbb{R}$ such that if $(w,\lambda) \in \omega$ solves (3), then we have either $\Sigma_w = O(n)$ or $\Sigma_w = \Sigma^\nu$ for some $\nu \in S^{n-1}$, i.e. if there is symmetry-breaking, then the symmetry breaks by axisymmetric solutions.

Theorem 2. If $\tilde{w}_r(1;p_0,\lambda_0) = 0$ and if moreover

$$a := \tilde{w}_{r\lambda}(1;p_0,\lambda_0)\tilde{w}_p(1;p_0,\lambda_0)-\tilde{w}_{rp}(1;p_0,\lambda_0)\tilde{w}_\lambda(1;p_0,\lambda_0) \neq 0 , \qquad (10)$$

then there is symmetry-breaking at (w_0,λ_0). More precisely, the neighborhood ω of theorem 1 can be chosen such that $F^{-1}(0)\cap\omega = C_s \cup C_a$, where

(i) C_s is a smooth line of spherically symmetric solutions, divided by (w_0,λ_0) in two connected components; along one of these components the solutions are positive, along the other one they are nonpositive.

(ii) C_a is an n-dimensional manifold of axisymmetric solutions, bifurcating from C_s at (w_0,λ_0), and invariant under the O(n)-action.

Remarks. (a) The condition $a\neq 0$ means that the sets $\{(p,\lambda)\in\mathbb{R}^2 | \tilde{w}(1;p,\lambda)=0\}$ and $\{(p,\lambda)\in\mathbb{R}^2 | \tilde{w}_r(1;p,\lambda)=0\}$ form in a neighborhood of (p_0,λ_0) two smooth curves which intersect transversally in the point (p_0,λ_0). (Compare with the paper of BUDD (1986)).

(b) If $f(w,\lambda) = \lambda\tilde{f}(w)$ then the condition $a\neq 0$ reduces to $\tilde{w}_p(1;p_0,\lambda_0) \neq 0$. This is the case studied by SMOLLER and WASSERMAN, who found symmetry-breaking under a further condition on $\tilde{f}$.

(c) One of the possibilities which can arise under the conditions of theorem 2 is that C_s has a turning point at (w_0,λ_0), with C_a bifurcating from that turning point.

There is still another way to formulate the transversality condition $a\neq 0$ of theorem 2; loosely speaking it means that along the branch C_s of spherically symmetric solutions the gradient at the boundary $\partial\Omega$ passes transversally through zero. A more precise statement is given in the next corollary to theorem 2. We assume that the nonlinearity $f(w,\lambda)$ is such that

$$(\tilde{w}(1;p,\lambda),\tilde{w}_r(1;p,\lambda),\tilde{w}_p(1;p,\lambda),\tilde{w}_\lambda(1;p,\lambda)) \neq (0,0,0,0) \qquad (11)$$

for all $(p,\lambda)\in\mathbb{R}^2$; this condition will be satisfied for generic f. Then the following holds.

Corollary. Let, for some interval $I\subset\mathbb{R}$, $\{(\tilde{p}(s),\tilde{\lambda}(s)) | s\in I\}$ be a smooth solution branch of the equation (9). Let $s_0 \in I$ be such that

(i) $\tilde{w}(r;\tilde{p}(s_0),\tilde{\lambda}(s_0)) > 0 \quad , \quad \forall r \in [0,1)$;

(ii) $\tilde{w}_r(1;\tilde{p}(s_0),\tilde{\lambda}(s_0)) = 0$;

(iii) $\frac{d}{ds} \tilde{w}_r(1;\tilde{p}(s),\tilde{\lambda}(s))\Big|_{s=s_0} \neq 0$.

Then there is symmetry-breaking at $(\tilde{w}(|x|;\tilde{p}(s_0),\tilde{\lambda}(s_0)),\tilde{\lambda}(s_0))$ and the conclusion of theorem 2 holds.

We conclude with a few comments on the proof, which is based on an equivariant Liapunov-Schmidt reduction. The crucial point is that the null space involved in this reduction is built from zeroth and first order spherical harmonics, a result due to SMOLLER and WASSERMAN (1986). Therefore the positivity condition in the foregoing results can be replaced by any other condition which ensures that no higher order spherical harmonics appear in the null space.

References

Budd, C.J. (1986) Symmetry breaking in elliptic systems. Preprint Oxford University.

Cerami, G. (1986) Symmetry breaking for a class of semi-linear elliptic problems. Nonlin. Anal., Th. Meth. & Appl. 10, 1-14.

Gidas, B., W.M. Ni and L. Nirenberg (1979) Symmetry and related properties via the maximum principle. Comm. Math. Phys. 68, 209-243.

Pospiech, C. (1986) Global bifurcation with symmetry breaking. Preprint Universität Heidelberg.

Smoller J. and A. Wasserman (1984) Existence, uniqueness and nondegeneracy of positive solutions of semi-linear elliptic equations. Comm. Math. Phys. 95, 129-159.

Smoller J. and A. Wasserman (1986) Symmetry-breaking for positive solutions of semilinear elliptic equations. Arch. Rat. Mech. Anal., to appear.

Vanderbauwhede, A. (1986) Symmetry-breaking at positive solutions. Z. Angew. Math. Phys., to appear.

A. Vanderbauwhede, Institute for Theoretical Mechanics, Krijgslaan 281, B-9000 Gent, Belgium.

International Series of
Numerical Mathematics, Vol. 79

AVERAGING OVER ANGLES WITH AN APPLICATION TO ROTOR DYNAMICS

F. Verhulst
Mathematisch Instituut, Rijksuniversiteit Utrecht
Netherlands

1. *Introduction.*

In many problems in engineering and physics small parameters arise in a natural way: weak coupling between modes, small damping, slow variation of coefficients in the equations corresponding with quantities in the system as mass, lenght etc. A large number of such problems can be reformulated in the following form

$$\begin{aligned} \dot{x} &= \varepsilon X(\phi,x) + \varepsilon^2 \ldots \quad , \quad x \in D \subset \mathbb{R}^n \\ \dot{\phi} &= \Omega(x) + \varepsilon \ldots \quad , \quad \phi \in T^m \end{aligned} \tag{1}$$

with $n,m \in \mathbb{N}$, ε a small parameter and initial values for x and ϕ added. We assume that the vector fields X and Ω are sufficiently smooth to carry out the operations which we need.

Example.

Consider the nonlinear one-dimensional oscillator

$$\ddot{x} + \varepsilon\mu\dot{x} + \omega^2(\varepsilon t)\, x + \varepsilon f(x) = 0$$

with μ constant and $f(x)$ a smooth function. Coordinate transformation $x = r\sin\phi$, $\dot{x} = \omega r\cos\phi$ and putting $\tau = \varepsilon t$ produces

$$\begin{aligned} \dot{r} &= \varepsilon[-\mu r\cos^2\phi - \frac{r}{\omega}\cos^2\phi\frac{d\omega}{d\tau} - \frac{\cos\phi}{\omega} f(r\sin\phi)] \\ \dot{t} &= \varepsilon \\ \dot{\phi} &= \omega + \varepsilon\ldots \end{aligned} \tag{2}$$

System (2) is of the form (1) with n=2, m=1. Assuming that $|\omega(\varepsilon t)|$ is bounded away from zero by constants independent of ε we average the first equation of (2) over ϕ to obtain

$$\dot{\tilde{r}} = -\frac{1}{2}\varepsilon[\mu\tilde{r} + \frac{\tilde{r}}{\omega}\frac{d\omega}{d\tau}]$$

Integration produces

$$\tilde{r}(t)\ \omega^{\frac{1}{2}}(\varepsilon t)\ e^{\frac{1}{2}\varepsilon\mu t} = \text{constant} \tag{3}$$

If $\mu=0$ this expression corresponds with a well-known adiabatic invariant; see for instance [5] § 32 and also [2].

Natural questions are: what is the asymptotic validity of the result, what happens if ω has zeros, which new phenomena arise if there are more angles $(m > 1)$?

2. *Asymptotic validity.*

Theorem 1.

Consider system (1) with m=1 and

$$0 < a \leqslant \inf_{x\in D} |\Omega(x)| \leqslant \sup_{x\in D} |\Omega(x)| \leqslant b < \infty$$

where a, b are ε-independent constants. Let (y,Ψ) be the solution of

$$\dot{y} = \varepsilon X^{\circ}(y)\ ,\ y(0) = x(0)$$
$$\dot{\Psi} = \Omega(y)\ ,\ \Psi(0) = \phi(0)$$

with

$$X^{\circ}(y) = \int_{S^1} X(\phi,y)d\phi$$

then, if $y(t)$ remains in $D_o \subset D$

$$x(t) = y(t) + O(\varepsilon) \text{ on the time-scale } 1/\varepsilon\ ,$$
$$\phi(t) = \Psi(t) + O(\varepsilon t\ e^{\varepsilon t})\ .$$

Proof

see [6] chapter 5.

Theorem 1 establishes the asymptotic validity of the adiabatic invariant (3) to $O(\varepsilon)$ on the time-scale $1/\varepsilon$. Higher order calculations involve $O(\varepsilon^2)$ approximations on the time-scale $1/\varepsilon$ for x and an $O(\varepsilon)$ approximation for ϕ. It is interesting to note that we can extend the time-scale of validity of the approximation for x to $[0,\infty)$ in the case that the averaged equation

$$\dot{y} = \varepsilon X^{\circ}(y)$$

contains a critical point which is asymptotically stable in the linear approximation. To see this we are using the *near-identity* or *normalising* transformation

$$x = z + \varepsilon u(\phi,z) = z + \varepsilon \frac{1}{\Omega(z)} \int^{\phi} [X(s,z) - X^{\circ}(z)]ds$$

Substitution into system (1) produces

$$\dot{z} = \varepsilon X^{\circ}(z) + \varepsilon^2 R_1(\phi,z;\varepsilon)$$

$$\dot{\phi} = \Omega(z) + \varepsilon\, R_2(\phi, z; \varepsilon)$$

From this stage on we can use the reasoning of the Eckhaus/Sanchez-Palençia Theorem (4.2.1 in [6]).

Theorem 2 (informal presentation).

Consider system (1) with $m=1$ and the assumptions of theorem 1. Moreover the averaged equation for y contains a critical point, asymptotically stable in the linear approximation. If $y(0) = x(0)$ is in the interior of the domain of attraction, we have

$$x(t) = y(t) + O(\varepsilon) \text{ on } [0,\infty).$$

A separate question is whether adiabatic invariants which can be found sometimes in a first-order calculation are valid on longer time-scales. In general this is not the case; an example from mechanics is the two-body problem with variable mass where to first-order the eccentricity of the orbit and the semi-major axis are adiabatic invariants, on a longer time-scale these quantities change drastically. See [8] and further references there.

The situation contrasts with the case of Hamiltonian systems where normalisation or averaging produces invariants, the existence of which are predicted by the KAM-theorem.

Theorem 1 and 2 are valid in domains which do not contain zeros of $\Omega(x)$. The equation $\Omega(x) = 0$ defines the *resonance manifold(s)* in $\mathbb{R}^n$. In and near a resonance manifold we have to apply a local analysis as in boundary layer theory. There are several interesting phenomena associated with resonance manifolds. Sometimes solutions pass through it but a certain scattering of orbits takes place (see [1]); in other cases solutions can be caught into resonance which is undesirable from the point of view of mechanical engineering.

3. *More than one angle.*

In the case of averaging over the m-torus, $m > 1$, there are many unsolved problems. The basic problems arise already if $m = 2$:

$$\begin{aligned} \dot{x} &= \varepsilon\, X(\phi_1, \phi_2, x) && , \; x \in \mathbb{R}^n \\ \dot{\phi}_1 &= \Omega_1(x) && , \; (\phi_1, \phi_2) \in T^2 \\ \dot{\phi}_2 &= \Omega_2(x) && \end{aligned} \tag{4}$$

with X periodic in ϕ_1 and ϕ_2. Suppose that Fourier expansion yields

$$X(\phi_1, \phi_2, x) = \sum_{k,l=-\infty}^{\infty} c_{kl}(x) e^{i(k\phi_1 + l\phi_2)}$$

Averaging over the angles yields the averaged equation

$$\dot{y} = \varepsilon \, c_{oo} \, (y) \tag{5}$$

but this only makes sense outside the resonances

$$k\phi_1 + l\phi_2 = 0 \quad , k,l \in \mathbb{Z}.$$

Resonance becomes important if system (4) admits that

$$k\Omega_1(x) + l\Omega_2(x) = 0 \quad , k,l \in \mathbb{Z} \tag{6}$$

Equation (6) defines an, in principle infinite, set of resonance manifolds sometimes called the Arnold web. Outside the Arnold web we may use the averaged equation (5) to obtain approximations of x as in the case of theorem 1 and 2. However, the presence of the set of resonance manifolds defined by (6) may complicate the dynamics of the system very much. Also, very little is known about adiabatic invariants in the case of more than one angle.

If we put the preceding observations in the context of Hamiltonian systems with two or more degrees of freedom we note that, each degree of freedom producing an action and an angle, we expect the presence of an infinite set of resonance manifolds. The normalisation process of the Hamiltonian (see [6]) introduces a natural hierarchy among these manifolds with very prominent ones and an infinite number of very small resonance domains. This makes the quantitative treatment of Hamiltonian systems possible in practice.

4. *The oscillator-flywheel problem.*

One of the prototype problems of mechanics is a slightly excentric rotor system (flywheel) mounted on an elastic support, see [3] or [4]. The equations are of the form

$$\begin{aligned} \ddot{x} + x &= \varepsilon f(x, \dot{x}, \phi, \dot{\phi}, \varepsilon) \\ \ddot{\phi} &= \varepsilon g(x, \dot{x}, \phi, \dot{\phi}, \varepsilon) \end{aligned} \tag{7}$$

We summarize the results as the technical details will be published elsewhere [7].

In the notation of our theorems: n=2, m=2.

a. A resonance manifold exists if the terms to order ε or order ε^2 are retained; the resonance corresponds with the case that the linearized elastic support frequency and the angular velocity $\dot{\phi}$ of the flywheel are equal.

b. Averaging outside the resonance manifold leads to simple expressions. The resonance domain is of typical size $O(\sqrt{\varepsilon})$ and can be analysed using the resonant combination angle. The equations contain two critical points: a saddle and a center; the approximation equations to first order are conservative which is not a structurally stable situation. Note that system (7) contains damping and forcing. This first-order result makes a second-order calculation necessary where we expect bifurcation of the centre to a generic case.

c. The saddle point corresponds with an unstable periodic solution and we expect that passage through the resonance domain is possible. A numerical test shows that this may happen indeed but that there are three sets of initial conditions outside the resonance manifold which lead to solutions caught in the resonance domain.

d. A second order calculation produces in addition to the purely imaginary eigenvalues small negative parts. The centre becomes an attracting spiral corresponding with a stable periodic solution in the resonance domain.

e. A technique inspired by work of Haberman and Melnikov is used to connect the solutions outside the resonance domain with the attracting set in and near the resonance manifold. The asymptotic approximations and the numerical results agree to the expected order of accuracy.

References

[1] Arnold, V.I., Conditions for the applicability, and estimate of the error, of an averaging method for systems which pass through states of resonance during the course of their evolution, Soviet Math. 6, 331-334 (1965).

[2] Bakaj, A.S., Integral manifolds and adiabatic invariants of systems in slow evolution; in Asymptotic Analysis II (F. Verhulst ed.), Lecture Notes Math. 985, Springer-Verlag (1983).

[3] Evan-Iwanowski, R.M., Resonance Oscillations in Mechanical Systems, Elsevier Publ. Co., Amsterdam (1976).

[4] Goloskokow, E.G. and Filippow, A.P., Instationäre Schwingungen Mechanischer Systeme, Akademie Verlag, Berlin (1971).

[5] Landau, L.D. and Lifschitz, E.M., Theoretische Physik Bd.1, Akademie Verlag, Berlin (1973).

[6] Sanders, J.A. and Verhulst, F., Averaging methods in nonlinear dynamical systems, Appl. Math. Sciences 59, Springer-Verlag (1985).

[7] Van den Broek, B., Domain of attraction in the oscillator-flywheel problem, Preprint 371, Rijksuniversiteit Utrecht (1985).

[8] Verhulst, F., Asymptotic expansions in the perturbed two-body problem with application to systems with variable mass, Celestial Mechanics 11, 95-129 (1975).

F. Verhulst, Mathematisch Instituut van de Rijksuniversiteit Utrecht, Postbus 80.010, 3508 TA Utrecht, The Netherlands.